Computational Microelectronics

Edited by S. Selberherr

The Drift Diffusion Equation and Its Applications in MOSFET Modeling

W. Hänsch

Springer-Verlag Wien New York

Dr. Dipl.-Phys. Ing. Wilfried Hänsch
RR 2 Box 2078
Charlotte, VT, USA

Typeset by Macmillan India Ltd., Bangalore, India

With 95 Figures

ISSN 0179-0307
ISBN-13:978-3-7091-9097-5   e-ISBN-13:978-3-7091-9095-1
DOI: 10.1007/978-3-7091-9095-1

Preface

To be perfect does not mean that there is nothing to add,
but rather there is nothing to take away

Antoine de Saint-Exupery

The drift–diffusion approximation has served for more than two decades as
the cornerstone for the numerical simulation of semiconductor devices.
However, the tremendous speed in the development of the semiconductor
industry demands numerical simulation tools that are efficient and provide
reliable results. This makes the development of a simulation tool an
interdisciplinary task in which physics, numerical algorithms, and device
technology merge. For the sake of an efficient code there are trade-offs
between the different influencing factors. The numerical performance of a
program that is highly flexible in device types and the geometries it covers
certainly cannot compare with a program that is optimized for one type of
device only. Very often the device is sufficiently described by a two-
dimensional geometry. This is the case in a MOSFET, for example, if the
gate length is small compared with the gate width. In these cases the
geometry reduces to the specification of a two-dimensional device. Here
again the simplest geometries, which are planar or at least rectangular
surfaces, will give the most efficient numerical codes. The device engineer
has to decide whether this reduced description of the real device is still
suitable for his purposes.

Also, it is impossible to incorporate in a simulation program which will be
used in an engineering environment the exact physics of the transport
phenomena in the device. The exact physical formulation is very often not
available in a mathematical formulation which is necessary to include it in a
numerical code. If such a formulation is known, it is usually very complicated and incompatible with the requirement of an efficient program. Therefore the general way is to describe the physical transport phenomena with
an empirical expression which contains the essential features and has a
simple enough algebraic structure to permit its fast evaluation on a

computer. In this respect modeling requires as much physics as necessary to describe the device behavior correctly. This contrasts with pure physics, where as much physics as possible is considered to get a better understanding of the basic mechanisms. The problem for modeling is then to decide which are the important features one has to keep. Very often there is clear experimental evidence as to which these might be as with, for instance, velocity saturation in high electric fields or the gate field reduced channel mobility in a MOSFET. Other phenomena are known to exist, such as the population of high energetic states in the carrier distribution, but they are hard to model because the experiments are only very indirectly connected to them. Here other sources for comparison have to be developed. Monte Carlo methods are very powerful in this respect. They allow one to run an artificial experiment and to study the physical scenario in detail. All in all, providing the correct physics for a simulation code is a formidable task that requires the knowledge of the essential experimental device data and also how to map it on to the carrier transport physics of a semiconductor device. But what happens if the device in question does not yet exist and no experimental material is available? If the models developed so far would describe the cause, and not only the effect, they should also cover these unknown structures. It is therefore of considerable interest to find the underlying physics of the models utilized in device modeling from a first principle formulation. The aim of this work is to find such a connection. Our guiding device will be the MOSFET, which is one of the most important components in today's integrated circuits. It belongs to the family of the field effect controlled devices and is essentially an unipolar device, which means that for its principal operation it is sufficient to consider only one type carrier. Naturally, we will especially focus on problems related to the physical phenomena in connection with this device. This will exclude some other interesting physics which is more related to the bipolar devices. The basic theme of the book will be to provide the physics in a form that can be used in a numerical code for an engineering environment. We will therefore always compromise between rigour and simplicity in the physical description under consideration. We will walk the long way from the basic principles of carrier transport in many-particle systems towards how to apply the drift–diffusion approximation to model the hot-carrier stability of a MOSFET device. In this respect we provide a bridge from the realm of pure physics to its application in the real device with relevant questions that require an answer. It is clear that the more we are removed from the classical approach of a physicist, the more we will encounter the way an engineer works on his problems: the scenario will be less sharply defined, and physical intuition and knowledge drawn from experience are more important than an exact formulation. We therefore especially invite the scientists to join us on the way and to find out how exciting physics still can be, even for a device that seems to have been completely understood for years. On the other hand, we challenge the engineers to find out that there is an underlying physics for the models they have so successfully invented and that this underlying physics might open the way for an extension of the conventional drift–diffusion approximation to cover further phenomena.

The book is organized into five chapters. Chapter 1 is a brief introduction of the physics of many-particle system in equilibrium and will derive the Boltzmann equation from first principle non-equilibrium quantum statistics. It will discuss what a particle in a many-body system is and under what circumstances it will be equivalent to the well-established classical concept of a particle. It will be shown that the Boltzmann equation is the correct classical limit $\hbar \to 0$ of the Wigner equation, including the collision terms. In Chapter 3, where we will discuss the carrier transport in the inversion channel of a MOSFET, we will come back to the generalized formulation of particle transport for systems with bound states. In Chapter 2 the underlying transport equation is the Boltzmann equation for an electron gas subjected to electron-impurity and electron–phonon scattering. We will present a detailed discussion of the linear response regime which includes the relaxation time approximation for the Boltzmann equation and the low field mobility in bulk material. Then we will move to the non-linear response regime where the system of carriers is no longer in equilibrium with the environment, and energy dissipation to the systems of phonons becomes an issue. Different approaches to obtain a modified drift–diffusion formulation for particle transport based upon momentum space averages of the Boltzmann equation are discussed. We will find the connection between the carrier temperature, velocity overshoot, and the proper parameterization of the momentum relaxation time. The end of this chapter is a generalization of the results to a two band system appropriate to describe carrier transport in III–V compounds.

In Chapter 3 we discuss the charge transport in the inversion layer of a MOSFET. The generalized quantum kinetic equation derived in Chapter 1 is used to rigorously find the correct transport equation for a system with bound states. This equation is evaluated in the $\lim \hbar \to 0$ to obtain a modified drift–diffusion approximation for the current density in an inversion channel. As a result of such a limit we derive an expression for the gate field reduced channel mobility from first principles. In Chapter 4 we investigate the impact ionization cross-section and the population of the high energetic states in the particle distribution function. We address the question whether these states are in a stationary state with the field or whether their character is predominantly ballistic. To have a clear understanding of the physics we compare the results of the analytical approach with Monte Carlo calculations. The chapter is closed by presenting suitable expressions for the impact ionization coefficient and gate oxide injection based on our findings. Finally, Chapter 5 is devoted to the specific problems of hot-carrier degradation of MOSFET devices. The results of the previous chapters are used to model the effects of hot-carrier-induced oxide damage on the device characteristics. This is done in two steps. In the first step we discuss in detail how modeling can be used for a detailed analysis of hot-carrier degraded devices. Here we discuss in detail what is the nature of the possible oxide damage and how can we verify it experimentally. In the second step we outline an approach how the degradation process itself can be modeled.

There is a reference list after each chapter. Because of the large number of publications in the field of modeling I have not even attempted to provide a complete bibliography. The references are chosen according to their useful-ness for further study. There are some original works which are worth consulting and review articles in which further bibliographical material can be found. Material of this book is, in part, also covered in the monographs by Siegfried Selberherr on modeling in general and Carlo Jacoboni and Paolo Lugli on Monte Carlo calculations, which are also published in this series. Both contain extensive reference lists.

This wide range of material from basic questions of transport physics to the complicated problem of MOSFET hot-carrier degradation cannot be covered by one person alone. Therefore I acknowledge the many discus-sions and help I have had from my colleges at the research laboratories at Siemens. It is impossible to name them all but without Christoph Werner and Werner Weber it would have been more difficult to discover the exciting physics hidden in the MOSFET device. I have to acknowledge the work done by Andreas Schwerin. In his PhD work he laid the foundation for most of the material presented in Chapter 5. In his master's work Thomas Vogelsang, a computer whiz, implemented the results of Chapter 3 into the MINIMOS code and did the quantum mechanical calculations for the equilibriums case. He also is responsible for the Monte Carlo calcu-lations presented in Chapter 4, which are a part of his PhD work. Sebastian Käsweber worked on the generalization of the extended drift–diffusion equation to the III–V compounds. Franz Hofman, Udo Schwalke, and Carlos Mazure provided experimental material to compare the modeling effort with experimental material. I must not forget Ulf Bürker, Hartwig Bierhenke, and Reiner Tielert, the people behind the curtain who made it possible for me to accomplish most of the work presented in this book.

Marden Seavey and John Faricelli from Digital Inc. at Hudson, Ma, I thank for the many discussions concerning the mobility and impact ionization models. To Jan Slotboom from Philips Nat. Lab. in Eindhoven I am indebted for many interesting discussions on various modeling and device physics issues.

I like to thank Siegfried Selberherr for the many discussions we have had over the years about what is modeling all about, for his critical comments on my own work, and for the continuous support during the preparation of this book. A considerable part in debugging the manuscript was done by Predrag Habaš and Philipp Lindorter, both members of Siegfried Selberherr's group. Also Gerhard Punz, Karl Traar, Andreas Schwerin, and Dan Cole contributed to make the manuscript more readable.

Anyone involved in a project like this knows that all effort is in vain unless the family steps back for a certain period of time. My wife Nancy accomplished the miracle of keeping our three children Inge, Anna, and Johannes and herself in reasonable good spirit and still managed to be for me, at any time, an indispensable partner in successfully completing this book. I am blessed with her presence and words are not sufficient to express my gratitude for her support.

Charlotte, Vermont, April 1991   Wilfried Hänsch

Contents

1.1 Introduction

The degree of complexity of today's devices makes numerical methods for evaluating their electrical behavior mandatory. The special demands of modeling therefore require a simple formulation of carrier transport containing the essential physics in a way expressible in numerical code. It is often impossible to derive special physical features from first principles, because in the complete system these must be combined with a number of different effects in a nontrivial setting. Figure 1 shows a selected number of physical phenomena that must be included in any realistic simulation of a MOSFET device.

The physics in this figure concentrates on the following phenomena: carriers in an inversion channel, surface roughness scattering, ohmic bulk-carrier mobility, velocity saturation, carrier heating, and the population of the high-energy tail of the distribution function. All these effects are very complicated and demand extensive treatment in their own right. However, they all must be considered simultaneously in a successful solution to the modeling problem: we cannot simply neglect factors such as surface scattering in order to obtain a satisfactory result. Up to now, the most convenient approach to modelling complex devices has been the drift–diffusion approximation, which refers to the current-density equation in the set of conventional semiconductor equations. The drift–diffusion approximation is so powerful for practical purposes that the questions of its range of validity is essential. At present, the degree of success is still uncertain, although the literature contains serious arguments that challenge its usefulness (Barker and Ferry 1980a, b, Ferry and Barker 1980). From a practical point of view, however, there is currently no reasonable alternative to the drift–diffusion approximation which does not involve enormous computer resources or a very high degree of complexity that restricts its use to very simple device structures. For this reason, we will investigate the physics underlying the simple drift–diffusion approximation.

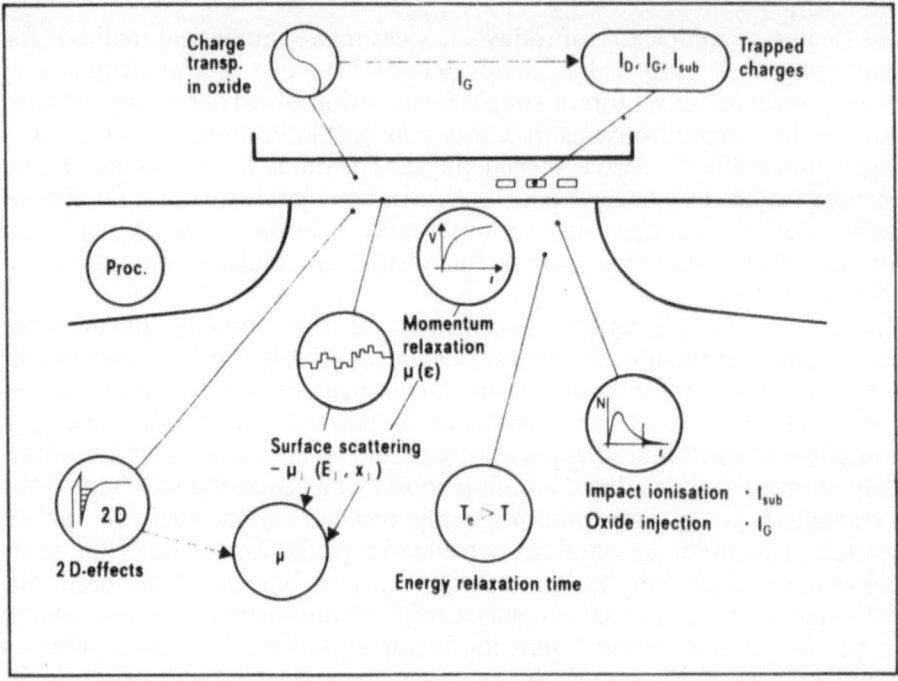

Fig. 1 Physical phenomena to be considered for the MOSFET under operating conditions. The symbol in the circle represents the fundamental physical question. The doping profiles have to be provided by process simulation (Proc). The charge transport in the oxide has to consider both the distribution of mobile charges that are injected from the active device area and their trapping in the oxide

Our approach will be to derive the drift–diffusion approximation to determine which assumptions are mandatory and what are their possible extensions. We cannot, of course, offer a general solution of the problem. As so often in modeling, half the answer comes from addressing the problem from the correct angle. In concrete terms, this means that the drift–diffusion approximation and/or possible extensions will give satisfactory results for some questions (possibly including the calculation of the terminal currents, which are integrated quantities of the device-internal distributions) but poor answers to another set of problems. The decision cannot be made *a priori*. A verification of the models by experimental means is mandatory. In seeking the roots of the drift–diffusion approximation, we will inevitably find the Boltzmann equation underlying it as the master equation. This is an integro-differential equation in seven dimensions of space (x, y, z), momentum (p_x, p_y, p_z), and time t. The solution of this complex equation will yield the distribution of particles in real space and momentum space at a given instant of time. The device-relevant quantities are calculated as appropriate momentum–space averages over this distribution function, which reduces the numerical variables to four. The analysis of the Boltzmann equation is therefore essential for a critical evaluation of the drift–diffusion approximation. We consequently decided to present a rigorous derivation of the Boltzmann equation in this chapter. This will of course initially lead us away from the modeling track. On the other hand, by following the derivation of the Boltzmann equation from the first principles of non-equilibrium quantum statistics we can see at what stage of the approximation certain phenomena (such as tunneling) will disappear and how they could possibly be included again.

We will not perform our analysis of transport in a many-body system in the most general way. To keep the algebra simple, we will use the continuum model of matter which disregards the periodicity of lattice material. Although this is a shortcoming, inclusion of the lattice structure will not change the essential results. We will include a brief discussion about where the lattice structure will have a significant influence. We will address the important question of the nature of a particle and how far this attractive concept is valid. We all associate carrier transport with drifting particles in a force field but hardly question the concept of a particle. To minimize referencing, we will include a short introduction to the Green's function method for a many-body system in equilibrium. This will provide the foundation for a proper derivation of the Boltzmann equation from first principles. At the end of this chapter, the Boltzmann equation will be derived in the form found in many textbooks.

1.2 Many-Body System in Equilibrium

This section provides the material allowing us to pursue a rigorous approach to deriving the Boltzmann equation, which will be given in the subsequent sections. It therefore serves to introduce the language used in many-particle physics. To cover the entire field of many-body systems would go beyond the scope of this book and would mean unnecessary competition to some excellent textbooks on this subject. Although the interested reader can choose between many monographs, the material presented here is drawn from the following textbooks. (i) *Quantum Theory of Many-Particle Systems* (Fetter and Walecka 1971) gives a very good introduction to the formalism; the presentation is very detailed and easy to follow. (ii) *The Many-Body Problem* (Parry 1973) is a concise text focusing primarily on the Matsubara finite-temperature formalism but also dealing with the equation of motion method. (iii) *Many-Particle Physics* (Mahan 1981) is an unlimited resource for examples and serves as a reference for the application of Green's functions in solid-state physics in general. The basic

material is presented very briefly in this book, making it somewhat in-
convenient for the newcomer to the field. Several other textbooks will be
listed in the references, particular mention should be made of the classic of
the Russian school: *Methods of Quantum Field Theory in Statistical Physics*
(Abricosov *et al.* 1961).

1.2.1 Quantum Mechanics of Many-Body Systems

Very few problems in quantum mechanics have exact solutions (Landau
and Lifshitz 1979, Davydov 1965, Merzbacher 1961). Even the one-particle
problem in an arbitrary potential is no longer generally solvable in closed
form. Things do not get any easier for two-body problems, such as that of
the helium atom. In a many-body system we are dealing with a very large
number of particles and this seems like a desperate case from the standpoint
of simple quantum mechanics. However, although it is in general difficult to
find the exact solution, often the problem under consideration allows a
solvable part to be separated. The remaining term then acts as a perturbation of the exactly known result. Perturbation theory was formalized to
calculate approximative solutions of the complete problem. The starting
point may for instance be the complete set of eigenfunctions for the zero-
order problem. In a many-body system, these are the completely sym-
metrized (bosons) or anti-symmetrized (fermions) wave functions. For non-
interacting particles, these are special linear combinations of products of
plane wave states (in a crystal, Bloch states (Kittel 1967)). The wave
functions depend on $3N$ space coordinates and time, where N is the number
of particles in the system. A one-particle eigenstate is assigned to each
particle. This notation for the wave functions is rather complex, so a
different notation is used in many-body physics. It makes use of the fact that
a many-body system of free and non-interacting identical particles is
sufficiently characterized by the number of particles n_s in an appropriately
chosen one-particle state s. For that state vector we use the symbol
$|n_1, n_2, \ldots, n_s, \ldots \rangle$. We now define the particle number operator N_s
whose eigenvalues are the number of particles in the state s.

$$N_s|n_1, \ldots n_s, \ldots \rangle = a_s^+ a_s|n_1, \ldots n_s, \ldots \rangle = n_s|n_1, \ldots n_s, \ldots \rangle \qquad (1.1)$$

The operators a_s^+ and a_s can be interpreted as adding and removing one
particle to the state s, respectively.

$$a_s^+ |n_1, \ldots, n_s, \ldots \rangle \sim |n_s, \ldots, n_s + 1, \ldots \rangle$$
$$a_s|n_1, \ldots, n_s, \ldots \rangle \sim |n_s, \ldots, n_s - 1, \ldots \rangle \qquad (1.2)$$

Physics tells us that there are two different kinds of identical particles,
which differ in their possible occupation of the one-particle states. Bosons

allow for an unlimited number of particles in the same one-particle state; we therefore have $n_s = 0, 1, 2, 3, 4, \ldots$. Fermions have to obey the Pauli exclusion principle which states that a one-particle state s can only be singly occupied or not at all; therefore the occupation number is: $n_s = 0, 1$. The former group includes all particles with integral spin numbers (photons, phonons, etc.) the latter those with half-integral spin (electrons, etc.). Applying the number operator N_s to the state vector $a_s^+ | \ldots n_s \ldots \rangle$ we therefore have only two options: in the case of bosons the particle number in the state s is increased by one and in the case of fermions either the state s is occupied, that is the case if $| \ldots n_s = 0 \ldots \rangle$, or N_s maps the new state on to 0, that happens if $| \ldots n_s = 1 \ldots \rangle$. This can be expressed by:

$$N_s a_s^+ |n_1, \ldots n_s, \ldots \rangle = (1 \pm n_s) a_s^+ |n_1, \ldots n_s, \ldots \rangle \qquad (1.3)$$

Here, the upper sign stands for bosons and the lower sign for fermions. From Eq. (1.3) we can derive commutation rules for the particle operators a_s and a_s^+ which contain the symmetry information of the wave function. Moving the c-number n_s on the right-hand side of a_s^+ and replacing it by N_s we obtain the operator identities

$$a_s a_s^+ - a_s^+ a_s = 1 \qquad \text{(bosons)}$$
$$a_s a_s^+ + a_s^+ a_s = 1 \qquad \text{(fermions)} \qquad (1.4)$$

These two commutation rules are the cornerstone in the formalism of many-body physics. They can easily be generalized to the case where creation and destruction operators refer to different state labels, then by the same argument as above, we have the following,

$$a_s a_{s'}^+ - a_{s'}^+ a_s = \delta_{ss'} \qquad \text{(bosons)}$$
$$a_s a_{s'}^+ + a_{s'}^+ a_s = \delta_{ss'} \qquad \text{(fermions)} \qquad (1.4)$$

Here $\delta_{ss'}$ is the Kronecker delta, which is one if both indices are the same and zero otherwise. Since the number states form a complete orthonormal basis set in occupation number space (sometimes referred to as Fock space), we have, for the matrix elements of a particle operator

$$|\langle \ldots n_s + 1 \ldots | a_s^+ | \ldots n_s \ldots \rangle|^2 = 1 \pm n_s \qquad (1.5)$$

An arbitrary state in occupation number formalism is now created by simply adding particles to the vacuum state $|0\rangle$ into their appropriate one-particle states s.

Once we have established this space we have to say how we do the physics. In the first place this means: what does the Hamiltonian look like? We will not go into the details of deriving the particle space Hamiltonian from the many-particle Schrödinger equation. We will simply state some general results (Fetter and Walecka 1971) and try to justify them from the physics they describe. To this end, we have to introduce the concept of field operators. This concept provides the link between the occupation number space we discussed previously and the wave mechanics. If we could separate an exactly solvable one-particle problem from the many-particle Schrödinger equation (for instance the free particle with wave number k), we can build up a basis in number space by filling up the vacuum state by placing particles in these states. The field operators Ψ and Ψ^+ are defined as follows

$$\Psi(x, t) = \sum_k \varphi_k(x) a_k(t)$$

(1.6)

$$\Psi^+(x, t) = \sum_k \varphi_k^*(x)^+ a_k(t)$$

Here the φ_k are the stationary solutions of the one-particle Schrödinger equation with eigenvalue ε_k. The time-dependence of the particle and field operators is given by the Heisenberg equation

$$ih \frac{\partial}{\partial t} O = [O, H]$$

(1.7)

H is the Hamiltonian of the system and the brackets stand for the commutator of the two operators O and H. To obtain the representation of the Hamiltonian in number space, the following simple procedure has to be followed:

(i) In the Schrödinger equation, separate the contributions that involve only particles of one kind, for instance only electrons or only phonons.
(ii) Now in these subsets distinguish contributions in which simultaneously only coordinates from one particle (interaction with external fields) or two (interaction among like particles) appear. We have now separated the one- and two-particle contributions.
(iii) The rest of the Hamiltonian describes the interaction between different kinds of particle.

Having identified all these different contributions, we can express them in the occupation number space by multiplying from the left side with the Ψ^+s at the corresponding coordinates and from the right side with the Ψs. This order of multiplication seems a bit arbitrary, but only by keeping this order can we guarantee consistency of the equation of motion for particle and field operators. The integration is performed over all space variables. To illustrate this procedure, we assume a many-body Hamiltonian that is a superposition of N one-particle contributions, for instance free particles interacting with randomly distributed impurities located at positions R_1 with potential V_{imp}. The total Hamiltonian of such an N-particle system is then:

$$H(x_1, p_1, \ldots, n_N, p_N) = \sum_{i=1}^{i=N} h(x_i, p_i) \tag{1.8}$$

$$h(x, p) = -\frac{\hbar^2}{2m} \nabla^2 + \Sigma_1 V_{\text{imp}}(R_1 - x) \tag{1.9}$$

The sum runs over all impurities. The representation in occupation number space is then, according to the rules stated above:

$$H = \int dx\, \Psi^+(x) h(x, p) \Psi(x)$$

$$= \sum_{kk'} \int dx\, \varphi_k^* h(x, p) \varphi_{k'} a_k^+ a_{k'}$$

where we used Eqs. (1.6) for the representation of the field operators. From the properties of the wave functions φ_k we can express H as

$$H = \sum_k \varepsilon_k a_k^+ a_k + \sum_{kk'} V_{kk'} a_k^+ a_{k'} \tag{1.10}$$

The first term is diagonal in k and is proportional to the occupation number in this state. The second term also contains non-diagonal terms and causes the unperturbed eigenvalue ε_k to change. In particular, we have for the matrix elements:

$$\varepsilon_k = \frac{\hbar^2 k^2}{2m} \tag{1.11}$$

$$V_{kk'} = \sum_1 \int dx\, \varphi_k^*(x) V_{\text{imp}}(R_1 - x) \varphi_{k'}(x) \tag{1.12}$$

If we had considered an electron in a crystal, the wave functions would have been Bloch functions and the band index n would have been added to the state label k. Furthermore, the k-summations would have been restricted to the first Brillouin zone. But the format of Eq. (1.10) would not have changed.

In the following, we will always assume the occupation number representation of the Hamiltonian. We will work in the electron–phonon system in a semiconductor. We must therefore identify the following contributions in the Hamiltonian:

$$H = H_0^e + H_0^{ph} + H^{e-ext} + H^{e-ph} + H^{e-e} + H^{ph-ph} \tag{1.13}$$

H_0^e and H_0^{ph} are independent electrons and phonons with no mutual interactions.

$$H_0^e = \sum_k \varepsilon_k a_k^+ a_k \tag{1.14}$$

$$H_0^{phe} = \sum_s \omega_s b_s^+ b_s \tag{1.15}$$

Here we have introduced ω_s for the energy dispersion of the phonons, and the phonon creation and destruction operators, b_s^+ and b_s, respectively. The state label s contains the wave number and polarization of the phonon. H^{e-ph} includes the electron–phonon interaction which will be of special importance in the later chapters.

$$H^{e-ph} = \sum_{kq} m_q a_k^+ a_{k-q}(b_q + b_{-q}^+) \tag{1.16}$$

Each term in the sum describes a possible event of scattering an electron from state k to a state $k-q$ by emission or absorption of a phonon. This is illustrated in Fig. 2.

The matrix element m_q depends on the kind of phonons the electrons interact with (acoustic, optical) and on the kind of coupling (deformational potential, piezoelectric, polar). If required, we will specify this matrix element in the context where it is used. H^{e-ext} is that part of the Hamiltonian which we evaluated in Eq. (1.12). It will be of special interest when we consider the influence of an external electric field. H^{e-e} and

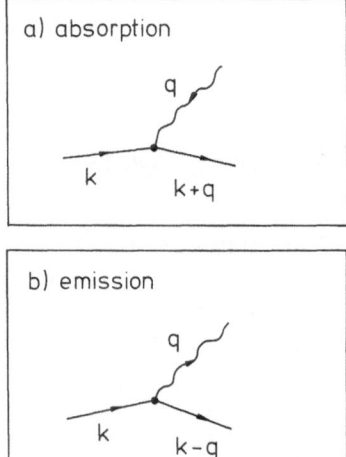

Fig. 2 Fundamental vertex for electron–phonon scattering. (a) An electron with momentum k and energy ε_k absorbs a phonon with momentum q and energy ω_q. (b) An electron with momentum k and energy ε_k emits a phonon with momentum q and energy ω_q

H^{ph-ph} stand for the interaction between electrons and between phonons, treated separately. In the subsequent treatment, we will neglect the phonon–phonon interaction and most of the electron–electron interaction. However, when we address the problem of Auger recombination or impact ionization we must be aware that these are direct consequences of the electron–electron interaction.

1.2.2 Green's Functions for Electrons

Once the formal language of the occupation number space has been established, the next step will be to find the link to measurable quantities. This is provided by calculating the expectation values of the quantum operators in a convenient thermodynamical ensemble. In equilibrium we are working with systems that have a constant temperature T. However, we have to allow for systems with a different but fixed particle number N. Therefore the expectation values are most conveniently calculated within the grand canonical ensemble. Readers who are not familiar with this approach should consult R. K. Pathria's *Statistical Mechanics* (Pathria 1972), for instance. The thermodynamic expectation value for any operator O in the grand canonical ensemble is

$$\langle O \rangle = \{\exp(-K\beta)O\} = \exp(\Omega\beta) \sum_{N,n_s} \langle \ldots | \exp(-K\beta)O | \ldots \rangle$$

$$(1.16)$$

Here we used $\beta = 1/k_B T$, where Ω is the thermodynamic potential which is given by,

$$\exp(-\Omega\beta) = \sum_{N,n_s} \langle \ldots | \exp(-K\beta) | \ldots \rangle \qquad (1.17)$$

and assures the correct normalization. The operator K is related to the Hamiltonian of the system,

$$K = H - \mu N \qquad (1.18)$$

with the number operator $N = \Sigma N_s$, with N_s from Eq. (1.1). The chemical or Fermi potential μ measures the energy required to remove one particle from an N-particle system. In systems with a variable particle number, such as the phonon system, we have $\mu = 0$. Applying K on an arbitrary state vector in occupation number space yields, for a state containing N particles:

$$K|n_1, \ldots n_N\rangle = (H - \mu N)|n_1, \ldots n_N\rangle = (E_N - \mu N)|n_1 \ldots n_N\rangle$$

$$E_N = \sum_k \varepsilon_k n_k \qquad (1.19)$$

$$N = \sum_k n_k$$

The trace in Eqs. (1.16) and (1.17) is calculated over all states belonging to a fixed number of particles N and over all N. The equation of motion for K is also the Heisenberg equation (1.7). The Green's function is defined as the thermodynamic expectation value of the time-ordered product of the field operators Ψ and Ψ^+:

$$G(xt, x't') = - i \langle T_\tau \Psi(xt) \Psi^+(x't') \rangle \qquad (1.20)$$

The time-ordering operator T_τ arranges the field operators such that the one with the earlier time is to the right of that having the later time. The physical interpretation of the Green's function is most conveniently given by separating its two components:

$$
\begin{aligned}
G^<(xt, x't') &= i \langle \Psi^+(x't') \Psi(xt) \rangle & t' > t \\
G^>(xt, x't') &= - i \langle \Psi(xt) \Psi^+(x't') \rangle & t > t'
\end{aligned}
\qquad (1.21)
$$

The first component $G^<$ describes the evolution of a one-particle state in an $N - 1$ particle system if at time t one particle was removed at position x from an N-particle system and was monitored again at a later time t' at position x'. In other words, a hole is created in an $N - 1$ particle system at time t and what is left of that hole is tested at a later time t'. We note that in the limit $t' \to t$ and $x' \to x$ $G^<$ coincides with the expectation value of the number operator, which is the particle density $n(x, t)$. Similar to $G^<$, the second component $G^>$ describes the correlation of a particle created at t' and monitored at a later instant t in a $N + 1$ particle system. The information contained in the Green's function is therefore related to one-particle properties of $N + 1$ particle systems and hole properties of $N - 1$ particle system.

Because N is usually a very large number, we are simply saying that the Green's function contains information about the one-particle (hole) properties of a many-particle system. The method using the Green's function is therefore an adequate tool to study the concept of a "particle" in such systems. Starting from a well defined set of one-particle states, we will see that including the coupling to impurities or phonons will destroy these particle states. However, if this coupling is weak enough we will then detect properties in the Green's function which are very similar to those of particles, thus permitting us to introduce the concept of quasi-particles. At the end of this section we will be able to provide a formal justification for the quasi-particle concept.

To continue, we insert the field operators into Eqs. (1.20) and (1.21)

$$G^<(xt, x't') = i \sum_{kk'} \varphi_{k'}^*(x') \varphi_k(x) \langle a_{k'}^+(t') a_k(t) \rangle \qquad (1.22)$$

$$G^>(xt, x't') = - i \sum_{kk'} \varphi_{k'}^*(x') \varphi_k(x) \langle a_k(t) a_{k'}^+(t') \rangle$$

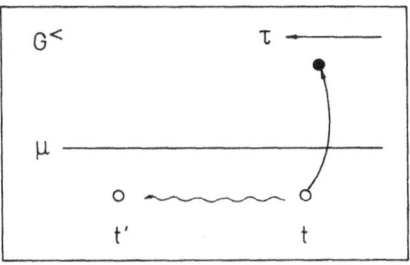

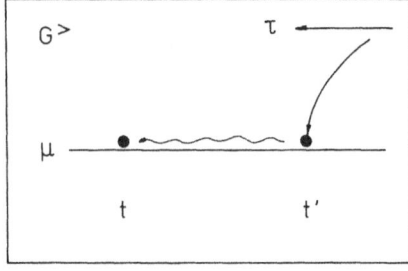

Fig. 3 The components of the Green's function G. All states below Fermi's energy are occupied with electrons at $T = 0$. Top $t' > t - G^<$: A hole is created (electron removed) at time t and evolves to time t'. Bottom $t > t' - G^>$: An electron is added at time t' and observed at time t

which allows us to obtain an alternative set of Green's functions which is more convenient for our further discussion

$$G_{kk'}^<(t, t') = i\langle a_{k'}^+(t')a_k(t)\rangle$$
$$G_{kk'}^>(t, t') = -i\langle a_k(t)a_{k'}^+(t')\rangle$$
(1.23)

Inserting the time-dependence of the number operators following from Eq. (1.7),

$$a_k(t) = \exp(itK/\hbar)a_k(0)\exp(-itK/\hbar)$$
(1.24)

we find that $G_{kk'}$ depends only on the time difference $\tau = t - t'$. In a spatially homogeneous system, we have $G_{kk'} \sim \delta_{kk'}$. In this case we can express Eqs. (1.23) as (for convenience we will from now on use units such that $\hbar = 1$):

$$G_k^<(\tau) = i\exp(\Omega\beta) \sum_{m,n} \exp(-K_m\beta)$$
$$\cdot \exp(-i(K_m - K_n)\tau)|\langle m|a_k^+|n\rangle|^2$$
$$G_k^>(\tau) = -i\exp(\Omega\beta) \sum_{m,n} \exp(-K_m\beta)$$
$$\cdot \exp(i(K_m - K_n)\tau)|\langle m|a_k|n\rangle|^2$$
(1.25)

The summation index $n(m)$ stands for a state vector with $N_{n(m)}$ particles occupying the single particle states k. The only non-zero matrix elements that we have are between states with $N_m = N_n - 1$ for the first equation and $N_m = N_n + 1$ for the second. Swapping the summation labels n and m in the second equation we have:

$$
G_k^<(\tau) = i\exp(\Omega\beta) \sum_{m,n} \exp(-K_m\beta)
$$

$$
\cdot\exp(-i(K_m - K_n)\tau)|\langle m|a_k^+|n\rangle|^2, \quad \tau < 0 \tag{1.26}
$$

$$
G_k^>(\tau) = -i\exp(\Omega\beta) \sum_{m,n} \exp(-K_n\beta)
$$

$$
\cdot\exp(-i(K_m - K_n)\tau)|\langle m|a_k^+|n\rangle|^2, \quad \tau > 0
$$

Here we have used

$$
|\langle m|a_k|n\rangle|^2 = \langle m|a_k|n\rangle\langle n|a_k^+|m\rangle
$$

$$
= \langle n|a_k^+|m\rangle\langle m|a_k|n\rangle = |\langle n|a_k^+|m\rangle|^2
$$

The Green's function is now

$$
G_k(\tau) = \theta(-\tau)G_k^<(\tau) + \theta(\tau)G_k^>(\tau) \tag{1.27}
$$

A Fourier transform with respect to time will reveal some very interesting analytical properties. For the first component we obtain

$$
G_k^<(\omega - i\delta) = \int_{-\infty}^{0} d\tau\exp(i(\omega - i\delta)\tau)G_k^<(\tau)
$$

$$
= \exp(\Omega\beta) \sum_{m,n} \frac{\exp(-K_m\beta)|\langle m|a_k^+|n\rangle|^2}{\omega - K_m + K_n - i\delta} \tag{1.28}
$$

and for the second

$$
G_k^>(\omega + i\delta) = \int_{0}^{\infty} d\tau\exp(i(\omega + i\delta)\tau)G_k^>(\tau)
$$

$$
= \exp(\Omega\beta) \sum_{m,n} \frac{\exp(-K_n\beta)|\langle m|a_k^+|n\rangle|^2}{\omega - K_m + K_n + i\delta}
$$

$$
= \exp(\Omega\beta) \sum_{m,n} \frac{\exp(-K_m\beta)|\langle m|a_k^+|n\rangle|^2}{\omega - K_m + K_n + i\delta}
$$

$$
\cdot\exp((K_m - K_n)\beta) \tag{1.29}
$$

The convergence factor $\delta > 0$ indicates that $G^>$ is an analytical function in the upper complex ω plane and $G^<$ in the lower one, and has to be taken to zero at the end of the calculation. To proceed, we have to use the identity:

$$\frac{1}{x \pm i\delta} = \frac{P}{x} \mp i\delta(x)$$

where P indicates the principal value and $\delta(x)$ is the Dirac δ-function. Comparing the imaginary part of Eqs. (1.28) and (1.29) yields:

$$\mathrm{Im}\, G_k^< (\omega) = \tfrac{1}{2} n_F(\omega) A(k, \omega) \tag{1.30}$$

$$\mathrm{Im}\, G_k^> (\omega) = -\tfrac{1}{2}(1 - n_F(\omega)) A(k, \omega) \tag{1.31}$$

with

$$A(k, \omega) = 2\pi(1 + \exp(\omega\beta)) \exp(\Omega\beta) \sum_{n,m}$$
$$\cdot \exp(-K_m\beta)|\langle m|a_k^+|n\rangle|^2 \delta(\omega - K_m + K_n) \tag{1.32}$$

and the Fermi distribution function

$$n_F(\omega) = (1 + \exp(\omega\beta))^{-1}$$

Equation (1.32) is the main result of this section: it is called the spectral function of the Green's function G. It contains all the information necessary to calculate the one-particle properties of a many-particle system. The calculation of this function is the central task of many-body solid-state physics and is directly related to optical experiments in particular. Before we demonstrate that its knowledge is sufficient to calculate Green's function itself, we have to point out some properties which will be important in the next section, where some results of this section will be generalized to nonequilibrium quantum mechanics. First of all we can show from Eq. (1.32) that $A(k, \omega) \geq 0$. Integration over ω gives

$$\int_{-\infty}^{\infty} \frac{d\omega}{2\pi} A(k, \omega) = \exp(\Omega\beta) \sum (\exp(-K_m\beta)$$
$$+ \exp(-K_n\beta)|\langle m|a_k^+|n\rangle|^2$$
$$= \exp(\Omega\beta) \sum \exp(-K_m\beta)\langle m|a_k^+ a_k + a_k a_k^+|m\rangle$$
$$= \exp(\Omega\beta) \sum \exp(-K_m\beta) = 1 \tag{1.33}$$

where we used the completeness relation of the basis set $\sum |n\rangle\langle n| = 1$ in the second line and the commutator relation (1.4) for fermion number operators in the third line. Equation (1.33) tells us that whatever happens in the system, the integration over ω always gives the same constant value. The physics behind this sum rule is very closely correlated with the law of particle conservation. Another important relationship is obtained by combining Eqs. (1.30) and (1.31):

$$\mathrm{Im}\, G_k^< (\omega) - \mathrm{Im}\, G_k^> (\omega) = \tfrac{1}{2} A(k, \omega) \tag{1.34}$$

It is instructive to calculate the spectral function directly from Eq. (1.31), which is possible for a non-interacting electron gas. For the grand canonical Hamiltonian K we have with Eq. (1.14),

$$K = \sum_k (\varepsilon_k - \mu) a_k^+ a_k \tag{1.35}$$

The matrix element is, according to Eq. (1.5)

$$|\langle m|a_k^+|n\rangle|^2 = 1 - n_k = m_k \tag{1.36}$$

and

$$K_m - K_n = \varepsilon_k - \mu$$

where we used the fact that $|m\rangle$ is an $N + 1$-particle state if $|n\rangle$ is an N-particle state. Due to Eq. (1.35), the trace over the particle number N separates into a product of sums over the occupation numbers which is also true for the thermodynamic potential Eq. (1.17):

$$A(k, \omega) = (1 + \exp(\omega\beta))$$
$$\cdot \frac{\sum m_k \exp(-(\varepsilon_k - \mu)\beta m_k \cdot \sum \exp(-(\varepsilon_{k1} - \mu) m_{k1} \cdots}{\sum \exp(-(\varepsilon_{k1} - \mu)\beta m_k \cdot \sum \exp(-(\varepsilon_{k1} - \mu) m_{k1} \cdots}$$
$$\cdot 2\pi \cdot \delta(\omega - \varepsilon_k + \mu)$$

In the case of fermions, the occupation number is only 0 or 1 and so the final result is:

$$A(k, \omega) = 2\pi\delta(\omega - \varepsilon_k + \mu) \tag{1.37}$$

The spectral function for a non-interacting electron gas is a delta function. We will show later that interactions have the tendency to broaden this delta function.
To find the relationship between the spectral function and the Green's function G or its components $G^<$ and $G^>$, we need to investigate some surprising properties of their Fourier transforms. For the Fourier transform of $G^<$ we have

$$G^<(\omega - i\delta) = \int_{-\infty}^0 d\tau \exp(i(\omega - i\delta)\tau) G^<(\tau)$$
$$= \int_{-\infty}^\infty \frac{d\omega'}{2\pi} G^<(\omega') \int_{-\infty}^0 d\tau \exp(i(\omega - \omega' - i\delta)\tau)$$
$$= i \int_{-\infty}^\infty \frac{d\omega'}{2\pi} \frac{G^<(\omega')}{\omega' - \omega + i\delta} \tag{1.38}$$

and similarly, we obtain for $G^>$,

$$G^>(\omega + i\delta) = i \int_{-\infty}^{\infty} \frac{d\omega'}{2\pi} \frac{G^>(\omega')}{\omega' - \omega - i\delta} \tag{1.39}$$

Here, we have omitted the state labels as they are not important for the present analysis. Separating real and imaginary parts we obtain from Eqs. (1.38) and (1.39):

$$\mathrm{Re}\, G^<(\omega) = - \int_{-\infty}^{\infty} \frac{d\omega'}{2\pi} P \frac{2\,\mathrm{Im}\, G^<(\omega')}{\omega' - \omega} \tag{1.40}$$

$$\mathrm{Re}\, G^>(\omega) = \int_{-\infty}^{\infty} \frac{d\omega'}{2\pi} P \frac{\mathrm{Im}\, G^>(\omega')}{\omega' - \omega} \tag{1.41}$$

Using this relationship in Eqs. (1.38) and (1.39), we finally obtain:

$$G^<(\omega) = \int_{-\infty}^{\infty} \frac{d\omega'}{2\pi} \frac{2\,\mathrm{Im}\, G^<(\omega')}{\omega - \omega' - i\delta} \tag{1.42}$$

$$G^>(\omega) = - \int_{-\infty}^{\infty} \frac{d\omega'}{2\pi} \frac{2\,\mathrm{Im}\, G^>(\omega')}{\omega - \omega' + i\delta} \tag{1.43}$$

Now we can use Eqs. (1.30) and (1.31) to find the desired result which links the Green's function to the spectral function.

$$G_k^<(\omega) = \int_{-\infty}^{\infty} \frac{d\omega'}{2\pi} \frac{n_F(\omega') A(k, \omega')}{\omega - \omega' - i\delta} \tag{1.44}$$

$$G_k^>(\omega) = \int_{-\infty}^{\infty} \frac{d\omega'}{2\pi} \frac{(1 - n_F(\omega')) A(k, \omega')}{\omega - \omega' + i\delta} \tag{1.45}$$

The physical meaning of these equations becomes clearer if we transform them back into the time domain.

$$G_k^<(\tau) = i \int_{-\infty}^{\infty} \frac{d\omega}{2\pi} n_F(\omega) A(k, \omega) \exp(-i\omega\tau)\theta(\tau) \tag{1.46}$$

$$G_k^>(\tau) = - i \int_{-\infty}^{\infty} \frac{d\omega}{2\pi} (1 - n_F(\omega)) A(k, \omega) \exp(-i\omega\tau)\theta(-\tau) \tag{1.47}$$

From the first of these Eqs (1.23) we can conclude that in the limit $\tau \to 0$ it coincides with the expectation value of the number operator for a system in equilibrium. Performing this limit we obtain

$$n_k = \int_{-\infty}^{\infty} \frac{d\omega}{2\pi} n_F(\omega) A(k, \omega) \tag{1.48}$$

For the case of an electron gas, we recover the well known Fermi distribution function if we use the spectral function from Eq. (1.37). We also see that if the spectral function is sufficiently broadened, the electron gas result is no longer valid. This is a first indication of the failure of the particle concept. Another conclusion that we can draw from Eqs. (1.46) and (1.47) is an interpretation of the Fourier transforms, which tells us that the following should hold:

$$G^{<}(\omega) = i n_F(\omega) A(k, \omega) \tag{1.49}$$

$$G^{>}(\omega) = - i(1 - n_F(\omega)) A(k, \omega) \tag{1.50}$$

This seems to be in contradiction with Eqs. (1.44) and (1.45). These can in fact be derived from the above equations by applying the Cauchy residuum theorem, which assumes that certain analytical properties of $A(k, \omega)$ and $n_F(\omega)$ are fulfilled. Equations (1.49) and (1.50) are very important in the context of the non-equilibrium case that we shall discuss later. Before we turn to that subject, let us briefly discuss how the spectral function is calculated.

1.2.3 Self-Energy

Perturbation theory is a very important method for calculating approximate solutions in wave mechanics. A perturbation theory also exists for the many-particle systems. The central expression will be the self-energy Σ, which is also necessary for calculating the spectral function. The self-energy is most conveniently introduced by studying the equation of motion for the Green's function Eq. (1.20). To accomplish this, we calculate the time derivative of G:

$$\frac{\partial}{\partial \tau} G(xx', \tau) = - i\delta(\tau)\delta(x - x') - i\left\langle \frac{\partial}{\partial t} \Psi(x\tau)\Psi^{+}(x'0) \right\rangle \theta(\tau)$$

$$+ i\left\langle \Psi^{+}(x'0) \frac{\partial}{\partial \tau} \Psi(x\tau) \right\rangle \theta(-\tau) \tag{1.51}$$

If we replace the time derivative of the field operators by their equation of motion Eq. (1.7) we get

$$\frac{\partial}{\partial \tau} G(xx', \tau) + i\langle [\Psi(x\tau), H] \Psi^{+}(x'0) \rangle \theta(\tau)$$

$$- i\langle \Psi^{+}(x'0)[\Psi(x\tau), H] \rangle \theta(-\tau)$$

$$= - i\delta(\tau)\delta(x - x') \tag{1.52}$$

Let us assume that we can separate a zero-order solution in the Hamiltonian, which is the essential idea of any perturbative approach, then we have

$$H = H_0 + H_{\text{int}} \tag{1.53}$$

H_0 represents the free-particle Hamiltonian, for instance, and we put all the interactions in H_{int}. With Eq. (1.9a) and the commutation rules for the field operators

$$\Psi(xt\,)\Psi^+(x't) + \Psi^+(x't)\Psi(xt) = \delta(x - x')$$

we obtain the following for the commutator in Eq. (1.52)

$$[\Psi(xt), H] = h(x, p, t)\Psi(xt) + [\Psi(xt), H_{\text{int}}] \tag{1.54}$$

Inserting Eq. (1.54) into Eq. (1.52) and re-establishing the time-order operator T_τ again we finally obtain the equation of motion for the Green's function G

$$\left(i\frac{\partial}{\partial \tau} - h(x, p, \tau) \right) G(xx', \tau) + i\langle T_\tau[\Psi(x\tau), H_{\text{int}}]\Psi^+(x'0)\rangle$$

$$= \delta(\tau)\delta(x - x') \tag{1.55}$$

It seems that we have not gained anything so far: to calculate the time-ordered expectation value G we have to solve an equation which itself contains an even more complicated time-ordered product. Although this expectation value is very similar to the Green's function, it is in fact very different. But to solve this equation approximately, the complicated expectation value is replaced by a functional identity which defines the self-energy operator Σ

$$i\langle T_\tau[\Psi(x\tau), H_{\text{int}}]\Psi^+(x'0)\rangle \equiv \int d\underline{x} \int_{-\infty}^{\infty} d\underline{\tau}\,\Sigma(x\underline{x}, \tau - \underline{\tau})G(\underline{x}x', \underline{\tau})$$

$$\tag{1.56}$$

To provide a suitable expression for this self-energy is the task of many-body perturbation theory. A systematic expansion scheme of the self-energy in the interaction strength is, for example, formalized by the Feynman diagram technique which was developed in quantum field theory. We will not give a detailed account on this formalism but refer to the textbooks mentioned above. However, a brief outline of the basic ideas will be given in the next section to understand the non-equilibrium case. Putting Eqs. (1.56) into Eq. (1.55), we derive the Dyson equation for the Green's function G

$$\left(i\frac{\partial}{\partial \tau} - h(x, p, \tau) \right) G(xx', \tau)$$

$$+ \int d\underline{x} \int_{-\infty}^{\infty} d\underline{\tau}\,\Sigma(x\underline{x}, \tau - \underline{\tau})G(\underline{x}x', \underline{\tau}) = \delta(\tau)\delta(x - x') \tag{1.57}$$

In this equation we can now identify a zero-order solution g_0 by its inverse,

$$g_0^{-1} = i\frac{\partial}{\partial\tau} - h \tag{1.58}$$

which permits us to write Eq. (1.57) symbolically as:

$$G = g_0 + g_0 \Sigma G \tag{1.59}$$

This equation has to be interpreted as an operator identity. Products are meant to be integrals, as shown explicitly in Eq. (1.57), for instance. We find the formal solution of Eq. (1.57) to be

$$G = \frac{1}{g_0^{-1} - \Sigma} \tag{1.60}$$

In real space variables $(x\tau)$, such an algebraic operation is not permitted. However, if we have a spatially homogeneous system, the Fourier transform of Eq. (1.57) with respect to x and τ will turn this equation into an algebraic one in the variables $(k\omega)$.

We have already learned that the Green's function G has some remarkable analytical properties in ω-space. In general, it has poles on both sides of the real ω axes in the complex ω-plane. This complex analytical behavior is a disadvantage. It is more convenient to consider functions with simpler structures. Therefore two different kinds of Green's functions that do not have a simple physical interpretation are considered: the retarded and advanced Green's functions G^r and G^a

$$G^r(xx',\tau) = -i\langle[\Psi(x\tau),\Psi^+(x'0)]_+\rangle\theta(\tau) \tag{1.61}$$

$$G^a(xx',\tau) = i\langle[\Psi(x\tau),\Psi^+(x'0)]_+\rangle\theta(-\tau) \tag{1.62}$$

The plus sign means the anti-commutator $[\Psi,\Psi^+]_+ = \Psi\Psi^+ + \Psi^+\Psi$ of the operators Ψ and Ψ^+. From the properties of their Fourier transforms it is clear, as we shall show below, that the poles of G^r and G^a are restricted to either above or below the real ω-axes in the complex ω-plane. This fact will be of special importance for connecting the finite temperature perturbation approaches with the Green's functions in the real-time domain. As we did with the Green's function G, it is convenient for the discussion of the analytical properties to consider the Green's functions $G_k^r(\tau)$ and $G_k^a(\tau)$, which is done similarly to Eqs. (1.22) and (1.23). Calculating the Fourier transform of G^r with respect to τ we obtain

$$G_k^r(\omega) = \exp(\Omega\beta) \sum_{n,m} \frac{(\exp(-K_m\beta) + \exp(-K_n\beta))|\langle m|a_k^+|n\rangle|^2}{\omega + K_m - K_n + i\delta} \tag{1.63}$$

which shows that G^r has its poles in the lower half of the ω-plane. The imaginary part of G^r is given by

$$\operatorname{Im} G_k^r(\omega) = - \pi \exp(\Omega\beta) \sum_{n,m} (\exp(-K_m\beta)$$

$$+ \exp(-K_n\beta))|\langle m|a_k^+|n\rangle|^2 \delta(\omega + K_m - K_n)$$

which, after swapping the summation indices in the second summand, yields

$$\operatorname{Im} G_k^r(\omega) = -\tfrac{1}{2} A(k,\omega) \tag{1.64}$$

where A is the spectral function defined in Eq. (1.32). We are now able to calculate the spectral function once we know the retarded Green's function G^r. This function in turn also obeys a Dyson equation, which can easily be shown by following exactly the same steps as we used before to derive Eq. (1.59). G^r is then given by

$$G^r = \frac{1}{(g_0^r)^{-1} - \Sigma^r} \tag{1.65}$$

In comparison to the solution for G, we now also have a retarded self-energy, which is obtained with an analogous decoupling scheme as Σ. However, there is no direct access for calculating Σ^r by perturbation theory. It is possible to obtain Σ^r by an analytic continuation of the finite temperature perturbation approach, as will be discussed in the next section. In a spatially homogeneous system, Eq. (1.65) turns into an algebraic identity in Fourier space. Taking the free particle as a zero-order solution and restoring the variables, we obtain

$$G_k^r(\omega) = \frac{1}{\omega - \varepsilon_k + \mu - \Sigma_k^r(\omega)} \tag{1.66}$$

With Eq. (1.64), we can now express the spectral function with the self-energy Σ^r

$$A(k,\omega) = \frac{-2\operatorname{Im}\Sigma_k^r(\omega)}{(\omega - \varepsilon_k + \mu - \operatorname{Re}\Sigma_k^r(\omega))^2 + (\operatorname{Im}\Sigma_k^r(\omega))^2} \tag{1.67}$$

From the fact that the spectral function is positive, we can conclude $\operatorname{Im}\Sigma^r \leq 0$! If we take the limit $\Sigma \to 0$ in Eq. (1.67) we discover a representation of the Dirac δ-function

$$\lim_{\Sigma^r \to 0} A(k,\omega) = 2\pi \lim_{\eta \to 0} \frac{1}{\pi} \frac{\eta}{(\omega - \varepsilon_k + \mu)^2 + \eta^2} = 2\pi\delta(\omega - \varepsilon_k + \mu) \tag{1.68}$$

The case $\Sigma^r = 0$ disregards all interactions and takes us back to the zero-order solution, the free particle Eq. (1.37). Together with Eqs. (1.46) and (1.47), we can now investigate the influence of the interactions on the particle concept. Inserting the free particle spectral function into these equations, we obtain the following for $g^<$, for instance, if we consider the limit $T \to 0$ in which the Fermi distribution function degenerates to a simple step function:

$$g_k^<(\tau) = i \exp(-i(\varepsilon_k - \mu)\tau)\theta(\mu - \varepsilon_k) \quad \tau < 0 \tag{1.69}$$

This result means that the hole that we have created at $\tau = 0$ propagates in time with constant amplitude. The correlation to find the hole after its creation is independent of τ. This means we have a perfect hole with an infinite life. Let us now consider the case of interactions in the system. For simplicity, we will only consider a self-energy with a constant imaginary part $-\Gamma > 0$. The real part is zero, which is no serious restriction. To monitor the propagation of the hole, we must now look at the following expression, again for $T \to 0$:

$$G_k^<(\tau) = i \exp(-i(\varepsilon_k - \mu)\tau) \int_{-\infty}^{\mu - \varepsilon_k} \frac{d\omega}{2\pi} \frac{2\Gamma}{\omega^2 + \Gamma^2} \exp(-i\omega\tau) \tag{1.70}$$

This integral cannot be represented in a closed form. However, a meaningful approximation can be obtained if we consider the limit $|\mu - \varepsilon_k| \gg \Gamma$. The upper integration limit can then be replaced by $\pm \infty$. If the lower sign holds, the integral is zero and if the upper sign holds the integral can be calculated with Cauchy's integral theorem to give,

$$G_k^<(\tau) = i \exp(-i(\varepsilon_k - \mu)\tau) \exp(\Gamma\tau)\theta(\mu - \varepsilon_k) \quad \tau < 0 \tag{1.71}$$

We notice that the interactions cause the hole to decay. It is no longer a stable state. Only if Γ is small enough can we identify $G^<$ with a propagating hole for a period of time $1/\Gamma$. A similar calculation could have been performed with the propagator $G^>$. In that case we would have monitored the time development of a particle placed on a free state outside the Fermi sphere. In conclusion, we can say that the imaginary part of the retarded self-energy Σ^r corresponds to the life time of a particle excitation of a many-body system. Only if this life time is sufficiently long compared with the excitation energy $|\varepsilon_k - \mu|$ or $|\varepsilon_k - \mu|/\Gamma \gg 1$ is the particle concept valid. An intuitive interpretation of this concept is given by the spectral function. We have seen that for an interaction-free system it is a δ-function with zero width. Interactions will smear the δ-function out and it will have a half-width proportional to the imaginary part of Σ^r. It is important, however, to remember that its ω-integration is always constant. This means that the maximum of the spectral function decreases with increasing interaction strength. This behavior is illustrated in Fig. 4.

Owing to the finite width, we will find a number of possible ε_k values for each energy parameter ω and vice versa. Considering the scattering of particles in which energy and momentum are exchanged, this means that energy and momentum conservation are no longer sharply defined as in

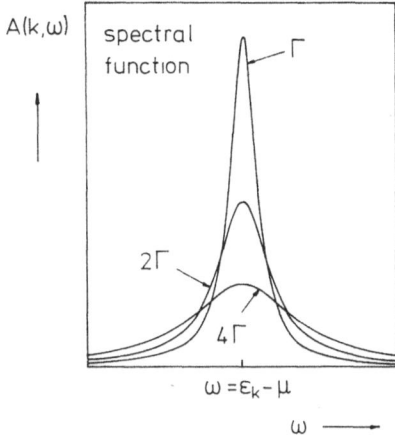

Fig. 4 The spectral function for different imaginary parts of the electrons self energy $\Gamma = -2\Sigma'$. The area enclosed is normalized to one. The spectral function has a maximum at the quasi-particle energy $\omega = \varepsilon_k - \mu$

classical systems. We have to consider each separately. The important feature we have learned is that in interacting many-particle systems, the energy ω and momentum k have to be considered as independent variables. Only if the line width of the spectral function is small do we have $\omega = \varepsilon_k$.

1.2.4 Perturbation Theory

In the previous section we established the language of many-body physics and studied some important properties of Green's function. The self-energy was especially important in that discussion. We did not yet look into the problem of calculating Σ. A rigorous approach is beyond the scope of this work. But to understand the non-equilibrium case, which is more important for our purposes, we have to present the principal ideas which lead to a systematic calculation scheme for Σ. The dynamics of the system are described by the solution of the Heisenberg equation (1.7) for the operators. We have

$$O_H(t) = \exp(iHt)O(t = 0)\exp(-iHt) \tag{1.72}$$

and the state vectors are constant in time in the Heisenberg picture, which is indicated by the subscript H.

$$\frac{\partial}{\partial t}\,|\,,t\rangle_H = 0 \tag{1.73}$$

For a perturbative approach, it is more convenient to describe the time

development in the interaction picture because it separates the contributions of the interaction term in the Hamiltonian from the zero-order
problem, which has a known solution. In the interaction picture, indicated
by the index I, the operators evolve according to,

$$i \frac{\partial}{\partial t} O_I(t) = [O_I, H_0] \tag{1.74}$$

and the state vectors

$$i \frac{\partial}{\partial t} |,t\rangle_I = H_{I,\text{int}}(t) |,t\rangle_I \tag{1.75}$$

where the time dependence of $H_{I,\text{int}}$ is given by solving Eq. (1.74). For an
arbitrary operator O_I, this is

$$O_I(t) = \exp(iH_0 t) O_H(t=0) \exp(-iH_0 t) \tag{1.76}$$

where the subscript H means to take the solution of the Heisenberg
equation at $t = 0$. A general solution of Eq. (1.75) defines the unitary time
evaluation operator $U(t, t')$:

$$|,t\rangle_I = U(t,t') |,t'\rangle_I = T_\tau \exp\left(-i \int_{t'}^{t} d\tau' H_{I,\text{int}}(\tau')\right) |,t'\rangle_I \tag{1.77}$$

The expectation value $\langle O \rangle$ is independent of the representation of the
system's time development because different representations are linked by
unitary transformations.

$$\langle O \rangle_t = \langle 0, |\exp(iHt)O(t=0)\exp(-iHt)|, 0\rangle$$
$$= \langle 0, |U(0,t)\exp(iH_0 t)O(t=0)\exp(-iH_0 t)U(t,0)|, 0\rangle$$

This allows the identification

$$\exp(-iHt) = \exp(-iH_0 t) U(t,0) \tag{1.78}$$

In the evaluation of the expectation values for the Green's function, Eq.
(1.16), we have to distinguish two different approaches. For zero temperature, it reduces to the matrix element in the many-body ground state. For
this limit, an analysis in the real-time domain is possible. The time-ordering
operator T_τ is now defined on a straight line coming from very early times
$-\infty$ and proceeding to very remote times toward $+\infty$.
It is now assumed that all interactions are switched on adiabatically from
$-\infty$ to full strength at $t = 0$ and then turn adiabatically off again when
reaching $t \to \infty$. In this sense, adiabatic switching means that the ground
state for the very early time $t \to -\infty$ $|0, -\infty\rangle$ is connected smoothly with
the ground state at $t \to \infty$ $|0, \infty\rangle$. It can be proved that both ground states
are identical up to a phase factor $\exp(i\alpha)$ (Fetter and Walecka 1971). The
Hamiltonian of the system now has an artificial time-dependence of the

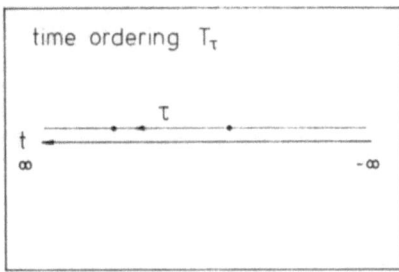

Fig. 5 The time-ordering operator T_τ acts on the physical time path that connects $-\infty$ and $+\infty$. Points to the left are considered later than points to the right

form:

$$H = H_0 + H_{int}\exp(-|\tau|\varepsilon) \tag{1.79}$$

where the limit $\varepsilon \to 0$ is understood at the end of the calculation. In the interaction picture, we have the following for Green's function:

$$G(x, x', \tau) = -i\,\frac{\langle 0,\ \infty|\,T_\tau S\Psi_I(x\tau)\Psi_I^+(x'0)|-\infty, 0\rangle}{\langle 0,\ \infty|S|-\infty, 0\rangle} \tag{1.80}$$

Here S is the S-matrix which is defined through the time development operator that connects the ground states $|,\ -\infty\rangle$ and $|,\ \infty\rangle$

$$S = \lim_{\substack{t\to\infty \\ t'\to-\infty}} U(t, t') = T_\tau\exp\left(-i\int_{-\infty}^{\infty} d\tau'H_{I,\,int}(\tau')\right) \tag{1.81}$$

The denominator in Eq. (1.80) accounts for the phase difference between the ground states $|,\ -\infty\rangle$ and $|,\ \infty\rangle$. Due to Wick's theorem, which is used to evaluate the expectation value of time-ordered products for $T = 0$, there is a systematic way of calculating a perturbation series of Eq. (1.80) with respect to H_{int}. This in turn leads to a Dyson equation equivalent to Eq. (1.57) with a specific self-energy Σ. For further details, we refer to the textbooks quoted above. A perturbation approach in connection with the non-equilibrium situation, which is very similar to the $T \to 0$ approach, will be briefly discussed in Appendix 1.

The situation is different for $T \neq 0$. Here the expectation value also involves matrix elements of the excited many-body states and not only the ground state. A thermodynamic perturbation theory is applied in this case. Starting again from the interaction representation we replace

$$i\tau \to \tau$$

which means that we are in the imaginary time domain.

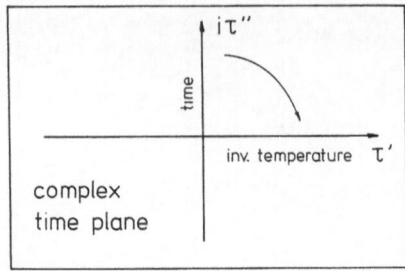

Fig. 6 In the thermodynamic perturbation theory the imaginary time (physical time) axis is mapped onto the real valued time (inverse temperature) axis

This imaginary time τ can be interpreted as an inverse temperature. A generalization of Eqs. (1.76), (1.77), and (1.78) gives

$$O_I(\tau) = \exp(K_0\tau)\exp(-K\tau)O_K(\tau)\exp(K\tau)\exp(-K_0\tau) \qquad (1.82)$$

$$U(\tau, \tau') = T_\tau \exp\left(-\int_{\tau'}^{\tau} d\tau'' K_{I,\text{int}}(\tau'')\right) \qquad (1.83)$$

$$\exp(-K\tau) = \exp(-K_0\tau)U(\tau, 0) \qquad (1.84)$$

which turns Eq. (1.82) into

$$O_K(\tau) = U(0, \tau)O_I(\tau)U(\tau, 0) \qquad (1.85)$$

The time-ordering operator is now acting on $-i\infty < it = \tau < i\infty$. The density operator for the grand canonical ensemble is obtained for $\tau = \beta$ in Eq. (1.84). We can now define an imaginary time or temperature Green's function analogously to Eq. (1.20)

$$G(x\tau, x'\tau') = -\frac{\Sigma\langle \ldots |\exp(-K\beta)T_\tau\Psi_K(x\tau)\Psi_{K+}(x'\tau')| \ldots\rangle}{\Sigma\langle \ldots |\exp(-K\beta))| \ldots\rangle} \qquad (1.86)$$

With Eqs. (1.84) and (1.85), this becomes the following in the interaction representation

$$G(x\tau, x'\tau') = -\frac{\Sigma\langle \ldots |\exp(-K_0\beta)T_\tau U(\beta, 0)\Psi_I(x\tau)\Psi_{I+}(x'\tau')| \ldots\rangle}{\Sigma\langle \ldots |\exp(-K_0\beta))U(\beta, 0)| \ldots\rangle} \qquad (1.87)$$

Like Eq. (1.80) for the zero temperature case, this equation will be the starting point for the perturbation approach. A remarkable property of the temperature Green's function is its anti-periodicity (for fermions) in the time arguments τ, τ',

$$G(x0, x'\tau') = -G(x\beta, x'\tau') \qquad (1.88)$$

which follows directly from Eq. (1.86) by using the invariance of the trace under cyclic permutation. The time arguments can be restricted to the interval $0 < \tau, \tau' < \beta$. G itself is a periodic function on an interval of length 2b. As a consequence of this periodicity, we can find a Fourier series representation of G with Fourier coefficients on a discrete set of points along the imaginary axis in the ω-plane,

$$G(x, x', \tau) = \beta^{-1} \Sigma \exp(-i\omega_n \tau) G(x, x', i\omega_n) \tag{1.89}$$

with $\omega_n = \pi \dfrac{(2n + 1)}{\beta}$, $n = \pm 1, \pm 2$, for fermions.

The thermodynamic perturbation theory for the Green's function for temperature again provides a self-energy Σ for the Dyson equation for G. Like G, Σ is also defined along the imaginary axis on the discrete points $i\omega_n$. We have learned that for the real-time Green's function, the retarded Green's function G^r and self-energy Σ^r are analytical functions in the upper ω-plane. The clue is now that we can identify G^r and Σ^r as analytical continuations of the temperature Green's function G and the self-energy Σ on to the upper branch of the real ω-axis

$$G(i\omega_n) \rightarrow G(\omega + i\delta) = G^r(\omega) \tag{1.90a}$$

$$\Sigma(i\omega_n) \rightarrow \Sigma(\omega + i\delta) = \Sigma^r(\omega) \tag{1.90b}$$

Equations (1.90) constitute the very important connection between real time and temperature Green's functions for thermodynamic equilibrium. To complete this discussion, we must mention that both temperature Green's function and the retarded Green's function can be expressed by the same spectral function A in Eq. (1.32)

$$F_k(\Omega) = \int_{-\infty}^{\infty} \frac{d\omega'}{2\pi} \frac{A(k, \omega')}{\Omega - \omega'} \tag{1.91}$$

$$G_k(i\omega_n) = F_k(i\omega_n), \qquad G_k^r(\omega) = F_k(\omega + i\delta)$$

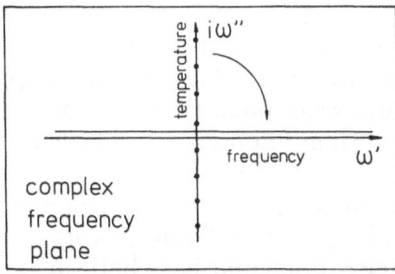

Fig. 7 The retarded Green's function G^r and self energy Σ^r are obtained by the analytical continuation of the temperature Green's function and self energy, that have poles $i\omega_n$ on the imaginary frequency axis, on to the real axis in the upper half-plane in frequency space

This whole section was devoted to studying the many-particle system in equilibrium. The last part in particular touched on the essential assumptions required for a perturbational analysis. In both the zero-temperature case and the finite-temperature formalism there is no explicit time dependence in the Hamiltonian, which is a feature of a system in equilibrium anyway. Equilibrium allows us to introduce a temperature Green's function for which a systematic perturbation theory can be developed. Once we know this function, we can obtain quantities that are important in the physical time domain by analytic continuation in Fourier space. For small deviations from equilibrium, it is therefore possible to expand the system's response to external forces around equilibrium. This is the task of linear response theory (Mahan 1981). Here, the external forces on the system are separated and the transport coefficients are calculated in equilibrium using the methods outlined above. In the following sections, we will generalize the concept of Green's functions to the case of Hamiltonians with an arbitrary explicit time-dependence which then allows a rigorous investigation of non-equilibrium phenomena in many-particle systems. This formulation will be formally very similar to our presentation so far and all the expressions we have discussed will be encountered again.

1.3 Non-Equilibrium Green's Functions

In trying to derive the Boltzmann equation, we set the stage in the preceding section for describing the particle concept in a many-particle system in terms of the spectral function. This analysis was restricted to the equilibrium situation. We will now proceed to generalize the methods we have learned to the non-equilibrium case, which is relevant for investigating transport phenomena. In the first place we must specify what we understand by non-equilibrium. We will allow external forces with an arbitrary time dependence and spatial variation to act on the electron–phonon system. Furthermore, we will consider the phonons as maintaining the equilibrium, which is characterized by a constant temperature T. Recent experimental and theoretical data justify this assumption for material with an indirect bandgap and weak polar coupling such as Si and Ge (Bordone *et al.* 1985). Non-equilibrium phonons are detected at low temperatures in the III–V compounds (Kocevar 1985) if the carrier density is sufficiently high to cause a significant energy transfer to the optical phonons. However, at temperatures exceeding that of liquid nitrogen, the phonon gas is in equilibrium with the environmental heat bath in this case too. However, this restriction is not necessary in principle. The analysis would then cover the coupled electron–phonon system in non-equilibrium, which is a very complicated problem, as even the equilibrium case has not been fully solved yet.

1.3.1 The Keldysh Formalism

As the non-equilibrium is defined by transient processes, it is obvious that we have to work in the physical time domain. Unfortunately, it seems that real time and finite temperature are mutually exclusive for perturbation theory. This is because we cannot apply the concept of adiabatically activating the interactions for an arbitrary time-dependent Hamiltonian. Therefore the S-matrix expansion, which is so useful for $T = 0$, does not apply. Let us recall that a central point in the S-matrix expansion was that the ground state for $t \to -\infty$ is identical with that of $t \to \infty$ by up to a phase factor, and this latter was precisely the ground-state expectation value of the S-matrix. The ingenious idea of Keldysh (Keldysh 1964, Rammer and Smith 1986) to overcome this problem was to alter the time path along which the time-ordering operator acts.

In the equilibrium case, this time path was the real-time axis extending from $-\infty$ to ∞. In the Keldysh formulation, the lowest time is again $-\infty$. Time then increases until a very long time t_0; this is long enough for all transient effects caused by introducing correlations among particles, due to activating the interactions, to die out and the systems response is only to the external perturbation. Now time increases further, but backwards with respect to physical time, until the starting point $-\infty$ is reached again. This time path guarantees that the start and end points are identical. The time-ordering operator T_τ^K will act on the upper branch $(+)$ of this path just like the conventional one. On the lower branch $(-)$, the time point closer to $-\infty$ will be further advanced than that closer to t_0. For all practical purposes, the long time t_0 is conveniently chosen to approach the limit $t_0 \to \infty$. We do not need the adiabatic switching of the interactions. The price we pay is a more complex time-ordering procedure which plays an important role in getting through the theory. The most important fact,

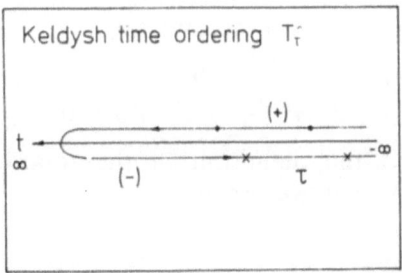

Fig. 8 The Keldysh time-ordering operator T_τ^K acts along a path that originates at $-\infty$ proceeds to a time t_0 and then returns to $-\infty$. For the upper branch of the contour $(+)$ times to the left are later than times to the right. For the lower branch $(-)$ times to the right are considered later than times to the left. A time on the lower branch is always later than a time on the upper branch

however, is that the perturbation analysis for the $T = 0$ case can be applied here immediately without major changes. That means that for a Green's function we can obtain a Dyson equation with a self-energy operator which is calculated with the use of a similar graphical analysis to that for the $T = 0$ case. A brief outline of this will be given in Appendix 1.

The Green's function is defined exactly as in the equilibrium case

$$G(xt, x't') = -i \langle T_\tau^K \Psi(xt)\Psi^+(x't') \rangle \tag{1.92}$$

According to the properties of the time-ordering operator, we can identify four different components

$$G^{++}(xt_+, x't'_+) = -i \langle T_\tau \Psi(xt_+)\Psi^+(x't'_+) \rangle \tag{1.93}$$

$$G^{+-}(xt_+, x't'_-) = i \langle \Psi^+(x't'_-)\Psi(xt_+) \rangle \tag{1.94}$$

$$G^{-+}(xt_-, x't'_+) = -i \langle \Psi(xt_-)\Psi^+(x't'_+) \rangle \tag{1.95}$$

$$G^{--}(xt_-, x't'_-) = -i \langle \underline{T}_\tau \Psi(xt_-)\Psi^+(x't'_-) \rangle \tag{1.96}$$

which differ in the positions of the time argument on the time path. For G^{++} and G^{--}, both time arguments lie on the same branch of the time path and T_τ^K therefore reduces to the ordinary T_τ and the anti-time ordering operator $\underline{T}_\tau$. G^{++} and G^{--} coincide with the conventional time or anti-time ordered Green's functions G and $\underline{G}$. For the other two components, the time arguments are located on different branches. Comparing Eqs. (1.94) and (1.95) with Eq. (1.23), we can identify G^{+-} with $G^<$ and G^{-+} with $G^>$. By omitting the variables, we can then make the following identification

$$G^{++} = G \tag{1.97}$$

$$G^{+-} = G^< \tag{1.98}$$

$$G^{-+} = G^> \tag{1.99}$$

$$G^{--} = \underline{G} \tag{1.100}$$

These four components are not independent of each other. Taking the difference of G and $G^<$, for instance, we get

$$G - G^< = \begin{cases} -i\langle \Psi\Psi^+ \rangle - i\langle \Psi^+\Psi \rangle = -i\langle [\Psi\Psi^+]_+ \rangle & t > t' \\ i\langle \Psi^+\Psi \rangle - i\langle \Psi^+\Psi \rangle = 0 & t < t' \end{cases} = G^r$$

where we have used the definition of the retarded Green's function Eq. (1.61). In the same way we obtain:

$$G - G^< = G^r \tag{1.101}$$

$$G - G^> = G^a \tag{1.102}$$

$$-\underline{G} + G^> = G^r \tag{1.103}$$

$$-\underline{G} + G^< = G^a \tag{1.104}$$

Combining these equations we have

$$G - G^> = -\underline{G} + G^<$$ (1.105)

and

$$G + \underline{G} = G^< + G^> = F$$ (1.106)

The four elements of the Keldysh Green's function can now be reduced to the three independent quantities G^r, G^a, and F. Representing the Keldysh G by a 2×2 matrix, the original components from Eqs. (1.97)–(1.100) can be transformed into the independent set G^r, G^a, and F with the aid of the unitary transformation

$$U = \tfrac{1}{2} \begin{vmatrix} 1 & -1 \\ 1 & 1 \end{vmatrix}$$ (1.107)

$$G \to U^+ G U = \begin{vmatrix} 0 & G^a \\ G^r & F \end{vmatrix}$$ (1.108)

The primary variables Σ, $\underline{\Sigma}$, $\Sigma^<$, and $\Sigma^>$ are calculated directly by the generalized perturbation theory and can be represented as Feynman diagrams, as discussed in Appendix 1. For our further analysis, however, it is more useful to utilize the independent quantities G^r, G^a, and F. In the primary variables, the Dyson equation for G is

$$\begin{vmatrix} G & G^< \\ G^> & \underline{G} \end{vmatrix} = \begin{vmatrix} g_0 & g_0^< \\ g_0^> & g_0 \end{vmatrix} + \begin{vmatrix} g_0 & g_0^< \\ g_0^> & g_0 \end{vmatrix} \begin{vmatrix} \Sigma & \Sigma^< \\ \Sigma^> & \underline{\Sigma} \end{vmatrix} \begin{vmatrix} G & G^< \\ G^> & \underline{G} \end{vmatrix}$$ (1.109)

and after applying the unitary transformation we obtain a system of three coupled equations for G^r, G^a, and F

$$G^r = g_0^r + g_0^r \Sigma^r G^r$$ (1.110)

$$G^a = g_0^a + g_0^a \Sigma^a G^a$$ (1.111)

$$F = G^r g_0^{r-1} f_0 g_0^{a-1} G^a + G^r \Omega G^a$$ (1.112)

Equations (1.110) and (1.111) are very similar to the corresponding Dyson equations for the equilibrium case. They primarily describe the influence of the interactions. The retarded self-energy Σ^r will again allow us to define a spectral function with finite width in a straightforward generalization of the equilibrium case. Equation (1.112) is a new relationship not yet familiar from the equilibrium formulation. The self-energies Σ^r, Σ^a, and Ω depend on all four primary self-energies Σ, $\underline{\Sigma}$, $\Sigma^<$, and $\Sigma^>$ in the following way:

$$\Sigma^r = \Sigma + \Sigma^< = -(\underline{\Sigma} + \Sigma^>)$$ (1.113)

$$\Sigma^a = \Sigma + \Sigma^> = -(\underline{\Sigma} + \Sigma^<)$$ (1.114)

$$\Sigma + \underline{\Sigma} = -(\Sigma^< + \Sigma^>) = \Omega$$ (1.115)

Equations (1.113) to (1.115) are obtained by applying the unitary transformation U to the Dyson equation (1.109). From the equilibrium case, we
know that the propagator $G^<$ is directly related to the particle density. By
inspecting Eqs. (1.94) and (1.98), we find that it has the same properties as
the equilibrium function. We will now concentrate our efforts on finding an
equation for this quantity. Replacing F by Eqs. (1.105) and (1.106) we
obtain

$$F = 2G^< - G^a + G^r \tag{1.116}$$

Inserting Eq. (1.116) into Eq. (1.112) and using Eqs. (1.110) and (1.111) for
the retarded and advanced Green's functions we find the following for $G^<$,

$$G^< = (1 + G^r \Sigma^r) g_0^< (1 + \Sigma^a G^a) - G^r \Sigma^< G^a \tag{1.117}$$

and a similar equation holds for $G^>$,

$$G^> = (1 + G^r \Sigma^r) g_0^> (1 + \Sigma^a G^a) - G^r \Sigma^> G^a \tag{1.118}$$

Equation (1.117) now gives the particle density $n(x, t)$ as a solution, if we
consider the limit $xt \to x't'$ in Eq. (1.94). The Boltzmann equation, however,
provides the distribution of particles in a 7-dimensional configuration space
(x, k, t). We will now show how this connection is made with $G^<$.

For the time being we will leave the analysis of Eq. (1.117), which, sur-
prisingly enough, is the generalized Boltzmann equation. To reveal this
connection we introduce the Wigner function $f(R, T, k, \omega)$, which is directly
related to Eq. (1.94). Here we use center-of-mass (R, T) and relative (r, τ)
coordinates

$$R = \tfrac{1}{2}(x + x'); \quad r = x - x' \tag{1.119}$$

$$T = \tfrac{1}{2}(t + t'); \quad \tau = t - t' \tag{1.120}$$

and calculate the Fourier transform with respect to (r, τ).

$$f(R, T, k, \omega) = -i \int dr \int_{-\infty}^{\infty} d\tau \exp(-ikr + i\omega\tau)$$

$$\cdot G^< (R + \tfrac{1}{2}r, T + \tfrac{1}{2}\tau; R - \tfrac{1}{2}r, T - \tfrac{1}{2}\tau) \tag{1.121}$$

or explicitly in the field operators Eq. (1.6)

$$f(R, T, k, \omega) = \int dr \int_{-\infty}^{\infty} d\tau \exp(-ikr + i\omega\tau)$$

$$\cdot \langle \Psi^+ (R - \tfrac{1}{2}r, T - \tfrac{1}{2}\tau) \Psi(R + \tfrac{1}{2}r, T + \tfrac{1}{2}\tau) \rangle \tag{1.122}$$

This function was first investigated by Wigner (1931) in a study of particle
correlation functions in many-particle systems. Although it is not a particle-
distribution function in a rigorous sense, it has many properties that allow

such an interpretation. The major drawback is that it cannot be shown for a general case that $f(R, T, k, \omega) \geq 0$, which is certainly a requirement for a density function. There are examples known for which $f(R, T, k, \omega) < 0$. The simplest example is the Wigner function for the harmonic oscillator. On the other hand $f(R, T, k, \omega)$ exhibits a feature absent in the classical distribution function $f(R, T, k)$, i.e. energy ω and momentum k are independent variables. We learned in the previous section that this is necessary in a system of particles subject to interactions. We will come back to this point at the end of this section. In the following, we will show the properties of $f(R, T, k, \omega)$ that justify its interpretation as a generalized distribution function.

(a) Particle density $n(R, T)$
To calculate the particle density we have to sum $f(R, T, k, \omega)$ over the variables k and ω

$$n(R, T) = \frac{1}{\Omega} \sum_k \int_{-\infty}^{\infty} \frac{d\omega}{2\pi} f(R, T, k, \omega) = \langle \Psi^+(RT)\Psi(RT) \rangle$$

(1.123)

with Ω as the volume considered.

(b) Particle continuity
The particle continuity allows us to define a proper way to calculate the particle-current density $j(r, t)$ from $f(R, T, k, \omega)$. To keep things simple, we will investigate a system which is described by the many-particle Hamiltonian similar to Eq. (1.8). This gives the following equation of motion for the field operator $\Psi(R, T)$

$$i \frac{\partial}{\partial T} \Psi(R, T) = h(R, p = -i\nabla, T)\Psi(R, T)$$

(1.124)

where we also allow for an explicit time dependence of the external potential in the one-particle Hamiltonian h which is similar to Eq. (1.9). Such a situation may be realized in a system of electrons in an arbitrary space and time-dependent electric field with potential $V(R, T)$ and scattering of the carriers by randomly distributed impurities. The results that follow do not depend on this restriction. We differentiate the particle density with respect to time T and use the equation of motion to express the time derivatives of the field operators

$$\frac{\partial}{\partial T} n(R, T) = -i \frac{1}{\Omega} \sum_k \int_{-\infty}^{\infty} \frac{d\omega}{2\pi} \int dr \int_{-\infty}^{\infty} d\tau \exp(-ikr + i\omega\tau)$$
$$\cdot \{-h(R - \tfrac{1}{2}r, T - \tfrac{1}{2}\tau) + h(R + \tfrac{1}{2}r, T + \tfrac{1}{2}\tau)\}$$
$$\cdot \langle \Psi^+(R - \tfrac{1}{2}r, T - \tfrac{1}{2}\tau)\Psi(R + \tfrac{1}{2}r, T + \tfrac{1}{2}\tau) \rangle \quad (1.125)$$

Separating the kinetic energy form the potential term gives for the former

$$\frac{1}{2m}\nabla'^2 - \nabla^2 \rightarrow \frac{1}{m}\nabla_R\nabla_r \tag{1.126}$$

which turns Eq. (1.125) into

$$\frac{\partial}{\partial T}n(R,T) = -i\frac{1}{\Omega}\sum\int_{-\infty}^{\infty}\frac{d\omega}{2\pi}\int dr\int_{-\infty}^{\infty} d\tau\exp(-ikr+i\omega\tau)$$

$$\left\{\frac{1}{m}\nabla_R\nabla_r + (V(R+\tfrac{1}{2}r, T+\tfrac{1}{2}\tau) - V(R-\tfrac{1}{2}r, T-\tfrac{1}{2}\tau))\right\}$$

$$\langle \Psi^+(R-\tfrac{1}{2}r, T-\tfrac{1}{2}\tau)\Psi(R+\tfrac{1}{2}r, T+\tfrac{1}{2}\tau)\rangle \tag{1.127}$$

Partial integration with respect to r finally allows ∇_r to act on the exponential function and we have

$$\frac{\partial}{\partial T}n(R,T) = -\frac{1}{\Omega}\nabla_R\sum_k\int_{-\infty}^{\infty}\frac{d\omega}{2\pi}\frac{k}{m}f(R,T,k,\omega) \tag{1.128}$$

This is the continuity equation which immediately allows the identification of the particle current density $j(R,T)$ as

$$j(R,T) = \frac{1}{\Omega}\sum_k\int_{-\infty}^{\infty}\frac{d\omega}{2\pi}\frac{k}{m}f(R,T,k,\omega) \tag{1.129}$$

Again we find that the macroscopic quantity j is calculated by summing f over the momentum k and the energy ω weighted by the particle group velocity $\nabla_k\varepsilon_k = k/m$. As can be seen by following the derivation, the kinetic term of the Hamiltonian is essential for obtaining the equation of continuity. The contribution from the potential term disappears. This will also be true for a more general interaction.

(c) Energy balance
Up to now, we have assigned to ω the meaning of an energy. We will now justify this assumption by showing that it is compatible with a macroscopic energy density E. To this end, we will investigate the stationary case first. We follow Eqs. (1.6) and (1.7) and assume we have found a basis set labelled s that diagonalizes the zero-order Hamiltonian H_0 and N is the numbers of states.

$$\Psi^+(xt) = \frac{1}{N}\sum_s \varphi_s^*(x)\exp(i\varepsilon_s t)a_s^+ \tag{1.130}$$

$$\Psi(xt) = \frac{1}{N}\sum_s \varphi_s(x)\exp(-i\varepsilon_s t)a_s \tag{1.131}$$

Inserting Eqs. (1.130) and (1.131) into Eq. (1.122) we get

$$f(R, k, \omega) = \frac{1}{N} \int dr \int_{-\infty}^{\infty} d\tau \exp(-ikr + i\omega\tau)$$

$$\cdot \sum_s \varphi_s^*(R - \tfrac{1}{2}r)\varphi_s(R + \tfrac{1}{2}r)\exp(-i\varepsilon_s\tau)n_s \qquad (1.132)$$

$$= \frac{1}{N} \sum_s \int dr \exp(-ikr)\, \varphi_s^*(R - \tfrac{1}{2}r)\varphi_s(R + \tfrac{1}{2}r)$$

$$\cdot 2\pi\delta(\omega - \varepsilon_s)n_s \qquad (1.133)$$

The energy density is calculated by

$$\mathscr{E}(R) = \frac{1}{\Omega} \sum_k \int_{-\infty}^{\infty} \frac{d\omega}{2\pi}\, \omega f(R, k, \omega) = \frac{1}{N} \sum_s |\varphi_s(R)|^2 \varepsilon_s n_s \qquad (1.134)$$

Equation (1.134) looks like the result expected from a rigorous particle picture. The quasi-particle density number n_s is given by Eq. (1.48) with zero width of the spectral function. This is because we started with a correct diagonalized Hamiltonian that gives us stable excitations. A generalization of Eq. (1.134) to the non-stationary case is now straightforward

$$\mathscr{E}(R, T) = \frac{1}{\Omega} \sum_k \int_{-\infty}^{\infty} \frac{d\omega}{2\pi}\, \omega f(R, T, k, \omega) \qquad (1.135)$$

in the same way that we derived the particle current density j, we can derive the energy current density by energy conservation

$$v_\varepsilon(R, T) = \frac{1}{\Omega} \sum_k \int_{-\infty}^{\infty} \frac{d\omega}{2\pi}\, \omega \frac{k}{m} f(R, T, k, \omega) \qquad (1.136)$$

Provided the Wigner function is known, we have shown that the macroscopic quantities $n, j, \mathscr{E}$, and v_ε can be calculated so that particle and energy conservation is satisfied. It is then permissible to identify $f(R, T, k, \omega)$ with the generalization of the classical distribution function $f(R, T, k)$.

We would like to close this section with a remark on a frequently found approximation of $f(R, T, k, \omega)$ (Kadanoff and Baym 1962) which is very closely linked to the quasi-particle concept. Let us recall that the Wigner function f is related to the Green's function $G^<$. In equilibrium we derived the important relationships Eqs. (1.49) and (1.50)

$$G^<(k, \omega) = in_F(\omega)A(k, \omega) \qquad (1.137)$$

comparing this with Eq. (1.121)

$$G^<(R, T, k, \omega) = if(R, T, k, \omega) \qquad (1.138)$$

we can look for a solution in the non-equilibrium case that has the following features

$$G^<(R, T, k, \omega) = i f(R, T, k) A(R, T, k, \omega) \tag{1.139}$$

A solution of this type is reasonably close to the exact solution if the non-equilibrium spectral function is sharply peaked at $\omega = \varepsilon_k$. In this case, it is meaningful to maintain only the momentum dependence in f. But if particle interactions cause a considerable width of the spectral function, k and ω must be kept independent. Because the non-equilibrium spectral function is itself unknown, it is hard to give an *a priori* justification of Eq. (1.139). A fully self-consistent calculation of the Keldysh G must be performed to derive its spectral function. We will not go into the analysis here but will return in the following section to Eqs. (1.117) and (1.118), which will lead us to the generalized Boltzmann equation.

1.3.2 Wigner and Boltzmann Equation

We will start our analysis of deriving the Boltzmann equation by re-writing Eqs. (1.117) and (1.118). The main objective is to obtain an integro-differential equation for $G^<$ and $G^>$. We will undertake a detailed investigation for $G^<$, which is proportional to the Wigner function $f(R, T, k, \omega)$ discussed in the previous section. A similar analysis holds for $G^>$, so we will merely quote the result at the end of this section. We now multiply Eq. (1.117) from the left with G^{r-1} and from the right with G^{a-1}. This results in the two equations.

$$(G^{r-1} + \Sigma^r)G^< = (G^{r-1} + \Sigma^r)g_0^< (1 + \Sigma^a G^a) - \Sigma^> G - \underline{\Sigma} G^> \tag{1.140}$$

$$G^<(G^{a-1} + \Sigma^a) = (1 + G^r \Sigma^r)g_0^< (G^{a-1} + \Sigma^a) - G\Sigma^< - G^< \underline{\Sigma} \tag{1.141}$$

with the aid of the Dyson equations for the retarded and advanced Green's functions, Eqs. (1.110) and (1.111), we further simplify the above equations to:

$$g_0^{r-1} G^< = g_0^{r-1}(g_0^<)g_0^{a-1} G^a - \Sigma^> \underline{G} - \underline{\Sigma} G^> \tag{1.142}$$

$$G^< g_0^{a-1} = G^r g_0^{r-1}(g_0^<)g_0^{a-1} - G\Sigma^< - G^< \underline{\Sigma} \tag{1.143}$$

For the inverse zero-order retarded and advanced Green's functions we find, as in Eq. (1.58), by re-installing the variables explicitly

$$g_0^{r-1}(x, t) = i\frac{\partial}{\partial t} - h(x, p = -i\nabla_x, t) \tag{1.144}$$

$$g_0^{a-1}(x, t) = -i\frac{\partial}{\partial t} - h(x, p = -i\nabla x, t) \tag{1.145}$$

where h is again that part of the Hamiltonian that is free of particle correlations due to interactions but contains the space and time-dependent external potential fields.

$$h(x, p = -i\nabla_x, t) = -\frac{1}{2m}\nabla^2 - qV(x, t) \tag{1.146}$$

In Eq. (1.467), we neglect the influence of a magnetic field so that the external field is adequately described by the scalar electrical potential obtained by solving the Poisson equation. The Green's function $G^<$ is a two-point function, which means that it depends on the two pairs (xt) and $(x't')$. The differential operator acting from the left acts on the first pair and that acting from the right on the second one. We continue by subtracting Eq. (1.143) from (1.142)

$$(g_0^{r-1}G^< - G^< g_0^{a-1}) = g_0^{r-1}g_0^< g_0^{a-1}G^a - G^r g_0^{r-1}g_0^< g_0^{a-1}$$
$$+ \Sigma^< \underline{G} + \Sigma G^< + G\Sigma^< + G^< \underline{\Sigma} \tag{1.147}$$

The first and second term on the right hand side contain the functional $g_0^{r-1}g_0^<$ which vanishes in an identical way. We can show this by writing the following, still omitting the variables

$$g_0^{r-1} = -i\frac{\partial}{\partial t} - h$$

$$g_0^< = -i\langle\Psi^+\Psi\rangle$$

$$g_0^{r-1}g_0^< = -i\left\{\left\langle\Psi^+ i\frac{\partial}{\partial t}\Psi\right\rangle - h\langle\Psi^+\Psi\rangle\right\} = 0 \tag{1.148}$$

The first summand is identical to the second by virtue of the equation of motion for the field operator, Eq. (1.7). Please note that the first pair of variables is assigned to Ψ, see Eq. (1.94). This result leaves Eq. (1.147) in the form

$$(g_0^{r-1}G^< - G^< g_0^{a-1}) = +\Sigma^< \underline{G} + \Sigma G^< + G\Sigma^< + G^< \underline{\Sigma} \tag{1.149}$$

For the further analysis, it is more convenient to re-write the right-hand side of Eq. (1.149) in a more symmetric form, utilizing Eqs. (1.101) to (1.104) for the various Green's functions and Eqs. (1.113) to (1.115) for the self-energy components.

$$(g_0^{r-1}G^< - G^< g_0^{a-1}) = \{G^>\Sigma^< - \Sigma^>G^<\}$$
$$+ \{\Sigma^<G^< - G^<\Sigma^<\} + \{\Sigma^aG^< - G^<\Sigma^a\}$$
$$- \{\Sigma^<G^a - G^a\Sigma^<\} \tag{1.150}$$

This is the generalized Boltzmann equation! Together with a similar equation for $G^>$, the Dyson equation for the retarded and advanced Green's functions, Eqs. (1.110) and (1.111), and the relationships between the different components of the Keldysh G and Σ which we find in Eqs. (1.101) to (1.106) and (1.113) to (1.115), respectively, we have a highly complex system whose general solution is impossible. The problem becomes clearer if we continue the analysis in the Wigner variables (R, T, k, ω). We first consider the left-hand side of Eq. (1.150). With Eqs. (1.144), (1.145), and (1.146) we get

$$g_0^{r-1} G^< - G^< g_0^{a-1} \rightarrow i \frac{\partial}{\partial T} G^<(R, T, r, \tau) + \frac{1}{m} \nabla_R \nabla_r G^<(R, T, r, \tau)$$

$$+ q(V(R + \tfrac{1}{2}r, T + \tfrac{1}{2}t) \tag{1.151}$$

$$- V(R - \tfrac{1}{2}r, T - \tfrac{1}{2}\tau)) G^<(R, T, r, \tau)$$

In going from (xt), $(x't')$ to the Wigner variables (RT) and $(r\tau)$ we have to replace the differential operators

$$\frac{\partial}{\partial t} \rightarrow \frac{1}{2} \frac{\partial}{\partial T} + \frac{\partial}{\partial \tau}$$

$$\frac{\partial}{\partial t'} \rightarrow \frac{1}{2} \frac{\partial}{\partial T} - \frac{\partial}{\partial \tau} \tag{1.152}$$

$$\nabla_x \rightarrow \tfrac{1}{2}\nabla_R + \nabla_r$$

$$\nabla_{x'} \rightarrow \tfrac{1}{2}\nabla_R - \nabla_r$$

The Fourier transform of Eq. (1.151) with respect to $(r\tau)$, Eq. (1.121), gives

$$g_0^{r-1} G^< - G^< g_0^{a-1} \rightarrow i \frac{\partial}{\partial T} G^<(R, T, k, \omega) + i \frac{k}{m} \nabla_R G^<(R, T, k, \omega)$$

$$+ q \left(V \left(R + i \frac{1}{2} \nabla_k, T - i \frac{1}{2} \frac{\partial}{\partial \omega} \right) \right. \tag{1.153}$$

$$\left. - V \left(R - i \frac{1}{2} \nabla_k, T + i \frac{1}{2} \frac{\partial}{\partial \omega} \right) \right) G^<(R, T, k, \omega)$$

If we neglect the left-hand side of Eq. (1.151), which as we will see soon contains the scattering terms, and replace $G^<$ by the Wigner function f, then Eq. (1.153) is the Wigner equation for a system of free particles in an external potential field $V(R, T)$

$$\frac{\partial}{\partial T} f(R, T, k, \omega) + \frac{\hbar k}{m} \nabla_R f(R, T, k, \omega)$$

$$+ \frac{q}{i\hbar} \left(V \left(R + i \frac{1}{2} \nabla_k, T - i \frac{1}{2} \frac{\partial}{\partial \omega} \right) \right.$$

$$\left. - V \left(R - i \frac{1}{2} \nabla_k, T + i \frac{1}{2} \frac{\partial}{\partial \omega} \right) \right) f(R, T, k, \omega) = 0 \tag{1.154}$$

In Eq. (1.154) we have re-established $\hbar$ at the proper places. The classical analogy to the scatter-free Wigner equation is the scatter-free Boltzmann equation, which is also called the Vlassow equation. This equation is obtained directly by taking the limit $\hbar \to 0$ in Eq. (1.154). To do this in a consistent way, we also have to take the classical limit

$$p = \hbar k$$

$$\varepsilon = \hbar \omega$$
(1.155)

where the momentum p is related to the wave number and the energy ε to the frequency. Furthermore, we have the following for a classical free particle

$$\varepsilon = \frac{p^2}{2m}$$
(1.156)

which then cancels the energy dependence in the distribution function f, which thus remains the classical distribution function $f(R, T, p)$.

$$\frac{\partial}{\partial T} f(R, T, p) + \frac{p}{m} \nabla_R f(R, T, p) - qE(R, T)\nabla_p f(R, T, p) = 0$$
(1.157)

Before we continue, let us compare the Wigner equation Eq. (1.154) and its classical counterpart Eq. (1.157). The first and most important observation we make is that the classical equation is local in the electric field. This means that the electric field enters the equation as a function in R and T. The field dependence in the Wigner equation is more complicated. It enters through a functional relationship in the potential field V. Through this relationship, the electric field influences the distribution function $f(R, T, k, \omega)$ in a very complex non-local way which is coupled directly to the momentum and energy dependence of f. In general, the potential field contains contributions from the electric potential, which is related to the solution of the Poisson equation, and the spatial variation of the band edge, which occurs if materials with different energy bandgaps are joined. This feature is easily included in the single-particle Hamiltonian h. The Wigner equation can deliver a solution which contains the penetration of particles in neighboring materials separated by a finite energy barrier. This is the tunnel effect known from quantum mechanics. In practical applications we find this when modeling contacts. In this respect, the Wigner equation, Eq. (1.154), or its generalization which also includes scattering (Eq. (1.163) below) could serve as a foundation for a rigorous self-consistent modeling of transport through contacts, which is not available at present. Numerical solutions of the Wigner equation for a finite energy barrier of finite width and for a potential well are known and demonstrate that it is able to describe quantum mechanical tunneling (Kluksdahl et al. 1989, Frensley 1986). In the classical equation, tunneling of particles through energy barriers is prohibited. Energy barriers serve as a hard wall and have to be treated with a reflective boundary condition. We will discuss this point further in Chapter 3. The Wigner equation does not account for scattering in the system. The scattering terms which we will now investigate are contained in the right-hand side of Eq. (1.150). The general form of the four different terms is

$$G\Sigma - \Sigma'G' \to \int dx''[G(x, x'')\Sigma(x'', x') - \Sigma'(x, x'')G'(x'', x')]$$

(1.158)

where the notation $x \equiv (x, t)$ has been used. This integral identity is now transformed into the Wigner coordinates.

$$\Delta_1 = \tfrac{1}{2}(x + x'') = \tfrac{1}{2}(R + \tfrac{1}{2}r + x'')$$
$$\Delta_2 = \tfrac{1}{2}(x'' + x') = \tfrac{1}{2}(R - \tfrac{1}{2}r + x'')$$
$$\delta_1 = x - x'' = R + \tfrac{1}{2}r - x''$$
$$\delta_2 = x'' - x' = x'' - R + \tfrac{1}{2}r$$

where the subscripts 1 and 2 mean the variables in the first and second factor, respectively. Introducing the new integration variable u

$$x'' = u + R + \tfrac{1}{2}r$$

we find for Eq. (1.158)

$$G\Sigma - \Sigma'G' \to \int du[G(R + \tfrac{1}{2}(r - u), u)\Sigma(R - \tfrac{1}{2}u, r - u)$$
$$-\Sigma'(R + \tfrac{1}{2}(r - u), u)G'(R - \tfrac{1}{2}u, r - u)]$$ (1.159)

So far we have not assigned any physical interpretation to the Wigner variables. But we have to do so if we are to continue. From our analysis of the properties of the Wigner function $f(R, T, k, \omega)$ we learn that the variables k and ω scale with the microscopic variations in the system because they are related to the energy and the wave number of the particles. On the other hand, the center-of-mass coordinates R and T scale with the variation of the external fields. If we only considered systems with very weakly varying external disturbances, a decoupling of microscopic and macroscopic scales would we possible. This could be simply achieved by assuming that the variation in the center of mass is slow compared to that in the relative coordinates r and τ. This is not always the case, especially if the system under consideration has potential steps. However, the numerical investigations of the Wigner equation show that the variation of the Wigner function with respect to the center-of-mass coordinates is always slow compared with rapid spatial and temporal changes in the external potential field. This is because the solution of a partial differential equation with stepwise continuous coefficient functions has a smoothing effect. This permits us to generalize the assumption that variations on the macroscopic and microscopic scales decouple. This assumption will fail if the external potential field is highly oscillatory in the time domain and can produce optical band-to-band transitions, or if its spatial variation is such that band-to-band tunneling becomes important. Granting the decoupling is valid, we can perform a Taylor expansion with respect to the center-of-mass coordinates in Eq. (1.159), which gives

$$G\Sigma - \Sigma'G' \rightarrow \int du[G(R,u)\Sigma(R,r-u) - \Sigma'(R,u)G'(R,r-u)]$$

$$+ \frac{1}{2}\int du[(r-u)\cdot\nabla_R G(R,u)\Sigma(R,r-u)$$

$$- (r-u)\cdot\nabla_R\Sigma'(R,u)G'(R,r-u)]$$

$$- \frac{1}{2}\int du[G(R,u)u\cdot\nabla_R\Sigma(R,r-u)$$

$$- \Sigma'(R,u)u\cdot\nabla_R G'(R,r-u)]$$

$$= \int du[G(R,u)\Sigma(R,r-u) - \Sigma'(R,u)G'(R,r-u)]$$

$$+ \frac{1}{2}\int du[(r-u)\cdot\nabla_R G(R,u)\Sigma(R,r-u)$$

$$- G(R,u)u\cdot\nabla_R\Sigma(R,r-u)]$$

$$- \frac{1}{2}\int du[(r-u)\cdot\nabla_R\Sigma'(R,u)G'(R,r-u)$$

$$- \Sigma'(R,u)u\cdot\nabla_R G'(R,r-u)]$$

A Fourier transform with respect to relative coordinates gives

$$G\Sigma - \Sigma G \rightarrow G(R,T,K,\omega)\Sigma(R,T,k,\omega)$$

$$- \Sigma(R,T,k,\omega)G(R,T,k,\omega)$$

$$\cdot i\tfrac{1}{2}[G(R,T,K,\omega), \Sigma(R,T,k,\omega)]$$

$$+ i\tfrac{1}{2}[G'(R,T,K,\omega), \Sigma'(R,T,k,\omega)] \qquad (1.160)$$

The first term is a simple algebraic product and $[\ ,\ ]$ is a generalized Poisson bracket

$$[A, B] = \nabla_R A \nabla_k B - \nabla_R B \nabla_k A - \frac{\partial}{\partial T} A \frac{\partial}{\partial \omega} B + \frac{\partial}{\partial T} B \frac{\partial}{\partial \omega} A \quad (1.161)$$

By collecting all four terms of the right-hand side of Eq. (1.150), we get together with Eq. (1.153)

$$i \frac{\partial}{\partial T} G^<(R, T, k, \omega) + i \frac{k}{m} \nabla_R G^<(R, T, k, \omega)$$

$$+ q \left(V \left(R + i \frac{1}{2} \nabla_k, T - i \frac{1}{2} \frac{\partial}{\partial \omega} \right) \right.$$

$$\left. - V \left(R - i \frac{1}{2} \nabla_k, T + i \frac{1}{2} \frac{\partial}{\partial \omega} \right) \right) G^<(R, T, k, \omega)$$

$$= G^>(R, T, k, \omega) \Sigma^<(R, T, k, \omega) - \Sigma^>(R, T, k, \omega) G^<(R, T, k, \omega)$$

$$+ i\tfrac{1}{2}[G^>(R, T, k, \omega), \Sigma^<(R, T, k, \omega)]$$

$$+ i\tfrac{1}{2}[G^<(R, T, k, \omega), \Sigma^>(R, T, k, \omega)]$$

$$- i[G^<(R, T, k, \omega), \Sigma^<(R, T, k, \omega)]$$

$$- i[G^<(R, T, k, \omega), \Sigma^a(R, T, k, \omega)]$$

$$+ i[G^a(R, T, k, \omega), \Sigma^<(R, T, k, \omega)] \quad (1.162)$$

This equation for $G^<$ is a generalization of the Wigner equation that includes scattering of particles. It has to be solved together with a similar one for $G^>$ and the Dyson equations for G^r and G^a. All these equations are coupled through the special form of the Keldysh self-energy that is provided by many-body perturbation theory. The Poisson brackets on the right-hand side of Eq. (1.162) make this coupled system very hard to solve. In deriving Eq. (1.162), we already assumed a slow variation of the center-of-mass coordinates in the Wigner function. An explicit analysis of the Poisson brackets for the case of a constant electric field which is suitable in metals can be found in the literature (Hänsch and Mahan 1983a, b). If we consider only contributions linear in the field, it can be shown that the left-hand side of Eq. (1.162) is renormalized by contributions from the retarded self-energy Σ^r in equilibrium. From our analysis in the previous sections we know that the imaginary part of Σ^r determines the spread of the spectral function A, Eq. (1.67) and therefore validates the quasi-particle concept. The correction terms are small if the spectral function is sufficiently sharply peaked, which is the case where interactions are weak. Granting that this is the case, we can neglect the contributions from the Poisson brackets and Eq. (1.162) becomes

$$i\frac{\partial}{\partial T}G^<(R, T, k, \omega) + i\frac{k}{m}\nabla_R G^<(R, T, k, \omega)$$

$$+ q\left(V\left(R + i\frac{1}{2}\nabla_k, T - i\frac{1}{2}\frac{\partial}{\partial\omega}\right)\right.$$

$$\left. - V\left(R - i\frac{1}{2}\nabla_k, T + i\frac{1}{2}\frac{\partial}{\partial\omega}\right)\right)G^<(R, T, k, \omega) \qquad (1.163)$$

$$= G^>(R, T, k, \omega)\Sigma^<(R, T, k, \omega) - \Sigma^>(R, T, k, \omega)$$

$$\cdot G^<(R, T, k, \omega)$$

Equation (1.163) still contains the quantum mechanical features of the Wigner equation but additionally has a right-hand side not equal to zero, which accounts for scattering. It was, however, assumed in deriving this equation that scattering is sufficiently weak so that the quasi-particle concept is applicable. On the other hand, it is worth while to consider Eq. (1.163) as an equation capable of unifying quantum and ensemble transport and also to apply it to systems where the quasi-particle concept is only weakly violated. As we will see shortly, information of the structure of the spectral function is still present. To continue, we now have to specify the self-energies $\Sigma^<$ and $\Sigma^>$. By following the rules discussed in Appendix 1, we obtain the following in an electron–phonon system:

$$\Sigma^{</>}(R, T, k, \omega) = -i\sum_q |m_q|^2 \int_{-\infty}^{\infty} \frac{d\Omega}{2\pi}$$

$$\cdot D^{</>}(q, \Omega)G^{</>}(R, T, k - q, \omega - \Omega) \qquad (1.164)$$

using the phonon Green's function in the lowest order we get

$$\Sigma^<(R, T, k, \omega) = -\sum_q |m_q|^2 (n_B(\omega_q)G^<(R, T, k - q, \omega - \omega_q)$$

$$\qquad (1.165)$$

$$+ [n_B(\omega_q) + 1]G^<(R, T, k - q, \omega + \omega_q))$$

$$\Sigma^>(R, T, k, \omega) = -\sum_q |m_q|^2 ([n_B(\omega_q) + 1]G^>(R, T, k - q, \omega - \omega_q)$$

$$+ n_B(\omega_q)G^>(R, T, k - q, \omega + \omega_q)) \qquad (1.166)$$

Here $n_B(\omega_q)$ is the occupation number for phonons with energy ω_q

$$n_B(\omega_q) = (\exp(\omega_q/k_B T) - 1)^{-1} \qquad (1.167)$$

For electrons scattered by randomly distributed impurities with density N_{imp} and interaction matrix element v_q we have:

$$\Sigma^{</>}(R, T, k, \omega) = -N_{imp}\sum_q |v_q|^2\, G^{</>}(R, T, k - q, \omega) \qquad (1.168)$$

The major difference in these two scattering mechanisms is that we have to account for an energy transfer to the phonon system while scattering to impurities is elastic. We will use impurity scattering and the coupling to optical phonons as representative elastic and inelastic scattering channels. To a good approximation, both are additive in the total self-energy $\Sigma^{</>}$,

$$\Sigma^{</>} = \Sigma_{ph}^{</>} + \Sigma_{imp}^{</>} \qquad (1.169)$$

Combining Eqs. (1.163), (1.165), (1.166), and (1.168), we now have

$$i\frac{\partial}{\partial T}G^{<}(R, T, k, \omega) + i\frac{k}{m}\nabla_R G^{<}(R, T, k, \omega)$$

$$+ q\left(V\left(R + i\frac{1}{2}\nabla_k, T - i\frac{1}{2}\frac{\partial}{\partial\omega}\right)\right.$$

$$\left. - V\left(R - i\frac{1}{2}\nabla_k, T + i\frac{1}{2}\frac{\partial}{\partial\omega}\right)\right)G^{<}(R, T, k, \omega)$$

$$= -G^{>}(R, T, k, \omega)\sum_q |m_q|^2(n_B(\omega_q)G^{<}(R, T, k - q, \omega - \omega_q)$$

$$+ [n_B(\omega_q) + 1]G^{<}(R, T, k - q, \omega + \omega q)) \qquad (1.170)$$

$$- G^{>}(R, T, k, \omega)N_{imp}\sum_q |v_q|^2 G^{<}(R, T, k - q, \omega)$$

$$+ \sum_q |m_q|^2([n_B(\omega_q) + 1]G^{>}(R, T, k - q, \omega - \omega_q)$$

$$+ n_B(\omega_q)G^{>}(R, T, k - q, \omega + \omega_q))G^{<}(R, T, k, \omega)$$

$$+ N_{imp}\sum_q |v_q|^2 G^{>}(R, T, k - q, \omega)G^{<}(R, T, k, \omega)$$

The step from Eq. (1.170) to the conventional Boltzmann equation is a short one. We take the limit $\hbar \to 0$, integrate the equation over the energy variable ω, and take the quasi-particle limit $\Gamma \to 0$. The limit $\hbar \to 0$ is calculated exactly like our approach in the Wigner equation Eq. (1.154) to obtain the Vlassow equation (1.157). To calculate the ω integration, we will use the quasi-particle limit approximations for $G^{<}$, Eq. (1.139), and a similar one for $G^{>}$

$$G^{<}(R, T, k, \omega) = if(R, T, k)A(R, T, k, \omega) \qquad (1.171)$$

$$G^{>}(R, T, k, \omega) = -i(1 - f(R, T, k))A(R, T, k, \omega) \qquad (1.172)$$

with the generalized spectral function A

$$A(R, T, k, \omega) =$$

$$\frac{-2\mathrm{Im}\,\Sigma^r(R, T, k, \omega)}{(\omega - \varepsilon(R, T, k) + \mu(R, T) + \mathrm{Re}\,\Sigma^r(R, T, k, \omega))^2 + \mathrm{Im}^2\Sigma^r(R, T, k, \omega)}$$

(1.173)

which also obeys the sum rule Eq. (1.33). The particle energy and the chemical potential are allowed to vary on the macroscopic scale in the center-of-mass variables R and T. A justification of Eq. (1.173) is given by a straightforward generalization of Eq. (1.64) to the retarded Green's function in the non-equilibrium case. However, in this case it is not sufficient to know the spectral function alone to obtain all components of the Keldysh Green's function. Its components $G^<$ and $G^>$ in particular require additional information provided by Eq. (1.150) which is not a Dyson-type equation.

$$\frac{\partial}{\partial T}\,f(R, T, k) + \frac{k}{m}\nabla_R f(R, T, k) - qE(R, T)\nabla_k f(R, T, k)$$

$$= [1 - f(R, T, k)]\sum_q |m_q|^2\,(n_B(\omega_q)f(R, T, k - q)$$

$$\cdot S(R, T, k, k - q, -\omega_q)$$

$$+ [n_B(\omega_q) + 1]f(R, T, k - q)S(R, T, k, k - q, \omega_q))$$

$$\cdot[(1 - f(R, T, k)]N_{\mathrm{imp}}\sum_q |v_q|^2\,f(R, T, k - q)$$

$$\cdot S(R, T, k, k - q, \omega_q = 0) \qquad (1.174)$$

$$- f(R, T, k)\sum_q |m_q|^2\,([n_B(\omega_q) + 1][1 - f(R, T, k - q)]$$

$$\cdot S(R, T, k, k - q, -\omega_q)$$

$$+ n_B(\omega_q)[1 - f(R, T, k - q)]S(R, T, k, k - q, \omega_q))$$

$$- f(R, T, k)N_{\mathrm{imp}}\sum_q |v_q|^2[1 - f(R, T, k - q)]$$

$$\cdot S(R, T, k, k - q, \omega_q = 0)$$

$$S(R, T, k, k', \omega') = \int_{-\infty}^{\infty} \frac{d\omega}{2\pi}\,A(R, T, k, \omega)A(R, T, k', \omega + \omega')$$

(1.175)

In the limit of free particles, the structure factor S degenerates to the energy conserving δ-function. We will now show that this is also true in the quasi-particle limit. To evaluate S, we will consider a spectral function with a finite width $\Gamma = -\text{Im}\,\Sigma^r$. We will neglect the real part of Σ^r, which can be included in a renormalized quasi-particle energy ε. For convenience, we omit the R and T dependence in the following as it is not essential. We must therefore evaluate the integral

$$S(k, k', \omega') = 4 \int_{-\infty}^{\infty} \frac{d\omega}{2\pi} \frac{\Gamma(k, \omega)}{(\omega - (\varepsilon_k - \mu))^2 + \Gamma^2(k, \omega)}$$
$$\cdot \frac{\Gamma(k', \omega + \omega')}{(\omega + \omega' - (\varepsilon'_k - \mu))^2 + \Gamma^2(k', \omega + \omega')}$$

using Cauchy's theorem. Without losing generality, we can assume $\Gamma > 0$ to have a definite situation. The denominator of the integrand has zeros in the complex plane at

$$\omega_{1/2} = \varepsilon k - \mu \pm i\Gamma(k, \omega_{1/2})$$

$$\omega_{3/4} = \varepsilon k - \mu - \Omega \pm i\Gamma(k', \omega_{3/4} + \Omega) \tag{1.176}$$

which means there are two poles in the upper and another two in the lower half-plane. Because of the sum rule for A, it has to vanish sufficiently fast in the limit $\omega \to \infty$. This allows the integration contour to be closed in either the upper or lower half-plane by a semi-circle of infinite radius. The integrand will not contribute on this path. Closing the integration path in the upper half-plane we get

$$S(k, k', \omega') =$$

$$= \frac{2\Gamma'}{(\omega' - (\varepsilon_{k'} - \varepsilon_k) + i\Gamma' + i\Gamma)(\omega' - (\varepsilon_{k'} - \varepsilon_k) + i\Gamma - i\Gamma')}$$

$$+ \frac{2\Gamma}{(\omega' - (\varepsilon_{k'} - \varepsilon_k) - i\Gamma' - i\Gamma)(\omega' - (\varepsilon_{k'} - \varepsilon_k) - i\Gamma + i\Gamma')} \tag{1.177}$$

Here we introduced the notation

$$\Gamma = \Gamma(k, \varepsilon_k - \mu)$$
$$\Gamma' = \Gamma(k', \varepsilon_{k'} - \mu)$$

In the quasi-particle limit $\Gamma \to 0$ we only consider the leading terms in Γ in Eq. (1.174) which gives

$$S(k, k', \omega') = \frac{2(\Gamma' + \Gamma)}{((\omega' - (\varepsilon_{k'} - \varepsilon_k))^2 + (\Gamma' + \Gamma)^2)}$$

$$\cdot \left(1 - \frac{(\Gamma - \Gamma')^2}{((\omega' - (\varepsilon_{k'} - \varepsilon_k))^2 + (\Gamma - \Gamma')^2)} \right)$$

In the quasi-particle limit $\Gamma \to 0$, the first factor will be very strongly peaked at $\omega' - (\varepsilon_{k'} - \varepsilon_k) = 0$ and will eventually turn into the Dirac δ-function (cf. also Eq. (1.68)). Under these conditions, the second factor will be of order one. We will therefore finally have an expression for the structure function S which looks similar to the spectral function, but with the difference that the width must be replaced by the joint width of the scattering partners

$$S(R, T, k, k', \omega') = \frac{2\Gamma(R, T, k, k', \omega')}{(\omega' - (\varepsilon_{k'} - \varepsilon_k))^2 + \Gamma^2(R, T, k, k', \omega')} \quad (1.178)$$

$$\Gamma(R, T, k, k', \omega') = \Gamma(R, T, k, \varepsilon(R, T, k) - \mu(R, T))$$
$$+ \Gamma(R, T, k', \varepsilon(R, T, k') - \mu(R, T))$$

or in the limit $\Gamma \to 0$

$$S(R, T, k, k', \omega') = 2\pi\delta(\omega' - \varepsilon_{k'} + \varepsilon_k) \quad (1.179)$$

Replacing the structure function S in Eq. (1.175) by Eq. (1.179), we obtain the classical Boltzmann equation for an electron–phonon–impurity system

$$\frac{\partial}{\partial T} f(R, T, k) + \frac{k}{m} \nabla_R f(R, T, k) - qE(R, T) \nabla_k f(R, T, k)$$

$$= [1 - f(R, T, k)] 2\pi \sum_q |m_q|^2 (n_B(\omega_q) f(R, T, k - q)$$
$$\cdot \delta(\varepsilon_k - \varepsilon_{k-q} - \omega_q) + [n_B(\omega_q) + 1]$$
$$\cdot f(R, T, k - q) \delta(\varepsilon_k - \varepsilon_{k-q} + \omega_q))$$
$$\cdot [(1 - f(R, T, k)] N_{\text{imp}} 2\pi \sum_q |v_q|^2$$
$$\cdot f(R, T, k - q) \delta(\varepsilon_k - \varepsilon_{k-q}) \quad (1.180)$$
$$- f(R, T, k) 2\pi \sum_q |m_q|^2 ([n_B(\omega_q) + 1]$$
$$\cdot [1 - f(R, T, k - q)] \delta(\varepsilon_k - \varepsilon_{k-q} - \omega_q)$$
$$+ n_B(\omega_q) [1 - f(R, T, k - q)] \delta(\varepsilon_k - \varepsilon_{k-q} + \omega_q))$$
$$- f(R, T, k) N_{\text{imp}} 2\pi \sum_q |v_q|^2 [1 - f(R, T, k - q)] \delta(\varepsilon_k - \varepsilon_{k-q})$$

This is the master equation for deriving the equations sufficient to describe the transport of electric charge in a classical device such as a MOSFET. On the path from the general equation Eq. (1.150) for the Green's function $G^<$ to the Boltzmann equation above some approximations were necessary.

— Heat Bath

The system of phonons is in equilibrium with the heat bath at a temperature T. If non-equilibrium effects of phonons are taken into account, there would be an additional transport equation for the phonons. The carrier and phonon transport equations are then coupled via the scattering terms

— Infinite Domain

Throughout the calculation, we assumed an infinite domain, which means that the wave numbers are quasi-continuous. Quantum size effects which reveal the discrete character of the spectrum are not included. How to include a discrete spectrum in the Boltzmann–Wigner equation is discussed in Chapter 3.

— Quasi Particle

The collision broadening of the electron's spectral function must be sufficiently small

$$\Gamma \ll \omega_q, k_B T$$

only then we can assign momentum and energy in such way to the quasi particle that it is well defined in the scattering event.

— Weakly Varying External Potential Field

The external potential field (solution of Poisson equation) has a slow variation on a microscopic scale

$$\frac{\partial}{V \partial R} V < \frac{1}{\lambda_{\text{th}}}, \frac{1}{\lambda_{\text{dB}}}$$

λ_{th} and λ_{dB} are the thermal and de Broglie wavelengths, respectively

$$\frac{\partial}{V \partial T} V < \frac{1}{\hbar} E_G$$

E_G is the optical bandgap of the semiconductor

— Moderate field strength

To avoid band-to-band tunneling the electrical field has to be of limited strength

$$E_G > \lambda_G \cdot \frac{\partial V}{\partial R}$$

Here the characteristic length is $\lambda_G = \sqrt{\hbar^2/(2mE_G)}$.

— Effective Mass Approximation
The effective mass approximation is not essential for the derivation of
Eq. (1.180). A generalization to a two-band system (coupled via scattering)
is possible.

References

Abricosov A. A., Gorkov L. P., Dzyaloshinski (1963): Methods of Quantum Field Theory in
     Statistical Physics (revised English translation). Dover Pns, New York.
Barker J. R., Ferry D. K. (1980): Solid State Electr. 23, 519.
Barker J. R., Ferry D. K. (1980): Solid State Electr. 23, 531.
Bordone P., Jacoboni C., Lugli P., Reggiani L., Kocevar P. (1985): Physica 134B, 169.
Davydov A. (1965): Quantum Mechanics. Pergamon, Oxford.
Ferry D. K., Barker J. R. (1980): Solid State Electr. 23, 545.
Fetter L. A., Walecka J. D. (1971): Quantum Theory of Many-Particle Systems. McGraw-
     Hill, New York.
Frensley W. R. (1986): Phys. Rev. B36, 1570.
Hänsch W. (1985): Phys. Rev. B31, 3504.
Hänsch W., Mahan G. D. (1983a): Phys. Rev. B28, 1902.
Hänsch W., Mahan G. D. (1983b): Phys. Rev. B28, 1886.
Kadanoff L. P., Baym G. (1962): Quantum Statistical Mechanics. Benjamin, New York.
Keldysh L. V. (1964): Zh. Eksp. Teor. Fiz 47, 1515 [Sov. Phys. JETP 20, 1018 (1965)].
Kittel C. (1967): Quantum Theory of Solids (4th printing). John Wiley, New York.
Kluksdahl N. C., Kirman A. M., Ferry D. K., Ringhofer C. (1989): Phys. Rev. B39, 7720.
Kocevar P. (1985): Physica 134B, 155.
Landau L. D., Lifshitz E. M. (1979): Lehrbuch der Theoretischen Physik, Vol. III—
     Quantenmechanik, (6 Auflage). Akademie Verlag, Berlin.
Mahan G. D. (1981): Many-Particle Physics. Plenum, New York.
Merzbacher E. (1961): Quantum Mechanics. John Wiley, New York.
Parry W. E. (1973): The Many-Body Problem. Clarendon Press, Oxford.
Pathria R. K. (1972): Statistical Mechanics. International Series in Natural Philosophy
     Vol. 45. Pergamon, Oxford.
Rammer J., Smith H. (1986): Rev. Modern Phys. 58, 323.
Wigner E. (1932): Phys. Rev. 40, 749.

Other Textbooks of General Interest

Ashcraft N. W., Mermin N. D. (1976): Solid State Physics. Holt, Rinehart and Winston,
     New York.
Callaway J. (1976): Quantum Theory of Solids (Student Edition). Academic Press, New
     York.
Haug A. (1972): Theoretical Solid State Physics Vols. I & II. International Series of
     Monographs in Natural Philosophy Vol. 36. Pergamon Press, Oxford.
Landau L. D., Lifshitz E. M. (1979): Lehrbuch der theoretischen Physik, Vol. V & IV—
     Statistische Physik (5. Auflage). Akademie Verlag, Berlin.
Seeger K. (1985): Semiconductor Physics. Springer Series in Solid State Science Vol. 40.
     Springer, Berlin.
Sze S. M. (1981): Physics of Semiconductor Devices (2nd edition). John Wiley, New York.

Hydrodynamic Model 2

2.1 Introduction

In the last chapter, we derived the Boltzmann equation from a calculation based on first principles. In reaching the desired result, we had to make approximations and simplifications in our general approach to obtain the equation that gives us the particle distribution f in classical phase space of the variables R, T, and k. There are other, more heuristic ways of deriving the Boltzmann equation based on the mechanics of classical particles in a slowly varying electric field (Madelung 1978). It is certainly possible to view the Boltzmann equation as a transport model for classical particles quite apart from its relationship to the more fundamental non-equilibrium statistical mechanics. This is the historical context in which it was derived and used for many years. As we shall see later, it is useful to go beyond the classical understanding of the equation towards possible extensions which are not contained in the purely classical picture.

The solution of the Boltzmann equation provides the distribution function f, thus making it possible to calculate all macroscopic variables such as the density n, current density j, energy density $\mathscr{E}$, energy current density v_ε, etc. All these quantities are of relevance for describing the performance of a device. The catch, however, is that solving the Boltzmann equation is a tedious task. It requires the solution of an integro-partial differential equation in 7 dimensions: 3 space, 3 momentum, and 1 time variable. In addition, we have the complex structure of the collision term. The solution of this complicated equation is feasible only for very simple situations. On the other hand, Monte Carlo methods are utilized to determine the distribution function directly (Jacoboni and Lugli 1989). A relevant example is the evolution of a system of classical particles subjected to scattering mechanisms, as discussed in the previous chapter. Although much progress has been made recently (Fischetti and Laux 1989), it is still a very costly procedure to obtain the performance of a device with this approach. It is therefore currently limited to research applications and is unsuitable for engineering purposes, where a fast and cheap evaluation of device performance is an important constraint. The hydrodynamic transport model does not solve the Boltzmann equation directly but rather a set of coupled equations representing the relationships between the different moments (momentum space averages) of the distribution function f. The mathematical problem is therefore to solve partial differential equations in four variables: 3 space and 1 time. This simplification must be paid for by a loss of the microscopic description of the scattering in the system. Average scattering rates appear in the mobility, for instance, which is a coefficient in the current-density equation. To obtain a useful result therefore, it is of utmost importance to model these scattering rates on the basis of the relevant physics in the system (Selberherr 1984). This is, of course, impossible without the help of experiments which provide information about the influence of the fundamental scattering processes. Here again there is a catch: the experiments are often performed under simple conditions (for instance constant electric field, no surface effects) to make an evaluation of the results possible. But in the device the situation is usually much more complex and it is often questionable whether such an experimentally verified relationship holds under these conditions. It is thus desirable also to include in the hydrodynamical model a self-consistent investigation of the transport coefficients. This will be the major concern of the sections that follow.

We will start our analysis with the linear response regime where only small deviations from equilibrium are considered. Here, we will focus especially on the ohmic mobility and the concept of the relaxation-time approximation of the scattering term in the Boltzmann equation. We will then study extensions, including the case of a strong perturbation of the system which can no longer be described by the linear response approach. Here, we will discuss the approach of Blotekjaer (1970), which is based on assuming a drifted Maxwellian for the distribution function as well as an alternative approach based on the expressability of the distribution function as an expansion of its moments. This latter approach also leads in a very natural way to include the scattering terms in the analysis in a form suitable for modeling. As in the previous chapter, we will assume the effective mass approximation here. We are fully aware that this approximation does not describe the transport sufficiently, especially the non-linear response, in Si. However, we believe that the physics extracted from this model will still be valid for more complex structures. After all, we do not aim to present an *ab initio* calculation of transport in Si, but rather focus on the principal relationships between the primary variables n, j, $\mathscr{E}$, v_ε, etc. involved. This leaves a number of coefficients which have to be fitted to the available experimental data. A third section will then investigate a possible extension of the hydrodynamic model for modelling III–V compound devices which can no longer be treated by the one-band effective mass approximation.

2.2 Linear Response and Relaxation-Time Approximation

We start this section with Eq. (1.180) from the previous chapter. This is the Boltzmann equation for electrons in a parabolic band subjected to impurity and phonon scattering. For simplicity, we will only consider the coupling to one branch of phonons with a phonon dispersion ω_q. More complicated situations can be considered by simply adding another label to the energy and performing the sum over all different kinds of phonons in the collision term. In a complete analysis, we must consider the coupling of the electrons to various acoustic and optical phonon branches. To keep the formulation as simple as possible, we will distinguish between a branch of acoustic and optical phonons whenever necessary. These two generic branches describe the essentials of the more exact models. In this chapter, we will consider bulk material and therefore not discuss the influence of an interface on the transport properties. This will be done in Chapter 3. If we isolate the scattering in the collision term of Eq. (1.180) we have,

$$\frac{\partial}{\partial T} f(R, T, k) + v(k)\nabla_R f(R, T, k) - qE(R, T)\nabla_k f(R, T, k) =$$

$$= \sum_{k'} [w(k, k')(1 - f(R, T, k))f(R, T, k') \qquad (2.1)$$

$$- w(k', k)f(R, T, k)(1 - f(R, T, k'))]$$

with the scattering rate $w(k, k')$ given by

$$w(k, k') = 2\pi |m_{k-k'}|^2 [n_B(\omega_{k-k'})\delta(\varepsilon_k - \varepsilon_{k'} - \omega_{k-k'})$$

$$+ (n_B(\omega_{k-k'}) + 1) \cdot \delta(\varepsilon_k - \varepsilon_{k'} + \omega_{k-k'})]$$

$$+ 2\pi |V_{k-k'}|^2 \delta(\varepsilon_k - \varepsilon_{k'}) \qquad (2.2)$$

This scattering rate contains the basic physics involved in the transport process. In the first place, we can differentiate elastic and inelastic scattering. In an inelastic scattering event, both momentum and energy are transferred during the scattering process. The coupling to phonons is typical. In elastic scattering, only momentum is transferred. A typical mechanism is scattering by ionized impurities. Sometimes the coupling to acoustical phonons is treated approximately as elastic scattering because the typical energy transferred $\omega_{k-k'}$ is small compared with a characteristic electron energy, for instance $k_B T$. Furthermore, it should be noted that through the occupation number n_B for the phonons, Eq. (1.167), there is an intrinsic temperature dependence in the scattering term. There are two contributions on the right-hand side of Eq. (2.1). The first one describes the scattering events that bring a particle into the phase–space element $\Delta R \Delta k$ centered around R and k. The second contains the scattering events that take a particle out of this region. In the lowest order, this contribution is related to the collision broadening discussed in the last chapter.

Very often, investigations of the Boltzmann equation start with Eq. (2.1) and appropriately chosen scattering rates. We will now show that there is a restriction on the possible choices of the scattering rate w. To this end, we will consider the equilibrium situation. Thermodynamic statistics tells us that equilibrium is described by a Fermi distribution function for the electrons with a spatial constant temperature T and a chemical (Fermi) potential $-\varepsilon_F$.

$$f_{equ}(R, k) = n_F(R, k) = \left(1 + \exp\left(\frac{\varepsilon_k - qV(R) + q\varepsilon_F}{k_B T}\right)\right)^{-1} \tag{2.3}$$

With Eq. (2.3), the left-hand side of Eq. (2.1) vanishes. Therefore the scattering rate has to satisfy the following equation in equilibrium,

$$\sum_{k'} [(w(k, k')(1 - n_F(R, k))n_F(R, k')$$
$$- w(k', k)n_F(R, k)(1 - n_F(R, k'))] = 0 \tag{2.4}$$

Using the identity

$$(1 - n_F(R, k))n_F(R, k') = \exp\left(-\frac{\varepsilon_{k'} - \varepsilon_k}{k_B T}\right) n_F(R, k)$$
$$\cdot (1 - n_F(R, k')) \tag{2.5}$$

we find that the scattering rate must have the property,

$$w(k', k) = \exp\left(-\frac{\varepsilon_{k'} - \varepsilon_k}{k_B T}\right) w(k, k') \tag{2.6}$$

This is known to be the requirement for detailed balance. Only if the scattering rates w obey Eq. (2.6) does the solution of Eq. (2.1) assume the correct equilibrium value. To show that Eq. (2.2) has the symmetry property of detailed balance is left as an exercise for the reader.

2.2.1 Linear Response

We now turn to the case where small external perturbations drive the carrier system out of equilibrium. This is usually realized by applying an external voltage to the device, for instance a *pn*-diode. An important question relates to the nature of the force responsible for driving the system out of equilibrium. The electric field would be a candidate of choice. But there can be considerable electric fields in the *pn* junctions of a diode without an external current flow. We find the answer by inspecting the equilibrium solution more closely. The characteristics of equilibrium are temporal and spatial constant temperatures and chemical (Fermi) potentials. To describe a system near equilibrium, we now introduce a weakly spatial and temporal dependence of T (in this case a small temperature gradient applied externally; see also discussion below) and ε_F and replace them in Eq. (2.3) by the following expressions.

$$\varepsilon_F \rightarrow \varepsilon_F(R, T) + \Lambda(k) \frac{\partial}{\partial R} \varepsilon_F(R, T) \tag{2.7}$$

$$T \rightarrow T(R, T) + \theta(k) \frac{\partial}{\partial R} T(R, T) \tag{2.8}$$

The vertex functions Λ and θ depend only on k and therefore describe the distortion of the distribution function from its equilibrium shape. In equilibrium, the gradients vanish and T and ε_F are constants. In non-equilibrium, the temporal and spatial dependence of T and ε_F must be found by solving the transport problem self-consistently. In a linear response, however, there is an important difference between the calculation of the Fermi potential and the temperature field: contributions from the heating of the carriers in the field are not included because energy dissipation is a process of higher order. The temperature field must therefore be considered as a given external field. We did not include a temporal gradient in Eqs. (2.7) and (2.8) because its contribution is negligible as long as the temporal variations of the perturbations are slow on the intrinsic time scale which is of the order of ps. The next step is to linearize Eq. (2.3) in Λ and θ. We then obtain the distribution function in the linear response regime

$$f = f_{equ} + q \frac{\partial f}{\partial \varepsilon} \Lambda \frac{\partial}{\partial R} \varepsilon_F - \frac{\partial f}{\partial \varepsilon} \frac{\varepsilon - qV + q\varepsilon_F}{k_B T} \theta \frac{\partial}{\partial R} k_B T \tag{2.9}$$

As a response to external perturbations, there will be a particle current and an energy or heat flow in the system. Both can be calculated with Eq. (2.9). The particle current

$$j(R, T) = -q \sum_k v_k f(R, T, k) \tag{2.10}$$

is now given by

$$j(R, T) = -q^2 \sum_k \frac{\partial f}{\partial \varepsilon} v_k \Lambda(k) \cdot \frac{\partial}{\partial R} \varepsilon_F(R, T)$$

$$+ q \sum_k \frac{\partial f}{\partial \varepsilon} v_k \frac{\varepsilon_k - qV(R, T) + q\varepsilon_F(R, T)}{k_B T(R, T)}$$

$$\cdot \theta \frac{\partial}{\partial R} k_B T \tag{2.11}$$

and the heat flow j_w

$$j_w(R, T) = \Sigma \, v_k(\varepsilon_k - qV(R, T) + q\varepsilon_k(R, T))f(R, T) \tag{2.12}$$

is, with Eq. (2.9)

$$j_w(R, T) = q\sum_k \frac{\partial f}{\partial \varepsilon} v_k(\varepsilon_k - qV(R, T)$$

$$+ q\varepsilon_F(R, T))\Lambda(k) \cdot \frac{\partial}{\partial R} \varepsilon_F(R, T)$$

$$- q\sum_k \frac{\partial f}{\partial \varepsilon} v_k(\varepsilon_k - qV(R, T) + q\varepsilon_F(R, T))$$

$$\cdot \frac{\varepsilon_k - qV(R, T) + q\varepsilon_F(R, T)}{k_B T(R, T)} \, \theta \, \frac{\partial}{\partial R} k_B T(R, T) \tag{2.13}$$

The particle and heat currents respond to both $\nabla \varepsilon_F$ and $\nabla k_B T$. This response can be expressed by a generalized current equation where each contribution is included in a component of a two-dimensional current vector J

$$J = \begin{vmatrix} j \\ j_w \end{vmatrix} = \begin{vmatrix} j_1 \\ j_2 \end{vmatrix} = \begin{vmatrix} K_{11} & K_{12} \\ K_{21} & K_{22} \end{vmatrix} \cdot \begin{vmatrix} F_1 \\ F_2 \end{vmatrix}, \begin{vmatrix} F_1 \\ F_2 \end{vmatrix} = -\frac{1}{k_B T} \begin{vmatrix} \nabla \varepsilon_F \\ \nabla \ln(k_B T) \end{vmatrix} \tag{2.14}$$

Following from the Onsager reciprocity theorem (Onsager 1931a,b, Yourgau et al. 1982) the cross-components obey $K_{12} = K_{21}$. This leaves us with

$$\theta(k) = -\Lambda(k)k_B T(R, T)/q \tag{2.15}$$

and the particle current is now

$$j(R, T) = -q\sum_k \frac{\partial f}{\partial \varepsilon} v_k \Lambda \cdot \left(q \frac{\partial}{\partial R} \varepsilon_F + (\varepsilon_k - qV(R, T) \right.$$

$$\left. + q\varepsilon_F) \frac{\partial}{\partial R} u_T(R, T) \right) \tag{2.16}$$

To calculate the vertex function Λ, which is still unknown, it is not necessary to include the effect of a small temperature gradient on the distribution function. The distribution function can therefore be expressed as

$$f(R, T, K) = f_{equ}(R, T, k) + q \frac{\partial f}{\partial \varepsilon} \Lambda(k) \cdot \nabla_R \varepsilon_F(R, T) \tag{2.17}$$

Before we turn to the determination of Λ, let us discuss some properties of the particle-current equation directly derived from Eq. (2.17). For this and the rest of this section we will assume the system to be at constant temperature. For the particle current we obtain the following from Eq. (2.17)

$$j(R, T) = - q\sum_k \frac{\partial f}{\partial \varepsilon} v_k \Lambda(k) \cdot \nabla_R \varepsilon_F(R, T) =$$

$$- q\mu n(R, T)\nabla_R \varepsilon_F(R, T) \tag{2.18}$$

and for the particle density n

$$n(R, T) = \sum_k f_{\text{equ}}(R, T, k)$$

$$= \int d\varepsilon \Omega(\varepsilon)\left(1 + \exp\left(\frac{\varepsilon - qV(R, T) + q\varepsilon_F(R, T)}{k_B T}\right)\right)^{-1} \tag{2.19}$$

Here $\Omega(\varepsilon)$ is the density of states.

In general, the mobility μ is a 3×3 matrix in space but reduces to a diagonal matrix in systems of sufficiently high symmetry. In the effective mass approximation, it reduces to a c-number which we will define later. Differentiating Eq. (2.19) with respect to R we have

$$\nabla n = \left[\frac{\partial}{\partial \varepsilon_F}n\right](E + \nabla \varepsilon_F) \tag{2.20}$$

and this gives the following for the gradient of the Fermi potential

$$\nabla \varepsilon_F = - E + \left[\frac{\partial}{\partial \varepsilon_F}n\right]^{-1}\nabla n \tag{2.21}$$

and the current equation is then

$$j(R, T) = q\mu n(R, T)E(R, T) - q\left[\frac{\partial}{\partial \varepsilon_F}n\right]^{-1}\mu n(R, T)\nabla_R n(R, T) \tag{2.22}$$

If we define the diffusion constant D

$$D = - \left[\frac{\partial}{\partial \varepsilon_F}n\right]^{-1}\mu n(R, T) \tag{2.23}$$

the current equation is cast into the well-known form

$$j(R, T) = q\mu n(R, T)E(R, T) + qD(R, T)\nabla_R n(R, T) \tag{2.24}$$

Equations (2.23) and (2.24) constitute the drift–diffusion approximation to the current equation. Equation (2.23) in particular is the generalized

Einstein relationship for a Fermi gas near the equilibrium. For a non-degenerate Fermi system, the distribution function turns into the Boltzmann distribution function. The functional derivative can now be explicitly calculated and gives the conventional Einstein relationship

$$D = \mu u_0; \quad u_0 = k_B T / q \tag{2.25}$$

and in the general case we have

$$D = \mu u_0 \frac{\int d\varepsilon \Omega(\varepsilon) f_{\text{equ}}(R, T, k)}{\int d\varepsilon \Omega(\varepsilon) f_{\text{equ}}(R, T, k)(1 - f_{\text{equ}}(R, T, k))} \tag{2.26}$$

It is obvious that for practical purposes it is much easier to deal with Eq. (2.25) than with Eq. (2.26). If, however, the effects of degeneracy are to be included, then the generalized Einstein relation Eq. (2.26) must be used. We will return to this question later in this section after having established a suitable expression for the mobility. To do this we must find Λ in Eq. (2.17). This will bring us to the relaxation time approximation of the scattering term in the Boltzmann equation, Eq. (2.1). From Eq. (2.17) we proceed to Eqs. (2.1) and (2.2). The left-hand side of Eq. (2.1) creates a number of derivatives with respect to R, T, and k. We collect only contributions that are linear in these derivatives, according to our assumption that the gradients in the system are small. It is thus consistent to neglect second-order contributions in derivatives with respect to R and T. A consequent linearization also neglects second-order contributions in ∇_k. In the scattering term we keep contributions linear in the vertex function Λ. This linearization scheme gives

$$\frac{\partial}{\partial T}(V(R, T) - \varepsilon_F(R, T)) + v_k \nabla_R \varepsilon_F(R, T)$$

$$= -\sum_{k'} w(k, k') \left[\Lambda_k \frac{f_{\text{equ}}(R, T, k')}{f_{\text{equ}}(R, T, k)} \right.$$

$$\left. - \Lambda_{k'} \frac{1 - f_{\text{equ}}(R, T, k')}{1 - f_{\text{equ}}(R, T, k)} \exp\left(-\frac{\varepsilon_{k'} - \varepsilon_k}{k_B T}\right) \right] \nabla_R \varepsilon_F(R, T)$$

$$= \sum_{k'} w(k, k') \frac{f_{\text{equ}}(R, T, k')}{f_{\text{equ}}(R, T, k)} [\Lambda_k - \Lambda_{k'}] \nabla_R \varepsilon_F(R, T) \tag{2.27}$$

To solve this equation, we introduce a relaxation time $\tau(\varepsilon_k)$ which is defined through the vertex function Λ_k

$$\Lambda_k = -v_k \tau(\varepsilon_k) \tag{2.28}$$

and depends on ε_k only! This time defines an intrinsic time and length scale for the problem. Assuming that the temporal variations in the system are small compared to this time scale, which will turn out to be in the ps regime, we can neglect the time derivatives in Eq. (2.27). An appropriate length scale is given by the average $\langle v\tau \rangle$. Inserting Eq. (2.28) into Eq. (2.27) the remaining vector identity is only solved if the relaxation $\tau(\varepsilon_k)$ time satisfies the relation

$$\frac{1}{\tau(\varepsilon_k)} = \sum_{k'} w(k, k') \frac{f_{\text{equ}}(R, T, k')}{f_{\text{equ}}(R, T, k)} \left[1 - \frac{\tau(\varepsilon_{k'})}{\tau(\varepsilon_k)} \cdot \frac{v_k v_{k'}}{v_{k^2}} \right] \tag{2.29}$$

In general, this equation cannot produce a solution that does not depend on R and T, as we assumed in Eqs. (2.7) and (2.8) and consequently also in Eq. (2.28). However, for the special case of elastic scattering, $\varepsilon_k = \varepsilon_{k'}$ holds and both the spatial and temporal dependence disappear in the summation, as does the ratio of the relaxation times. So we have finally derived for the relaxation time $\tau(\varepsilon_k)$

$$\frac{1}{\tau(\varepsilon_k)} = \sum_{k'} w(k, k') \left[1 - \frac{v_k v_{k'}}{v_{k^2}} \right] \tag{2.30}$$

an expression that depends only on the scattering properties of the system. Only in the linear response regime does the relaxation time $\tau(\varepsilon_k)$ allow us to rewrite the Boltzmann equation (2.1) in an equivalent form: the relaxation time approximation

$$\frac{\partial}{\partial T} f(R, T, k) + v_k \nabla_R f(R, T, k) - qE(R, T)\nabla_k f(R, T, k) =$$

$$- \frac{f(R, T, k) - f_{\text{equ}}(R, T, k)}{\tau(\varepsilon_k)} \tag{2.31}$$

The essential assumptions leading to the relaxation time approximation were:

  (i) elastic collisions only, thus limited to impurity scattering and to acoustic phonons if $\omega_q \ll k_B T$
  (ii) relaxation time depends only on ε_k
  (iii) linear response; ∇_R and $\partial/\partial T$ are small on the scale $\langle v_k \tau \rangle$ and $\langle \tau \rangle$, respectively. Restrictions to the field will be discussed later.

Together with Eqs. (2.17) and (2.28), the solution of Eq. (2.31) in the linear response regime is given by

$$f(R, T, k) = f_{\text{equ}}(R, T, k) - q \frac{\partial f}{\partial \varepsilon} \tau(\varepsilon_k) v_k \cdot \nabla_R \varepsilon_F(R, T) \tag{2.32}$$

Only in that particular limit will Eq. (2.31) produce physically meaningful results, as we shall demonstrate at the end of this section.

2.2.2 Low Field Mobility

We can now continue our discussion of the mobility, which is expressed according to Eqs. (2.18) and (2.32) as

$$\mu_{ij} = -q \frac{1}{n(R, T)} \sum_k \frac{\partial f}{\partial \varepsilon} \tau(\varepsilon_k) v_{ki} v_{kj} \qquad (2.33)$$

This is in general a nondiagonal tensor. If the system under consideration has sufficient symmetry, the tensor will become diagonal. This is the case in cubic systems, for example. Because f_{equ} and τ depend only on the energy, it is convenient to introduce an energy integration.

$$\mu_{ij} = -q \frac{1}{n(R, T)} \int d\varepsilon \frac{\partial f}{\partial \varepsilon} \tau(\varepsilon) \sum_k v_{ki} v_{kj} \delta(\varepsilon - \varepsilon_k) \qquad (2.34)$$

If the energy surface $\varepsilon = \varepsilon_k$ is a rotational paraboloid, it is possible to transform it into a sphere for which the effective mass approximation holds. We will not discuss this transformation here. Interested readers are referred to the monograph by Ester Conwell (1967), for instance. In the effective mass approximation, the mobility tensor is a c-number. For the remaining k-summation we get

$$\sum_k v_{ki} v_{kj} \delta(\varepsilon - \varepsilon_k) = \frac{2}{3} \Omega(\varepsilon) \frac{\varepsilon}{m} \delta_{ij} \qquad (2.35)$$

which finally leads to the following expression for the mobility

$$\mu = \frac{q \tau_{tr}}{m} \qquad (2.36)$$

with an effective transport lifetime $\tau_{tr} (\approx \langle \tau \rangle)$

$$\tau_{tr} = \frac{1}{k_B T} \cdot \frac{\int d\varepsilon \Omega(\varepsilon) f_{equ}(R, T, \varepsilon) (1 - f_{equ}(R, T, \varepsilon)) \varepsilon \tau(\varepsilon)}{\int d\varepsilon \Omega(\varepsilon) f_{equ}(R, T, \varepsilon)} \qquad (2.37)$$

In situations where the Fermi statistic has to be used for the carriers we find an explicit dependence on R and T in the mobility which comes from the spatial and temporal variations of the electric potential V and the Fermi potential ε_F. We find the same for the diffusion constant D. For practical applications, especially in numerical simulations, this gives rise to inconvenience because both μ and D are very complex functionals of V and ε_F. The situation is considerably relaxed in non-degenerate systems where only the Boltzmann limit of the Fermi function has to be considered. Here, the contributions from V and ε_F cancel and the transport lifetime reduces to

$$\tau_{tr} = \frac{4}{3\sqrt{\pi}} \cdot \int d\varepsilon \exp\left(-\frac{\varepsilon}{k_B T}\right) \varepsilon^{3/2} \tau(\varepsilon) \qquad (2.38)$$

For the rest of this section, we will study only non-degenerate systems. The analysis of the mobility will therefore be based on Eqs. (2.36) and (2.38). The relaxation time calculated in Eq. (2.29) has to be compared with the collision time τ_c

$$\frac{1}{\tau_c(k)} = \sum_{k'} w(k, k') \tag{2.39}$$

Both scattering times depend on the scattering rate $w(k, k')$. However, the rate responsible for particle transport carries extra weight with respect to the direction of the velocities of the two scattering partners. Scattering contributions where the velocity does not change direction, i.e. small-angle scattering, are suppressed relative to those with large-angle changes. This is due to the scattering term in the Boltzmann equation containing terms for both scattering in and scattering out of the phase space volume, resulting in cancellation. This does not appear in the collision time, which counts only the outwardly scattered events. As a result, both can have considerably different values. The relaxation time τ_{tr} is determined in transport experiments and is therefore often called the transport lifetime. The collision time τ_c is determined experimentally by optical measurements relating to the width of the spectral function and is therefore called the optical lifetime. As an example, we calculate the energy-dependent transport lifetime $\tau_{tr}(\varepsilon)$ from Eq. (2.30), and the collision lifetime $\tau_c(\varepsilon)$ from Eq. (2.39) for a screened Coulomb potential.

$$V(x) = \frac{q^2}{\kappa x} \exp(-k_c x) \tag{2.40}$$

This potential approximately describes the scattering of carriers by ionized impurities. In an effective mass approximation, the transport lifetime is

$$\frac{1}{\tau_{tr}(\varepsilon_k)} = 2\pi N_{imp} \sum_{k'} |V_{kk'}|^2 \delta(\varepsilon_k - \varepsilon_{k'})(1 - \cos\theta_{kk'}) \tag{2.41}$$

and the collision time

$$\frac{1}{\tau_c(\varepsilon_k)} = 2\pi N_{imp} \sum_{k'} |V_{kk'}|^2 \delta(\varepsilon_k - \varepsilon_{k'}) \tag{2.42}$$

Here N_{imp} is the density of randomly distributed impurities and $\cos\theta_{kk'}$ the angle between the initial and final state k-vector of the scattering event. The matrix element contains the Fourier transform of the potential Eq. (2.40)

$$V_{kk'} = \frac{q^2}{\kappa} \int d^3x \frac{\exp(-k_c x)}{x} \exp(-i(k - k')x) = 4\pi \frac{q^2}{\kappa} \frac{1}{|k - k'|^2 + k_c^2} \tag{2.43}$$

Inserting Eq. (2.43) into Eqs. (2.41) and (2.42), we obtain the following for the transport lifetime

$$\frac{1}{\tau_{tr}(\varepsilon)} = \frac{1}{2}\pi N_{imp}\left(\frac{q^2}{\kappa}\right)^2 \frac{1}{\sqrt{(2m)}}\frac{1}{\varepsilon\sqrt{\varepsilon}}\left\{\ln(1 + \varepsilon/\varepsilon_c) - \frac{4\varepsilon/\varepsilon_c}{1 + 4\varepsilon/\varepsilon_c}\right\}$$

(2.44)

and for the collision time

$$\frac{1}{\tau_c(\varepsilon)} = \frac{1}{4}\pi N_{imp}\left(\frac{q^2}{\kappa}\right)^2 \frac{1}{\sqrt{(2m)}}\frac{1}{\varepsilon\sqrt{\varepsilon}}\frac{(4\varepsilon/\varepsilon_c)^2}{1 + 4\varepsilon/\varepsilon_c}$$

(2.45)

with $\varepsilon_c = k_c^2/(2m)$. It is of interest to look at different limiting cases. Let us first consider the limit of a short-range interaction, i.e. where $\varepsilon_c \to \infty$. We then get

$$\lim_{\varepsilon_c \to \infty} \frac{1}{\tau_{tr}(\varepsilon)} = \lim_{\varepsilon_c \to \infty} \frac{1}{\tau_c(\varepsilon)} = \frac{1}{4}\pi N_{imp}\left(\frac{q^2}{\kappa}\right)^2 \frac{1}{\sqrt{(2m)}}\frac{1}{\varepsilon\sqrt{\varepsilon}}\left(4\frac{\varepsilon}{\varepsilon_c}\right)^2$$

(2.46)

In this limit, the transport and collision lifetimes are the same. This is not the case in the weak screening limit $\varepsilon_c \to 0$.

$$\lim_{\varepsilon_c \to 0} \frac{1}{\tau_{tr}(\varepsilon)} = \frac{1}{2}\pi N_{imp}\left(\frac{q^2}{\kappa}\right)^2 \frac{1}{\sqrt{(2m)}}\frac{1}{\varepsilon\sqrt{\varepsilon}}\left\{\ln(1 + 4\varepsilon/\varepsilon_c) - 1\right\} \quad (2.47)$$

$$\lim_{\varepsilon_c \to 0} \frac{1}{\tau_c(\varepsilon)} = \frac{1}{2}\pi N_{imp}\left(\frac{q^2}{\kappa}\right)^2 \frac{1}{\sqrt{(2m)}}\frac{1}{\varepsilon\sqrt{\varepsilon}}(4\varepsilon/\varepsilon_c) \qquad (2.48)$$

Taking the ratio of Eqs. (2.47) and (2.48), we obtain the surprising result

$$\lim_{\varepsilon_c \to 0} \frac{\tau_c(\varepsilon)}{\tau_{tr}(\varepsilon)} = \frac{1}{4}\frac{\varepsilon_c}{\varepsilon}\left\{\ln(1 + 4\varepsilon/\varepsilon_c) - 1\right\} \to 0$$

(2.49)

This means that for large elastic scatterers we have $\tau_{tr} \gg \tau_c$. The reason is that large scattering centers predominantly have small momentum transfers of order $k_c \approx 1/\lambda_c$, where λ_c is the characteristic range of the interaction, and therefore contribute significantly to forward scattering. These scattering events are suppressed in the transport lifetime. Statements about carrier lifetime and mean free paths could lead to incorrect conclusions if it is unclear whether they refer to transport lifetime or to optically determined collision broadening.

Scattering by randomly distributed impurities is one channel for elastic scattering. The other important channel is the elastic approximation of acoustic phonon scattering. In general, the transport lifetime due to acoustic phonon scattering is given by the first term in the scattering rate w of

Eq. (2.2)

$$\frac{1}{\tau_{tr}(\varepsilon)} = 2\pi \sum_{k'} |m_{k-k'}|^2 [n_B(\omega_{k-k'})\delta(\varepsilon_k - \varepsilon_{k'} - \omega_{k-k'})$$

$$+ (n_B(\omega_{k-k'}) + 1) \cdot \delta(\varepsilon_k - \varepsilon_{k'} + \omega_{k-k'})](1 - \cos\theta_{kk'}) \quad (2.50)$$

We obtain the elastic limit if $k_B T \gg \omega_q$. In this limit, the phonon occupation number is

$$\lim_{k_B T \gg \omega_q} n_B(\omega) = \frac{1}{\exp(\omega/k_B T) - 1} = \frac{k_B T}{\omega} - \frac{1}{2} \quad (2.51)$$

If we only keep the leading terms in the phonon energy ω_q, it can be dropped in the argument of the δ-function. This is justified for $k_B T \gg \omega_q$ because in the linear response regime the bulk of carriers will have an energy comparable to $k_B T$ and the energy transfer to the phonon system is very small. With the matrix element for deformation potential coupling (Mahan 1981)

$$|m_{kk'}|^2 = C^2 \frac{1}{2\rho\omega_{k-k'}} |k - k'|^2 \quad (2.52)$$

and an isotropic phonon dispersion relation $\omega_{kk'} = s|k - k'|$, we obtain the following for the transport lifetime

$$\frac{1}{\tau_{tr}(\varepsilon_k)} = 4\pi C^2 \frac{k_B T}{2\rho s^2} \sum_{k'} \delta(\varepsilon_k - \varepsilon_{k'})(1 - \cos\theta_{kk'}) \quad (2.53)$$

and a similar expression for the collision time. Here s, ρ, and C are the velocity of sound, the specific density, and the deformation energy constant of the material, respectively. All these quantities are measurable and can be found in the literature. Performing the k-sum in Eq. (2.53) gives a τ_{tr} for acoustic phonon scattering

$$\frac{1}{\tau_{tr}(\varepsilon)} = \frac{1}{\pi} C^2 \frac{k_B T}{2\rho s^2} (2m)^{2/3} \sqrt{\varepsilon} \quad (2.54)$$

which is proportional to the density of states $\Omega(\varepsilon) \sim \sqrt{\varepsilon}$.
We are now in a position to calculate the low-field mobility of the bulk material. The different contributions to the relaxation time are given by Eqs. (2.46) and (2.47) for ionized impurity scattering for its extreme limits, and Eq. (2.54) for scattering by acoustic phonons. Many more scattering channels are possible, such as scattering by neutral impurities or carrier–carrier scattering. The former is usually less important than ionized impurity scattering because the interaction is much weaker (dipole coupling). It might, however, be of importance for very high doping densities or very low temperatures when most of the dopands are only incompletely

ionized. In that case we have to consider the degenerate electron system which also involves a more complex structure of the transport coefficients μ and D. Carrier–carrier scattering is an inelastic process to a very high degree. If particles of comparable mass interact, a considerable transfer of energy and momentum occurs between them. In addition it is impossible to linearize this interaction because two particles are involved in the scattering process. We therefore restrict ourselves to the scattering rates mentioned above. The energy-dependent transport relaxation time is, by virtue of Eq. (2.2)

$$\frac{1}{\tau_{tr}(\varepsilon)} = \frac{1}{\tau_{tr}(\varepsilon)^{imp}} + \frac{1}{\tau_{tr}(\varepsilon)^{acph}} \tag{2.55}$$

and the mobility is obtained by inserting Eq. (2.50) into Eq. (2.38) and then using Eq. (2.36). If we consider only the limits of short ($i = 1$) and long ($i = -1$) range scattering for the ionized impurities for simplicity, the mobility can be expressed as

$$\mu = \mu_{max} \int d\varepsilon \, \exp(-\varepsilon) \frac{\varepsilon}{1 + (N_{imp}/n^i_{ref}) \cdot \varepsilon^{i-1}} \tag{2.56}$$

$$\alpha = \frac{1}{\pi} \left(\frac{C}{2\rho a_0^3 s^2} \right) \left(\frac{k_B T}{Ry} \right)^{3/2} \frac{C}{\hbar} \left(\frac{m_0}{m} \right)^{3/2} \tag{2.57}$$

$$\beta^1 = 16\pi \frac{a_0^3}{\kappa^2} \left(\frac{Ry}{k_B T} \right)^{3/2} \left(\frac{k_B T}{\varepsilon_c} \right)^2 \frac{Ry}{\hbar} \left(\frac{m_0}{m} \right)^{1/2} \tag{2.58}$$

$$\beta^{-1} = 2\pi \frac{a_0^3}{\kappa^2} \left(\frac{Ry}{k_B T} \right)^{3/2} \frac{Ry}{\hbar} \left(\frac{m_0}{m} \right)^{1/2} \tag{2.59}$$

Here m_0, a_0, and Ry are the free electron mass, the Bohr radius, and the Rydberg energy unit. In the constants α and β we have reinstalled Planck's constant $\hbar$ in its proper places.

$$\mu_{max} = \frac{4}{3} \frac{q}{m \sqrt{\pi \alpha}} \tag{2.60}$$

$$n^i_{ref} = \frac{\alpha}{\beta^i} \tag{2.61}$$

The integral can be performed analytically for both cases. For small scattering centers we get

$$\mu = \frac{\mu_{max}}{1 + N_{imp}/n^1_{ref}} \tag{2.62}$$

and for long-range scatterers

$$\mu = \mu_{max} \left\{ 1 - \frac{N_{imp}}{n_{ref}^{-1}} g(\sqrt{(N_{imp}/n_{ref}^{-1})}) \right\} \tag{2.63}$$

Here $g(x)$ is the auxiliary function

$$g(x) = - ci(x)\cos(x) - x^{-2} si(x)\sin(x) \tag{2.64}$$

Both Eqs. (2.62) and (2.63) show a monotonic decrease of the mobility with increasing impurity concentration. This is a well-known experimental fact. But it is also observed that a minimal saturation value exists for this decrease as a lower limit. Neither of the two expressions derived above can account for this saturation behavior. We will now direct our investigation to find a possible explanation. To this end, we will need some of the material considered in Chapter 1. The energy conserving δ-function in the scattering rate, Eq. (2.2), was produced by taking the quasi-particle limit of the structure function S, Eq. (1.175). In deriving this result, we disregarded the influence of the real part of the electron's retarded self-energy $\operatorname{Re}\Sigma^r$. If we repeat the calculation, but now explicitly retaining $\operatorname{Re}\Sigma^r$, we will derive the following for the structure function S in the limit $\Gamma = - \operatorname{Im}\Sigma^r \to 0$

$$S(k, k', \omega) = \frac{2\Gamma(k, k', \omega)}{(\omega - (\varepsilon_k - \varepsilon_{k'}) - \operatorname{Re}\Sigma^r(k, k', \omega))^2 + \Gamma^2(k, k', \omega)} \tag{2.65}$$

with

$$\Sigma^r(k, k', \omega) = \operatorname{Re}\Sigma^r(k', \varepsilon_{k'} + q\varepsilon_F) - \operatorname{Re}\Sigma^r(k, \varepsilon_k + q\varepsilon_F)$$

If we consider elastic scattering, we have to take the limit $\omega \to 0$. This gives us, in the quasi-particle limit $\Gamma \to 0$

$$S(k, k', 0) = 2\pi \delta(\varepsilon_{k'} - \varepsilon_k + \operatorname{Re}\Sigma^r(k', \varepsilon_{k'} + q\varepsilon_F)$$
$$- \operatorname{Re}\Sigma^r(k, \varepsilon_k + q\varepsilon_F))$$
$$= \frac{2\pi}{1 + \dfrac{\partial}{\partial \varepsilon_k} \Sigma^r(k, \varepsilon_k + q\varepsilon_F)} \delta(\varepsilon_{k'} - \varepsilon_k) \tag{2.66}$$

Equation (2.66) is a renormalization of the energy-conserving δ-function. The self-energy for impurity scattering is proportional to N_{imp} (Appendix 1). Therefore the renormalization factor of the δ-function has the tendency to decrease with increasing doping concentration. To quantify this, we have to calculate $\operatorname{Re}\Sigma^r$ for randomly distributed impurities. For this, we have according to Eq. (A1.30)

$$\operatorname{Re}\Sigma^r(k, \varepsilon_k + q\varepsilon_F) = N_{\text{imp}} \sum_{k'} |V_{kk'}|^2 \operatorname{Re} G^r(k', \varepsilon_{k'} + q\varepsilon_F)$$

$$= N_{\text{imp}} \sum_{k'} |V_{kk'}|^2 \frac{\varepsilon_k - \varepsilon_{k'}}{(\varepsilon_k - \varepsilon_{k'})^2 + \delta^2} \qquad (2.67)$$

Here, we have used the zero-order equilibrium Green's function as a first approximation and the limit $\delta \to 0$ is assumed. Together with the scattering matrix element, Eq. (2.43), we obtain the following after performing the angular integrations

$$\operatorname{Re}\Sigma^r(k, \varepsilon_k + q\varepsilon_F) = 2N_{\text{imp}} \left(\frac{q^2}{\kappa}\right)^2 \frac{m}{k} \int k' dk' \frac{1}{(k'^2 + k_c^2)}$$

$$\cdot \ln \frac{\left(1 - \frac{1}{2}\frac{k'}{k}\right)^2 + \delta^2}{\left(1 + \frac{1}{2}\frac{k'}{k}\right)^2 + \delta^2} \qquad (2.68)$$

Further analytical calculation is impossible. We are only interested in the qualitative behavior of Eq. (2.68), and will therefore continue by replacing the slowly varying ln-function as follows

$$\ln \frac{\left(1 - \frac{1}{2}\frac{k'}{k}\right)^2 + \delta^2}{\left(1 + \frac{1}{2}\frac{k'}{k}\right)^2 + \delta^2} = 2 \ln \left| \frac{1 - \frac{1}{2}\frac{k'}{k}}{1 + \frac{1}{2}\frac{k'}{k}} \right| = -2 \begin{cases} \dfrac{k'}{k}; \dfrac{k'}{k} < 2 \\[2mm] 4\dfrac{k}{k'}; \dfrac{k'}{k} > 2 \end{cases} \qquad (2.69)$$

This approximation cuts off the logarithmic divergences in the integral which are damped anyway for a small but finite δ. Preceeding with the integration gives

$$\operatorname{Re}\Sigma^r(k, \varepsilon_k + q\varepsilon_F) = -4N_{\text{imp}} \left(\frac{q^2}{\kappa}\right)^2 m \left\{ \left(\frac{1}{2k^2 k_c} - \frac{1}{k_c^3}\right) \operatorname{arctg} \frac{2k}{k_c} \right.$$

$$\left. - \frac{1}{2}\left(\frac{1}{k^2 k_c} - \frac{1}{k_c^3}\right) \frac{2k/k_c}{1 + (2k/k_c)^2} + \frac{\pi}{4k_c^3} \right\} \qquad (2.70)$$

As it turns out, the limit of long-range scattering fits the experimental findings better than that of short-range scattering. So in Eq. (2.70) we will consider only the limit $k_c \to 0$. The leading contributions in this case are

$$\operatorname{Re}\Sigma^r(k, \varepsilon_k + q\varepsilon_F) = -\frac{1}{2} N_{\text{imp}} \left(\frac{q^2}{\kappa}\right)^2 \left\{ \frac{\pi - 1}{\varepsilon_k k_c} + \frac{2m}{k_c^3} \right\} \qquad (2.71)$$

For the renormalization of the δ-function, we need the energy derivative of $\operatorname{Re}\Sigma^r$. With Eq. (2.71) in the leading contribution in $k_c \to 0$ we get the following

$$\frac{\partial}{\partial \varepsilon_k} \operatorname{Re} \Sigma^r(k, \varepsilon_k + q\varepsilon_F) = N_{\text{imp}} \gamma^{-1} \left(\frac{k_B T}{\varepsilon_k} \right)^2 \tag{2.72}$$

with

$$\gamma^{-1} = 4 \frac{a_0^3}{\kappa^2} \left(\frac{Ry}{k_B T} \right)^2 \frac{1}{a_0 k_c} \tag{2.73}$$

The renormalized transport lifetime is obtained by replacing the δ-function in Eqs (2.41) and (2.53) by Eq. (2.66) together with Eq. (2.71). For the mobility, we finally get

$$\mu = \mu_{\text{max}} + (\mu_{\text{min}} - \mu_{\text{max}}) \frac{N_{\text{imp}}}{n_{\text{ref}}^{-1}} g(\sqrt{(N_{\text{imp}}/n_{\text{ref}}^{-1})}) \tag{2.74}$$

$$\mu_{\text{min}} = \mu_{\text{max}} \gamma^{-1} n_{\text{ref}}^{-1} \tag{2.75}$$

In the limit $N_{\text{imp}} \to \infty$, equation (2.74) assumes the minimal mobility μ_{min}. To obtain an estimate for μ_{max}, μ_{min}, and the reference density n_{ref} we evaluate these quantities for typical numbers for silicon material. At room temperature the velocity of sound is $s = 9 \cdot 10^5$ cm/s. For the range of the interaction we assume several interatomic distances so that $1/k_c = 10 \cdot 10^{-8}$ cm. The deformation potential constant we adjust to be $C = 2.5$ eV to obtain the correct intrinsic mobility. We then find $\mu_{\text{max}} = 1300$ cm^2/Vs, $\mu_{\text{min}} = 25$ cm^2/Vs and $n_{\text{ref}}^{-1} = 8 \cdot 10^{17}$ cm^{-3}. These numbers compare reasonably with those found by fitting expressions similar to Eq. (2.74) to experimental data (Selberherr 1984).

2.2.3 Limits

We derived the relaxation-time approximation of the Boltzmann equation for a system responding to a small external perturbation. It is inherent to the relaxation-time approximation that only elastic collisions are considered. This means that no energy exchange occurs between the carrier system and the heat bath, which in general comprises the acoustic phonons. It is indeed puzzling that despite this we can calculate the resistivity of a sample, which is very closely linked to the energy dissipation into the environment. The catch is that energy dissipation is a second-order effect which must be very small to be neglected. Energy dissipation expressed as Joule heating, for instance, is proportional to the product of current density and electrical field. The current density scales in the linear response regime with the gradient of the Fermi potential ε_F, which is the force responsible for driving the system away from equilibrium. The electric field has not yet entered the theory as a prominent force. It has its place, however, if energy dissipation is considered, which, again, is not included in the relaxation-time approximation. To find the limits of the relaxation-time approximation we will now investigate the non-stationary solution of the Boltzmann equation, Eq. (2.31)

$$\frac{\partial}{\partial T} f(R, T, k) + v_k \nabla_k f(R, T, k) - qE(R, T)\nabla_R f(R, T, k)$$

$$= - \frac{f(R, T, k) - f_{\text{equ}}(R, T, k)}{\tau(\varepsilon_k)} \tag{2.31}$$

A formal solution of Eq. (2.31) can be found along the trajectories $\{R(T), k(T)\}$ in the classical phase space. These are defined by the equations of motion

$$\frac{\partial}{\partial T} R(T) = v(k(T)) \tag{2.76}$$

$$\frac{\partial}{\partial T} k(T) = - qE(R(T)) \tag{2.77}$$

Imposing the initial condition at $T = T_0$

$$f(R(T_0), T_0, k(T_0)) = f_{\text{equ}}(R(T_0), T_0, k(T_0)) = f_{\text{equ},0}$$
$$R(T_0) = R_0 \tag{2.78}$$
$$k(T_0) = k_0$$

the Boltzmann equation in the relaxation-time approximation is solved by:

$$f_T = \exp\left(- \int_{T_0}^{T} \frac{d\tau}{\tau_{\text{tr}}(\tau)}\right)\left\{f_{\text{equ},0} + \int_{T_0}^{T} \frac{d\tau}{\tau_{\text{tr}}(\tau)} f_{\text{equ},\tau}\right.$$

$$\left. \cdot \exp\left(\int_{T_0}^{\tau} \frac{d\tau'}{\tau_{\text{tr}}(\tau')}\right)\right\} \tag{2.79}$$

In the relaxation-time approximation scheme, $f_{\text{equ},\tau}$ is given by

$$f_{\text{equ},\tau} = \exp\left(- \frac{\varepsilon_{k(\tau)} - qV(R(\tau)) + q\varepsilon_F(R(\tau))}{k_B T}\right) \tag{2.80}$$

In general the solution of Eq. (2.79) cannot be represented in a simple closed form. So we will simplify the problem by assuming an energy-independent relaxation time and a constant electric field. However, these approximations will not invalidate the principal physical conclusions drawn from the end result. For the equations of motion, Eqs. (2.76) and (2.77), we now get

$$k(\tau) = - qE(\tau - T_0) + k_0 \tag{2.81}$$

$$R(\tau) = - \frac{qE}{2m}(\tau - T_0)^2 + \frac{k_0}{m}(\tau - T_0) + R_0 \tag{2.82}$$

and especially for the energy of the electron we have

$$\varepsilon_{k(\tau)} - qV(R(\tau)) = \varepsilon_{k_0} - qV(R_0) \tag{2.83}$$

Inserting Eq. (2.80) together with Eq. (2.83) into Eq. (2.79), we obtain the following for the distribution function

$$f_T = f_{\mathrm{equ},0} \exp\left(-\frac{T - T_0}{\tau_{\mathrm{tr}}}\right)$$

$$\cdot\left\{1 + \int_{T_0}^{T} \frac{d\tau}{\tau_{\mathrm{tr}}} \exp\left(\frac{\tau - T_0}{\tau_{\mathrm{tr}}} + \frac{q}{k_B T}\varepsilon_{F0} - \frac{q}{k_B T}\varepsilon_F(R(\tau))\right)\right\}$$

(2.84)

If we limit this investigation to homogeneous bulk material, the Fermi potential will be linear in R or $\varepsilon_F = V$ which gives

$$\varepsilon_{F0} - \varepsilon_F(R(\tau)) = \frac{k_0 E}{m}(\tau - T_0) - \frac{qE^2}{2m}(\tau - T_0)^2$$

(2.85)

and using this in Eq. (2.84) we get the explicit time dependence of the distribution function

$$f_T = f_{\mathrm{equ},0} \exp\left(-\frac{T - T_0}{\tau_{\mathrm{tr}}}\right)\left\{1 + \int_{T_0}^{T} \frac{d\tau}{\tau_{\mathrm{tr}}}\right.$$

$$\left.\cdot\exp\left(\frac{\tau - T_0}{\tau_{\mathrm{tr}}}\left[1 + \frac{qk_0\tau_{\mathrm{tr}}}{k_B T m}E\right] - \frac{q^2 E^2}{k_B T 2m}\tau_{\mathrm{tr}}\left(\frac{\tau - T_0}{\tau_{\mathrm{tr}}}\right)^2\right)\right\}$$

(2.86)

This expression does not show its asymptotic behavior for $T - T_0 \to \infty$ very clearly. With the aid of simple algebra, Eq. (2.86) is brought into the following form:

$$f_T = f_{\mathrm{equ},0}\left\{\frac{\exp\left(\frac{T - T_0}{\tau_{\mathrm{tr}}} \cdot \frac{k_0 v_d}{k_B T} - \varepsilon_d\left(\frac{T - T_0}{\tau_{\mathrm{tr}}}\right)^2\right)}{1 + \frac{k_0 v_d}{k_B T} - 2\varepsilon_d\left(\frac{T - T_0}{\tau_{\mathrm{tr}}}\right)^2} + \frac{1}{2}\left(\frac{\pi}{\varepsilon_d}\right)^{1/2}\right.$$

$$\cdot\exp\left(-\frac{T - T_0}{\tau_{\mathrm{tr}}}\right)\exp\left(\frac{\left[1 + \frac{k_0 v_d}{k_B T}\right]^2}{4\varepsilon_d}\right)$$

$$\left.\cdot\,\mathrm{erfc}\left(-\frac{\left[1 + \frac{k_0 v_d}{k_B T}\right]^2}{4\varepsilon_d}\right) + \exp\left(-\frac{T - T_0}{\tau_{\mathrm{tr}}}\right)\right\}$$

(2.87)

with

$$\varepsilon_d = \frac{mv_d^2}{2k_B T}; \quad v_d = \mu_0 E$$

Here erfc is the complementary error function and μ_0 the low-field mobility. In the limit $T - T_0 \to \infty$, only the first term in Eq. (2.87) contributes significantly. Re-installing the phase space variables R and k, the distribution functions reads as follows in the asymptotic limit

$$f(R, T, k) = \exp\left(-\frac{\varepsilon_k}{k_B T}\right) \cdot \frac{1}{1 + \mu_0 \dfrac{kE}{k_B T} - \dfrac{q^2 E^2}{k_B Tm}(T - T_0)^2} \qquad (2.88)$$

We can draw several conclusions from this equation
  (i) There is no stationary solution in the limit $T - T_0 \to \infty$; this is, of course, an artifact of the assumption that no energy-dissipating scattering process exists.
  (ii) Linearizing Eq. (2.88) in the electric field we obtain Eq. (2.32) for homogeneous bulk material ($\nabla \varepsilon_F = \nabla V = -E$))

$$f(k) = \exp\left(-\frac{\varepsilon_k}{k_B T}\right) \cdot \left\{1 - \frac{\mu_0}{k_B T} kE\right\} \qquad (2.89)$$

The second term in the denominator of Eq. (2.88) is of the order v_d/v_{TH}; here, $v_{th} = \sqrt{(k_B T/m)}$ is the thermal velocity of the carriers. The linearization is possible for small fields, which means that $v_d \ll v_{th}$, and if the third term in the denominator of Eq. (2.88) is small compared to the second one. This is always assured for sufficiently small time intervals $\Delta T = T - T_0$ for which

$$\frac{q^2 E^2}{m} \Delta T^2 = v_d^2 m \frac{\Delta T^2}{\tau_{tr}^2} \ll k_B T \qquad (2.90)$$

holds. This restricts the relaxation-time approximation to the time scale

$$\tau_{tr} \ll \Delta T \ll \frac{\sqrt{(k_B Tm)}}{qE} \qquad (2.91a)$$

The lower limit ensures that enough scattering events occur to maintain a stationary state with the field. Owing to the lack of dissipative scattering, the upper limit must restrict the energy gain of the carriers in the field. In a realistic physical situation, there will of course be a stationary state in the limit $\Delta T \to \infty$ because energy-dissipating scattering processes, for instance by acoustic phonons, always occur. Especially if $\Delta T/\tau_{tr} \gg 1$, enough scattering events can occur to carry the excess energy away. If, however, the field is increased so that during the time between collisions the energy gain from the field is comparable to $k_B T$, then even with a large number of collisions the energy cannot completely dissipate into the heat bath. In this case, we must leave the linear response regime and are forced to include the coupling to the heat bath explicitly in our analysis. The critical field strength is estimated from Eq. (2.90), with $\Delta T = \tau_{\text{tr}}$

$$\frac{q^2 E_{\text{crit}}^2 \tau_{\text{tr}}^2}{m} = k_B T \tag{2.91b}$$

or the equivalent expression $v_d = v_{\text{th}}$. For a rough estimate of the critical field, we assume m to be the free electron mass and $\tau_{\text{tr}} = 0.1$ ps, which gives $E_{\text{crit}} \approx 4 \cdot 10^4$ V/cm.

If this critical field is exceeded in homogeneous material, we leave the linear response regime and encounter nonlinear effects such as velocity saturation and carrier heating. This situation will be investigated in detail in the following section.

2.3 Nonlinear Response and the Moment Method

It is generally understood that the drift–diffusion approximation for carrier transport refers to Eq. (2.24). Carriers either drift in a force field or experience a diffusive motion driven by a density gradient. Both components together comprise the total particle flow. Equations of this kind are used in many different areas of transport physics. We derived Eq. (2.24) assuming a weakly perturbed system, which means that the transport coefficients μ and D do not depend on the driving force $\nabla \varepsilon_F$. The observation that μ and D are related by Einstein's relationship is especially important in this context (2.23). The mobility calculated for this case relies heavily on the transport relaxation time τ_{tr}. It follows from the collision term of the Boltzmann equation (2.1) if we disregard all dissipative scattering events. This introduces some ambiguities, as we have discussed at the end of the previous section. Because there are always dissipative scattering processes in an electron–phonon system, a rigorous analysis should contain them from the beginning. The problem we face is that a general analysis does not take us very far due to the complex structure of the collision term in the Boltzmann equation. There are various ways of dealing with this problem.

The most advanced approach to carrier transport leaves the realm of the drift–diffusion approximation completely. It only considers the microscopic scattering events based on the most accurate available band structure, phonon dispersion relations, and scattering matrix elements. Taking these as an input to a Monte Carlo calculation, we can hope to obtain an accurate description of the physics on the macroscopic level (Fischetti and Laux 1989). At present, the Monte Carlo approach works in bulk material with simple *pn*-structures. The situation is more complicated for the MOSFET because of the Si/SiO_2 interface.

The most successful approach is a pragmatic one: to modify the drift–diffusion approximation, Eq. (2.24) so that the experimentally observed features are correctly described. In bulk material, the most prominent effect to be included is the velocity saturation in high fields. It is observed that once a critical field E_{crit} is exceeded, the mobility is no longer a constant but decreases so that for a sufficient large field, the drift velocity is a constant. The situation is clear in homogeneous bulk material, where the mobility is simply parametrized with respect to the electric field E, $\mu \to \mu(E)$ to match the experimental finding. These models are then used in the more complex device situation. The problem involved with this "transfer" is whether the electric field remains the correct parameter responsible for velocity saturation. As we have learned in analyzing the linear response problem in a general non-homogeneous situation, the current was driven not by the field but by the gradient of the quasi-Fermi potential. Should we replace the electric field E by $\nabla \varepsilon_F$? This choice is by no means the only one. Nor is the functional form of $\mu(F)$ uniquely determined. An account of the various possibilities can be found in the monograph by Selberherr (Selberherr 1984). In short, the nonlinear response is incorporated by choosing a proper field-dependent mobility for the drift–diffusion equation, Eq. (2.24), and Einstein's relationship, Eq. (2.23). As a result, the current density is still driven in the direction of $\nabla \varepsilon_F$.

The third approach we will pursue in the following is the moment method for the Boltzmann equation (Blotekjaer 1970, Cook and Frey 1982, Fukuma and Übbing 1984, Hänsch and Selberherr 1986, Forghieri et al. 1986, Thoma et al. 1989). The device-relevant quantities are the particle density n and the current density j. Both are expressed as certain averages of the distribution function $f(R, T, k)$. The semiconductor equations constitute a system of partial differential equations which mutually relate these averages. Examples are the continuity equation and the current equation, which couple the density n and the current density j in the lowest order. Additionally, of course, there is the Poisson equation that relates the charge density to the electric field. The former two equations are transport equations and can be derived rigorously from the Boltzmann equation, as we did for the current equation in the previous section for the linear-response regime. In the general situation of arbitrary deviations from equilibrium, the Boltzmann equation cannot be solved rigorously, so we do not know what the distribution function looks like. On the other hand, we are not interested in the distribution function itself but rather in its averages. This feature is used in the moment method. We take the Boltzmann equation and calculate the device-relevant averages. This yields a hierarchy of partial differential equations that relates the various moments (or averages) to each other. In general, by this procedure we create a hierarchy of infinite order. From experience, and only from experience, we know that it is sufficient to keep only the first few moments. As we shall see below, for the linear response regime these are the first two, i.e. n and j. To go beyond the linear response, we must always successively add the next pair of moments. The first component then produces a scalar quantity and the second a generalized current density flow. This adds the energy density $\mathscr{E}$ and the energy flow, v_ε for instance, to the drift–diffusion equation as first-order corrections. Little is known about the influence of the higher-order moments on the solution of the transport problem. The very first pair describes the particle balance in the system and the second pair the energy balance. The latter is required to include the energy exchange between the carrier system and the lattice.

We start from the Boltzmann equation Eq. (2.1) of a non-degenerate electron gas in the effective mass approximation.

$$\frac{\partial}{\partial T} f(R, T, k) + v(k)\nabla_R f(R, T, k) - qE(R, T)\nabla_k f(R, T, k) =$$

$$\sum_{k'} [w(k, k')f(R, T, k') - w(k', k)f(R, T, k)] \quad (2.92)$$

The forthcoming analysis is easily generalized to a one-band system if the velocity $v(k)$ is maintained instead of k/m.

(i) zero order—particle density n

$$n(R, T) = \sum_k f(R, T, k) \quad (2.93)$$

Because we do not at the moment include particle exchange between bands, the collision term in Eq. (2.92) will disappear if the k summation is performed. So does the third term on the left-hand side because the distribution function vanishes at $k \to \pm \infty$. We therefore obtain the continuity equation

$$\frac{\partial}{\partial T} n(R, T) + \nabla_R v(R, T) = 0 \quad (2.94)$$

(ii) first order—particle current density v

$$v(R, T) = \sum_k v(k)f(R, T, k) = \sum_k \frac{k}{m} f(R, T, k) \quad (2.95)$$

We multiply Eq. (2.92) by one component of the velocity $v_i(k)$ ($i = x, y, z$)

and sum over k, which gives an equation for the current density.

$$\frac{\partial}{\partial T} v_i(R, T) + \frac{2}{m} \frac{\partial}{\partial R_j} T_{ij}(R, T) + q \frac{E_i(R, T)}{m} n(R, T)$$

$$= \sum_{kk'} (v_i(k) - v_i(k'))w(k, k')f(R, T, k') \qquad (2.96)$$

here T_{ij} is the energy tensor

$$T_{ij}(R, T) = \frac{1}{2} m \sum_k v_i(k)v_j(k)f(R, T, k) = \sum_k \frac{k_i k_j}{2m} f(R, T, k) \quad (2.97)$$

and we have adopted the convention that a summation has to be performed over repeated indices.

(iii) second order—energy density $\mathcal{E}$

$$\mathcal{E}(R, T) = \sum_k \varepsilon_k f(R, T, k) = \text{Tr}(T) \qquad (2.98)$$

In the effective mass approximation, the energy density $\mathcal{E}$ is related to the trace of the energy tensor $(\text{Tr}(T) = \Sigma T_{ii})$. Multiplying Eq. (2.92) by $\varepsilon_k = k^2/2m$ and summing again over k we get the energy conservation equation

$$\frac{\partial}{\partial T} \mathcal{E}(R, T) + \nabla_R v_\varepsilon(R, T) + qE(R, T)v(R, T)$$

$$= \sum_{kk'} (\varepsilon_k - \varepsilon_{k'})w(k, k')f(R, T, k') \qquad (2.99)$$

(iv) third order—energy current density v_ε

$$v_\varepsilon(R, T) = \sum_k \varepsilon_k v(k)f(R, T, k) \qquad (2.100)$$

We now multiply Eq. (2.92) by $\varepsilon_k v_i(k)$ and sum over k to obtain the equation for the energy-current component $v_{\varepsilon i}$

$$\frac{\partial}{\partial T} v_{\varepsilon i}(R, T) + \frac{2}{m} \frac{\partial}{\partial R_j} \theta_{ij}(R, T) + q \frac{2}{m} E_j(R, T)(T_{ij}(R, T)$$

$$+ \delta_{ij} \frac{1}{2} \text{Tr}(T)) = \sum_{kk'} (\varepsilon_k v_i(k) - \varepsilon_{k'} v_i(k'))w(k, k')$$

$$\cdot f(R, T, k') \qquad (2.101)$$

here θ_{ij} is the auxiliary tensor

$$\theta_{ij}(R, T) = \frac{1}{2} m \sum_k \varepsilon_k v_i(k)v_j(k)f(R, T, k) = \sum_k \frac{k^2}{2m} \frac{k_i k_j}{2m} f(R, T, k)$$

$$(2.102)$$

Inspecting Eqs. (2.96), (2.99), and (2.101) we find that without explicit knowledge of the distribution function $f(R, T, k)$ we are unable to deal with the scattering terms. Furthermore we notice that the current equations Eqs. (2.96) and (2.101) refer to moments of the next hierarchy of conservation laws. To close the hierarchy we have to assume a truncation scheme in the last current level that we consider. In our cases, we must approximate the auxiliary tensor θ in Eq. (2.101). The tensors T and θ introduce a complexity which is undesired in practical applications. Even in the effective mass approximation, these tensors are in general non-diagonal. In view of the fact that these tensor quantities are absent in the drift–diffusion approximation Eq. (2.24), which proved to be very successful, and that in the limiting case of linear response we have to reproduce that equation, we will replace the tensor components.

$$A_{ij} \rightarrow \frac{1}{3} \delta_{ij} Tr(A) \tag{2.103}$$

This substitution conserves the trace of the tensor A. With Eq. (2.103) the generalized set of transport equations, although still complicated enough, simplifies considerably.

$$\frac{\partial}{\partial T} n(R, T) + \nabla_R v(R, T) = 0 \tag{2.104}$$

$$\frac{\partial}{\partial T} v(R, T) + \frac{2}{3m} \nabla_R \varepsilon(R, T) + q \frac{E(R, T)}{m} n(R, T)$$
$$= \sum_{kk'} (v(k) - v(k'))w(k, k')f(R, T, k') \tag{2.105}$$

$$\frac{\partial}{\partial T} \mathscr{E}(R, T) + \nabla_R v_\varepsilon(R, T) + qE(R, T)v(R, T)$$
$$= \sum_{kk'} (\varepsilon_k - \varepsilon_{k'})w(k, k')f(R, T, k') \tag{2.106}$$

$$\frac{\partial}{\partial T} v_\varepsilon(R, T) + \frac{2}{3m} \nabla_R \mathscr{E}''(R, T) + q \frac{5}{3m} E(R, T)\mathscr{E}(R, T)$$
$$= \sum_{kk'} (\varepsilon_k v(k) - \varepsilon_{k'} v(k'))w(k, k')f(R, T, k') \tag{2.107}$$

In Eq. (2.107) we introduced the abbreviation $\mathscr{E}''$ for the trace of θ, which according to Eq. (2.102) is

$$\mathscr{E}''(R, T) = \sum_k \varepsilon_k^2 f(R, T, k) \tag{2.108}$$

Equations (2.104)–(2.107) serve as the basis for a generalization of the drift–diffusion equation (2.24) beyond the linear response regime. To be useful for practical purposes, we still have to solve the problem of truncating Eq. (2.107) and to approximate the collision terms. So far we have not assumed any special form of the distribution function f. For the scattering terms at least, we cannot continue without information concerning f.

2.3.1 Drifted Maxwellian

Historically, the drifted Maxwellian approximation was used first to generalize the drift–diffusion equation

$$f(R, T, k) = \exp\left(-\frac{\varepsilon_{k - v_d(R,T)m} + qV(R, T) - q\varepsilon_F(R, T)}{k_B T_c(R, T)}\right) \quad (2.109)$$

The objective for this approximation is that the scattering among the carriers is sufficiently strong for the carrier system to be in equilibrium at temperature T_c and to drift along with velocity v_d. To maintain such a situation, the momentum and energy transfer has to be much faster between carriers than to the phonon system. Because scattering between likely particles is a two-particle process, it will scale quadratically with the particle density. On the other hand, the scattering matrix element will scale with the density-dependent screening length of the screened Coulomb potential. Because the range of the screened Coulomb potential decreases with increasing density, the interaction strength will not scale quadratically with the density. The question is: will the Coulomb interaction be sufficiently strong to maintain equilibrium among the subset of carriers. An ultimate answer to this question has not yet been given. There are advocates of both extreme cases (Stratton 1957, Wingreen et al. 1986). Monte Carlo calculations are of little help because it is very difficult to incorporate carrier–carrier scattering and the results seem to be model-dependent. However, Monte Carlo calculations show that without carrier–carrier scattering the distribution function is not a drifted Maxwellian. We will come back to this question in Chapter 4, where we will investigate the problem of modeling impact ionization and oxide injection. There, the shape of the distribution function is of extreme importance.

In the transport equations (2.104)–(2.107) we are interested in the averages of the distribution function, as these will be less sensitive to the special form of this function. The drifted Maxwellian deserves some attention for this very reason. Utilizing Eq. (2.109), we obtain the following for the current density, the energy density, the energy-current density, and the auxiliary quantity $\mathscr{E}''$.

$$v(R, T) = v_d(R, T)n(R, T) \quad (2.110)$$

$$\mathscr{E}(R, T) = \frac{3}{2} k_B T_c(R, T)n(R, T) + \frac{1}{2} v_d(R, T)^2 mn(R, T) \quad (2.111)$$

$$= \theta_{tot}(R, T)n(R, T) \quad (2.112)$$

$$v_\varepsilon(R, T) = \frac{5}{3} v_d(R, T) \left[\theta_{\text{tot}}(R, T) - \frac{1}{5} v_d(R, T)^2 m \right] n(R, T) \quad (2.113)$$

$$\mathscr{E}''(R, T) = \frac{5}{3} \left[\theta_{\text{tot}}(R, T)^2 - \frac{2}{5} \left(\frac{1}{2} v_d(R, T)^2 m \right)^2 \right] n(R, T)$$

$$= \frac{5}{3} \left[\frac{\mathscr{E}(R, T)^2}{n(R, T)} - \frac{2}{5n(R, T)^3} \left(\frac{1}{2} v(R, T)^2 m \right)^2 \right] \quad (2.114)$$

Here we have introduced θ_{tot} which is the average energy per particle for the drifted Maxwellian. It comprises a temperature contribution and a kinetic energy term, Eq. (2.111). Equation (2.111) provides a definition of the carrier temperature T_c that distinguishes "thermal" and "kinetic" energies in the average energy per particle θ_{tot}.

$$\theta_{\text{tot}}(R, T) = \theta_{\text{the}}(R, T) + \theta_{\text{kin}}(R, T)$$

$$= \tfrac{3}{2} k_B T_c(R, T) + \tfrac{1}{2} v_d(R, T)^2 m \quad (2.115)$$

As we will see later, this is not the only possible way to define a carrier temperature. Equation (2.114) provides a truncation for the transport equations (2.104)–(2.107) in a very natural way. An argument very often found in the literature is that the drifted Maxwellian does not allow heat flow. It is based on the fact that the moment method was not utilized pairwise. Usually, the energy–current flow equation was not considered. With Eq. (2.114), the generalized semiconductor equations read

$$\frac{\partial}{\partial T} n(R, T) + \nabla_R v(R, T) = 0 \quad (2.116)$$

$$\frac{\partial}{\partial T} v(R, T) + \frac{2}{3m} \nabla_R \mathscr{E}(R, T) + q \frac{E(R, T)}{m} n(R, T)$$

$$= \sum_{kk'} (v(k) - v(k')) w(k, k') f(R, T, k') \quad (2.117)$$

$$\frac{\partial}{\partial T} \mathscr{E}(R, T) + \nabla_R v_\varepsilon(R, T) + q E(R, T) v(R, T)$$

$$= \sum_{kk'} (\varepsilon_k - \varepsilon_{k'}) w(k, k') f(R, T, k') \quad (2.118)$$

$$\frac{\partial}{\partial T} v_\varepsilon(R, T) + \frac{10}{9m} \nabla_R \left(\frac{\mathscr{E}(R, T)^2}{n(R, T)} - \frac{2}{5n(R, T)^3} \left(\frac{1}{2} v(R, T)^2 m \right)^2 \right)$$

$$+ q \frac{5}{3m} E(R, T) \mathscr{E}(R, T)$$

$$= \sum_{kk'} (\varepsilon_k v(k) - \varepsilon_{k'} v(k')) w(k, k') f(R, T, k') \quad (2.119)$$

Although we were able to truncate the hierarchy of equations, Eqs. (2.116)–(2.119) are not closed in the sense that they provide a self-consistent set of equations to calculate n, v, $\mathscr{E}$, and v_ε. Higher-order moments are still contained in the scattering terms. The truncation scheme is therefore not self-consistent. Within the drifted Maxwellian approach, it is still possible to express the scattering terms as functions of T_c and v_d (Blotekjaer and Lunde 1969) but it is difficult to expand them in the relevant lower moments n, v, E, and v_ε. They are therefore approximated by a generalized relaxation-time approximation.

$$\sum_{kk'}(v(k) - v(k'))w(k, k')f(R, T, k') \to \frac{v(R, T)}{\tau_v} \tag{2.120}$$

$$\sum_{kk'}(\varepsilon_k - \varepsilon_{k'})w(k, k')f(R, T, k') \to \frac{\mathscr{E}(R, T) - \mathscr{E}_0(R, T)}{\tau_\varepsilon} \tag{2.121}$$

$$\sum_{kk'}(\varepsilon_k v(k) - \varepsilon_{k'}v(k'))w(k, k')f(R, T, k') \to \frac{v_\varepsilon(R, T)}{\tau_{v\varepsilon}} \tag{2.122}$$

The relaxation times τ_v, τ_ε, and $\tau_{v\varepsilon}$ have to be adjusted to fit the experimental situation. In general they depend on n, v, $\mathscr{E}$, and v_ε to close the set of equations (2.116)–(2.119) self consistently. However, because their functional form is not given *a priori* different choices are possible. This can have an influence on the solution. $\mathscr{E}_0$ is the energy density for a carrier system with the phonon temperature, $T_c = T$. With Eqs. (2.120)–(2.122) the generalized semiconductor equations read

$$\frac{\partial}{\partial T}n(R, T) + \nabla_R v(R, T) = 0 \tag{2.123}$$

$$v(R, T) - \tau_v\frac{\partial}{\partial T}v(R, T) = n(R, T)\frac{q\tau_v}{m}E(R, T) + \frac{\tau_v}{m}\frac{2}{3}\nabla_R\mathscr{E}(R, T) \tag{2.124}$$

$$\frac{\partial}{\partial T}\mathscr{E}(R, T) + \nabla_R v_\varepsilon(R, T) = \frac{\mathscr{E}(R, T) - \mathscr{E}_0(R, T)}{\tau_\varepsilon}$$

$$- qE(R, T)v(R, T) \tag{2.125}$$

$$v_\varepsilon(R, T) - \tau_{v\varepsilon}\frac{\partial}{\partial T}v_\varepsilon(R, T) = \frac{5\tau_{v\varepsilon}}{3\tau_v}\left(v(R, T) - \tau_v\frac{\partial}{\partial T}v(R, T)\right)$$

$$\cdot\frac{\mathscr{E}(R, T)}{n(R, T)} + \frac{10}{9}\frac{\tau_{v\varepsilon}}{m}\mathscr{E}(R, T)\nabla_R\frac{\mathscr{E}(R, T)}{n(R, T)}$$

$$- \frac{1}{9}\tau_{v\varepsilon}m\nabla_R\left(\frac{v(R, T)^4}{n(R, T)^3}\right) \tag{2.126}$$

The relaxation times τ_v, τ_ε, and $\tau_{v\varepsilon}$ are of the order ps and smaller. Therefore the time derivatives in the current equations. Eqs. (2.124) and (2.126), will have a negligible influence as long as the external perturbations vary on a time scale which is much slower. Transients in a circuit are not faster than several hundred ps. In the continuity equations, Eqs. (2.123) and (2.125), the time derivatives are not scaled with the relaxation times here they are essential to describe the transient behavior of the device. To simplify the notation we introduce the thermal voltage u_T

$$u_T(R, T) = q^{-1} \frac{2\mathscr{E}(R, T)}{3n(R, T)} \tag{2.127}$$

and neglect the time derivatives in Eqs. (2.124) and (2.126)

$$\frac{\partial}{\partial T} n(R, T) + \nabla_R v(R, T) = 0 \tag{2.128}$$

$$v(R, T) = n(R, T)\mu E(R, T) + \mu \nabla_R(u_T(R, T)n(R, T)) \tag{2.129}$$

$$\frac{\partial}{\partial T}(u_T(R, T)n(R, T)) + \frac{2}{3q}\nabla_R(v_\varepsilon(R, T)) = n(R, T)\frac{u_T(R, T) - u_0}{\tau_\varepsilon}$$

$$- \frac{2}{3}E(R, T)v(R, T) \tag{2.130}$$

$$v_\varepsilon(R, T) = q\frac{5}{2}\text{æ}u_T(R, T)(v(R, T) + \mu n(R, T)\nabla_R(u_T(R, T)))$$

$$- \frac{1}{9}\tau_{v\varepsilon}m\nabla_R\left(\frac{v(R, T)^4}{n(R, T)^3}\right) \tag{2.131}$$

here μ is the mobility

$$\mu = \frac{q\tau_v}{m}$$

and æ is the ratio of energy and particle current relaxation time

$$\text{æ} = \frac{\tau_{v\varepsilon}}{\tau_v}$$

Equations (2.128)–(2.131) hold under very general circumstances and are in principle not tied to the drifted Maxwellian. Only the last term in the energy current equation, Eq. (2.131), is linked directly to the drifted Maxwellian. Usually it is neglected in a numerical solution because its very nonlinear behavior in the velocity gives a poorly conditioned discretized system. This can than be viewed as a linearization in v if one is willing to accept that the relaxation times are of weak variation in v. To solve the system, Eqs. (2.128)–(2.131), we still have to provide suitably parameterized relaxation

times. As mentioned above, their choice is not unique and will certainly
have an influence on the solution. To remove this ambiguity we will now
look at the problem from a different angle.

2.3.2 Moment Expansion of the Distribution Function

In the linear response regime we derived a distribution function of the form
(Eq. (2.32))

$$f(R, T, k) = f_{\mathrm{equ}}(R, T, k) + k \cdot g(R, T, k)$$

thus separating a even and odd contribution in k. This result will serve as an
objective for the generalized Ansatz (Hänsch and Miura-Mattausch 1986)

$$f(R, T, k) = \alpha(R, T, k) + k \cdot \beta(R, T, k) \tag{2.132}$$

Here α and β are yet unknown functions which are even in k. The function α
is related to the densities (n, $\mathscr{E}$, ...) or even moments of f and β is related to
the currents (v, v_ε, ...) or odd moments of f. Defining the even weight
function ϕ^i

$$\phi^i(k) = \varepsilon(k)^{i-1}; \quad i = 1, 2, 3, \ldots \tag{2.133}$$

we obtain for the density moments

$$n(R, T) = \sum_k \phi^1(k)\alpha(R, T, k) = \Psi^1(R, T)$$

$$\mathscr{E}(R, T) = \sum_k \phi^2(k)\alpha(R, T, k) = \Psi^2(R, T)$$

$$\mathscr{E}''(R, T) = \sum_k \phi^3(k)\alpha(R, T, k) = \Psi^3(R, T)$$

$$\vdots$$

With the functions φ_j from the dual space of $\{\phi_i\}$ we can expand α as
follows

$$\alpha(R, T, k) = \sum_j \varphi_j(k)\Psi^j(R, T) \tag{2.134}$$

because the orthogonality relation

$$\sum_k \varphi_j(k)\phi^i(k) = \delta_{ij}$$

With a similar argument an expansion of β is possible with respect to the
odd moments or currents

$$\beta(R, T, k) = \sum \varphi_{j'}(k)\Psi^{j'}(R, T)$$

here $\varphi_{j'}$ will be orthogonal to the odd or current weights $\phi^{i'}$

$$\phi^{i'}(k) = k\phi^i(k) \tag{2.135}$$

and we have

$$v(R, T) = \sum_k \phi^{1'}(k) k \cdot \beta(R, T, k) = \Psi^{1'}(R, T)$$

$$v_\varepsilon(R, T) = \sum_k \phi^{2'}(k) k \cdot \beta(R, T, k) = \Psi^{2'}(R, T)$$

$$\vdots$$

Maintaining in this expansion only the first four terms we get

$$f(R, T, k) = \varphi_1(k)\Psi^1(R, T) + \varphi_{1'}(k) k \cdot \Psi^{1'}(R, T)$$
$$+ \varphi_2(k)\Psi^2(R, T) + \varphi_{2'}(k) k \cdot \Psi^{2'}(R, T) + \cdots$$

$$(2.136)$$

The distribution function Eq. (2.136) contains information up to the energy current density v_ε. If we use this expansion in the scattering terms and if we find a suitable truncation for Eq. (2.107) then Eqs. (2.104)–(2.107) will be closed and self-consistent with respect to the first four moments of the distribution function. To truncate the hierarchy we again have to express $\mathscr{E}''$ in a convenient way. To neglect it altogether is a too rough approximation because it would not allow a heat current contribution to the energy flow. In thermal equilibrium we get for $\mathscr{E}''$

$$\mathscr{E}''_{equ}(R) = \frac{5\mathscr{E}_{equ}(R)^2}{3n(R)} \qquad (2.137)$$

Based on Eq. (2.137) we postulate

$$\mathscr{E}''(R, T) = \frac{5\mathscr{E}(R, T)^2}{3n(R, T)} \qquad (2.138)$$

to truncate the hierarchy. As we see later this will provide the correct form of the heat current to the energy current density. More important, however, are the scattering terms in Eqs. (2.104)–(2.107). We define the relaxation times τ_v, τ_ε, and τ_{v_ε} as in the case of the drifted Maxwellian, see Eqs. (2.120)–(2.122), but now we will insert Eq. (2.136) for the distribution function and calculate their correct parameterized form. Before we do this we like to summarize the two important specifics of the model

(i) matrix elements depend only on the magnitude of the momentum transfer $m_q = m_{|q|}$

(ii) phonon dispersion energies are isotropic

$$\omega_q = \omega_{|q|}$$

Nothing more is needed for what follows next.

We start with the momentum relaxation time τ_v. Inserting the distribution function from Eq. (2.136) into Eq. (2.120) we get

$$\frac{1}{\tau_v} = v(R, T) \sum_{kk'} \frac{(v(k) - v(k'))w(k, k')}{v(R, T)^2} [\varphi_1(k')\Psi^1(R, T)$$
$$+ \varphi_{1'}(k')k' \cdot \Psi^{1'}(R, T) + \varphi_2(k')\Psi^2(R, T)$$
$$+ \varphi_{2'}(k')k' \cdot \Psi^{2'}(R, T)] \tag{2.139}$$

The scattering matrix element $w(k, k')$ depends only on the magnitude of the k-vectors and their scalar product

$$w(k, k') = w(|k|, |k'|, k \cdot k')$$

and the φ-functions are even functions in k. By simple symmetry arguments we find that only the even moments do not contribute to the sum in Eq. (2.139). In particular we also find that the remaining part is diagonal. This brings us to the final result

$$\frac{1}{\tau_v} = A + B \frac{v(R, T) \cdot v_\varepsilon(R, T)}{v(R, T)^2} \tag{2.140}$$

The coefficients A and B do not depend on the moments $\Psi^{i(')}$ anymore. Therefore they are constant with respect to the moments of the distribution function. There is, however, a possible spatial dependence through a spatial dependence of the matrix elements. For the energy relaxation time τ_ε, defined in Eq. (2.121), we get the following same arguments

$$\frac{1}{\tau_\varepsilon} = \frac{C \cdot \mathscr{E}(R, T) - D \cdot n(R, T)}{\mathscr{E}(R, T) - \mathscr{E}_0(R, T)} \tag{2.141}$$

Again C and D do not depend on the moments of f. Their values are determined by considering a situation close to equilibrium. For this we rewrite Eq. (2.141) in the following way

$$\frac{1}{\tau_\varepsilon} = C \cdot \left[1 + \frac{\mathscr{E}_0(R, T) - D/C \cdot n(R, T)}{\mathscr{E}(R, T) - \mathscr{E}_0(R, T)} \right] = C \cdot \left[1 + \frac{u_0 - 2D/3C}{u_T - u_0} \right] \tag{2.142}$$

Unless $2D/3C$ equals u_0, Eq. (2.142) approaches infinity at equilibrium. Therefore the energy relaxation time is independent from the moments of f, in the approximation we are currently discussing.

$$\tau_\varepsilon = \tau_{\varepsilon 0} \tag{2.143}$$

The energy current relaxation time τ_{v_ε} is defined in Eq. (2.122) and is found to be expressed as

$$\frac{1}{\tau_{v_\varepsilon}} = A' + B' \frac{v_\varepsilon(R, T) \cdot v(R, T)}{v_\varepsilon(R, T)^2} \tag{2.144}$$

Equations (2.140), (2.143), and (2.144) are the relaxation times parametrized consistently within the four moment expansion. The transport problem is now formulated in a self-consistent closed form. With Eq. (2.138) for the truncation of the energy current equation and Eqs.. (2.140), (2.143), and (2.144) for the relaxation times we have for the generalized semiconductor equations

$$\frac{\partial}{\partial T} n(R, T) + \nabla_R v(R, T) = 0 \tag{2.145}$$

$$v(R, T) = n(R, T)\mu E(R, T) + \mu \nabla_R(u_T(R, T)n(R, T)) \tag{2.146}$$

$$\frac{\partial}{\partial T}(u_T(R, T)n(R, T)) + \frac{2}{3q}\nabla_R(v_\varepsilon(R, T)) = n(R, T)\frac{u_T(R, T) - u_0}{\tau_\varepsilon}$$

$$- \frac{2}{3}E(R, T)v(R, T) \tag{2.147}$$

$$v_\varepsilon(R, T) = q\frac{5}{2}æu_T(R, T)(v(R, T) + \mu n(R, T)\nabla_R(u_T(R, T))) \tag{2.148}$$

These are identical with those derived from the drifted Maxwellian apart from the last term in the energy current equation and the now provided form of the relaxation times, however.

In the drifted Maxwellian approach the carrier temperature T_c entered the formulation in a natural way through the form of the distribution function. However, it does not appear in the transport problem explicitly. Here the total energy density $\mathscr{E}$ is essential. The reason for this is that the temperature is not a moment of the distribution function itself. An alternative definition of the carrier temperature T_c can be obtained by using Eq. (2.127)

$$k_B T_c(R, T) = qu_T(R, T) = \frac{2}{3}\frac{\mathscr{E}(R, T)}{n(R, T)} \tag{2.149}$$

This definition is independent of the specific form of the distribution function and is more suitable for the transport problem. It has to be contrasted with Eqs. (2.112) and (2.115)

$$k_B T_c(R, T) = \frac{2}{3}\left[\frac{\mathscr{E}(R, T)}{n(R, T)} - \frac{1}{2}\frac{v(R, T)^2}{n(R, T)^2}m\right] \tag{2.150}$$

which is based on the drifted Maxwellian approach.

We are going to analyze the relaxation times in more detail. For this purpose we will in analogy with the linear response case define the force that is responsible for driving the system off equilibrium. This force F is simply found by inspection of the particle current equation (2.146) which reads

$$v(R, T) = n(R, T)\mu F(R, T) \tag{2.151}$$

with F

$$F(R, T) = E(R, T) + n(R, T)^{-1}\nabla_R(u_T(R, T)n(R, T)) \tag{2.152}$$

In the limit of only small deviations from equilibrium F is $\nabla\varepsilon_F$. In this case the transport coefficient μ and therefore τ_v is a constant. In the general case the momentum relaxation time found in Eq. (2.140) is

$$\frac{1}{\tau_v} = A + B\left[\frac{5}{2}\,æu_T(R, T)\left(1 + \frac{F(R, T)\cdot\nabla_R u_T(R, T)}{F(R, T)^2}\right)\right] \tag{2.153}$$

the energy current relaxation time is

$$\frac{1}{\tau_{v_\varepsilon}} = B' + \frac{A'}{\frac{5}{2}\,æu_T(R, T)}$$

$$\cdot \frac{1 + \dfrac{F(R, T)\cdot\nabla_R u_T(R, T)}{F(R, T)^2}}{1 + 2\dfrac{F(R, T)\cdot\nabla_R u_T(R, T)}{F(R, T)^2} + \left(\dfrac{\nabla_R u_T(R, T)}{F(R, T)}\right)^2} \tag{2.154}$$

and the energy relaxation time τ_ε is a constant. Equations (2.153) and (2.154) are not independent from each other. They are coupled through $æ = \tau_{v_\varepsilon}/\tau_v$. The consequences of this coupling are dramatic! For the sake of simplicity we will continue our discussion for the case where the spatial variation in the thermal voltage is small. Then Eqs. (2.153) and (2.154) simplify considerably

$$\frac{1}{\tau_v} = A + B\,æu_T(R, T) \tag{2.155}$$

$$\frac{1}{\tau_{v_\varepsilon}} = B' + \frac{A'}{æu_T(R, T)} \tag{2.156}$$

We have included a factor 2/5 in the constant A'. At this stage it is worthwhile to solve the coupled set of equations to study the behavior of the relaxation times for extreme cases. For æ we get

$$\text{æ} = u_T(R, T)^{-1}\frac{u_T(R, T)A - A'}{B' - Bu_T(R, T)} \tag{2.157}$$

As a result the momentum relaxation time τ_v will saturate in the limit $u_T \to \infty$ to a finite non zero value. This is in contradiction with the experimental findings which propose $\tau_v \to 0$ for $u_T \to \infty$. The explanation for this misbehavior is found in the very strong coupling of all current relaxation times which is only incompletely taken account of in our case. To be more explicit let us consider the influence of the next pair of moments in our expansion. This would be E'' and the corresponding current flow $v_{\varepsilon''}$. Proceeding in the same way as before we would obtain for the three current relaxation times τ_v, τ_{v_ε}, and $\tau_{\varepsilon''}$

$$\frac{1}{\tau_v} = A + B \cdot \frac{vv_\varepsilon}{v^2} + C \cdot \frac{vv_{\varepsilon''}}{v^2} + \cdots$$

$$= A + B\text{æ}u_T + C\text{æ}'u_T^2 + \cdots \tag{2.158}$$

$$\frac{1}{\tau_{v_\varepsilon}} = A' \cdot \frac{v_\varepsilon v}{v_\varepsilon^2} + B' + C' \cdot \frac{v_\varepsilon^2 v_{\varepsilon''}}{v_\varepsilon^2} + \cdots$$

$$= A'/(\text{æ}u_T) + B' + C'\text{æ}''u_T + \cdots \tag{2.159}$$

$$\frac{1}{\tau_{v_{\varepsilon''}}} = A'' \cdot \frac{v_{\varepsilon''}v}{v_{\varepsilon''}^2} + B'' \cdot \frac{v_{\varepsilon''}v_\varepsilon}{v_{\varepsilon''}^2} + C'' + \cdots$$

$$= A''/(\text{æ}'u_T^2) + B''/(\text{æ}''u_T) + C'' + \cdots \tag{2.160}$$

with

$$\text{æ} = \frac{\tau_{v_\varepsilon}}{\tau_v}; \quad \text{æ}' = \frac{\tau_{v_{\varepsilon''}}}{\tau_v}; \quad \text{æ}'' = \frac{\tau_{v_{\varepsilon''}}}{\tau_{v_\varepsilon}}$$

This clearly demonstrates that the four moment expansion is in general insufficient to give the correct behavior in the limit $u_T \to \infty$. A finite order truncation might, as shown above, produce unphysical results. Due to the fact that also the higher order velocity densities $v_\varepsilon, v_{\varepsilon''}, \ldots$ etc are based on particle motion in a force field, we expect their saturation if the force field is large enough. This is of course only true if the transport is scattering limited and we therefore can identify a strong scattering regime in contrast to the weak scattering case. Here particle motion is limited by the free flight of the carriers in the electrical field. A particular feature of the latter is the absence of a stationary state in an accelerating field. Especially the drift diffusion approximation can be considered as the strong coupling limit $\tau_\varepsilon = 0$. Which means that the carriers are tightly coupled to the phonon system that comprises the heat bath. To go one step beyond the drift diffusion equation means to consider the limit $\tau_\varepsilon \to 0$ in lowest order of the finite energy relaxation time. This will reduce the scattering rates τ_v, τ_{v_ε}, ... etc to a linear dependence in u_T:

$$\frac{1}{\tau_{v_\varepsilon}} = \gamma \frac{1}{\tau_v} \tag{2.161}$$

$$\frac{1}{\tau_v} = A + Bu_T \tag{2.162}$$

here γ is a number of order one at room temperature. Equation (2.162) is essentially the same as the originally derived Eq. (2.155) if we set $æ = 1$. Near equilibrium Eq. (2.161) is equivalent to the Wiedemann–Franz law and can be considered as its generalization.

Putting everything together again we end up with the following set of generalized semiconductor equations

$$\frac{\partial}{\partial T} n(R, T)n(R, T) + \nabla_R v(R, T) = 0 \tag{2.163}$$

$$v(R, T) = n(R, T)\mu F(R, T) \tag{2.164}$$

$$\frac{\partial}{\partial T}(u_T(R, T)n(R, T)) + \frac{2}{3q}\nabla_R(v_\varepsilon(R, T)) = n(R, T)\frac{u_T(R, T) - u_0}{\tau_{\varepsilon 0}}$$

$$-\frac{2}{3}E(R, T)v(R, T) \tag{2.165}$$

$$v_\varepsilon(R, T) = q\frac{5}{2\gamma}u_T(R, T)(v(R, T) + \mu n(R, T)\nabla_R(u_T(R, T))) \tag{2.166}$$

$$\mu = \frac{\mu_0}{1 + \gamma\left(u_T(R, T)\left[1 + \frac{F(R, T)\cdot\nabla_R u_T(R, T)}{F(R, T)^2}\right] - u_0\right)} \tag{2.167}$$

Here we used Eq. (2.162) in the equation for the mobility μ and introduced the two new variables μ_0 and γ instead of A and B. It is clear by inspection that μ_0 is the low field mobility which we considered in the previous section. The Poisson equation and Eqs. (2.163)–(2.167) are closed in the sense that also the parameterization of the transport coefficients μ and τ_ε with respect to the four moments n, v, E, and v_ε is known. Adjustable parameters are γ and $\tau_{\varepsilon 0}$. Because these parameters do not depend on the active variables their value is easily determined in the most simple situation: the homogeneous case.

For the linear response the $\nabla_R \varepsilon_F$ was the driving the system off equilibrium. A straightforward generalization to the non-linear response case is given by the quantity F, which also drives the current and assumes $\nabla_R \varepsilon_F$ in the limit $u_T \to u_0$. Therefore it is worthwhile to write the current equation Eq. (2.164) and the energy balance equations (2.165) and (2.166) as expansions with respect to F.

$$v(R, T) = n(R, T)\mu F(R, T) \tag{2.168}$$

$$-\frac{\partial}{\partial T} u_T(R, T) + \frac{u_T(R, T) - u_0}{\tau_{\varepsilon 0}} = \frac{2}{3} \mu F(R, T)^2$$

$$\cdot \left[1 + \frac{3}{2} \frac{F(R, T) \cdot \nabla_R u_T(R, T)}{F(R, T)^2} \right]$$

$$- \frac{2}{3} \mu u_T(R, T) \frac{1}{n(R, T)} \nabla_R n(R, T)$$

$$+ \frac{5}{3} \frac{u_T(R, T)}{n(R, T)} \nabla_R v(R, T) + O^2(u_T) \tag{2.169}$$

The last term on the right hand side of Eq. (2.169), which we did not write out explicitly, is of second order in u_T. We will neglect this order here in accordance with the expansion of τ_v. Considering this contribution in more detail we find that it contributes in regions of fast spatial variation of u_T. A more detailed mathematical analysis is found in the work of Hänsch and Schmeisser (1989). If we neglect impact ionization then we would have $\nabla_R v \sim 0$ and the third term on the right hand side of Eq. (2.169) would also vanish. In the further analysis we will find that the second term is of order F^0 in the limit $F \to \infty$ while the first term is of order F^1. Therefore we will only keep this contribution as the leading term for $F \to \infty$.

$$v(R, T) = n(R, T)\mu F(R, T) \tag{2.170}$$

$$-\frac{\partial}{\partial T} u_T(R, T) + \frac{u_T(R, T) - u_0}{\tau_{\varepsilon 0}}$$

$$= \frac{2}{3} \mu F(R, T)^2 \left[1 + \frac{3}{2} \frac{F(R, T) \cdot \nabla_R u_T(R, T)}{F(R, T)^2} \right] \tag{2.171}$$

Equations (2.170) and (2.171) are a reduced form of the generalized semiconductor equations Eqs. (2.163)–(2.166). They are correct in leading terms of the generalized driving force F if we neglect term of order $O^2(u_T)$ in the energy balance. If we would assume a very weak variation of u_T, on a yet undetermined scale, we can also neglect the second term in brackets. We then obtain

$$-\frac{\partial}{\partial T} u_T(R, T) + \frac{u_T(R, T) - u_0}{\tau_{\varepsilon 0}} = \frac{2}{3} \mu F(R, T)^2 \tag{2.172}$$

and for the mobility we have

$$\mu = \frac{\mu_0}{1 + \gamma(u_T(R, T) - u_0)} \qquad (2.173)$$

Because γ does not depend on the active variables n, v, and u_T we can determine its value for the stationary case. Inserting Eq. (2.173) into (2.172) we can solve the energy equation for u_T and going back with the result into Eq. (2.173) we derive for the mobility

$$\mu = \frac{2\mu_0}{1 + (1 + \frac{8}{3}\mu_0\tau_{\varepsilon 0}\gamma F(R, T)^2)^{1/2}} \qquad (2.174)$$

In deriving Eq. (2.174) we did not assume a constant electric field which is usually the input in empirically found mobility models. The mobility found in Eq. (2.174) reduces for the case of a constant electric field in homogeneous material to the model proposed by Jaggi (1969). The problem of finding the correct parameterization for non-homogeneous case is solved by considering from the beginning an expansion in the driving force F. However, for a homogeneous material and a constant field we have $F = E$. In the limit $E \to \infty$ the current equation (2.170) together with the mobility equation (2.174) gives a saturated velocity! To obtain the experimentally determined value v_{sat} we have

$$\gamma = \frac{3\mu_0}{2\tau_\varepsilon v_{\mathrm{sat}}^2} \qquad (2.175)$$

which gives a familiar expression for the mobility

$$\mu = \frac{2\mu_0}{1 + \left(1 + 4\dfrac{\mu_0^2}{v_{\mathrm{sat}}^2} F(R, T)^2\right)^{1/2}} \qquad (2.176)$$

and for the thermal voltage we obtain

$$u_T = u_0 + \frac{2}{3}\tau_\varepsilon v_{\mathrm{sat}}^2 \frac{\mu_0 - \mu}{\mu_0\mu} \qquad (2.177)$$

For the stationary case this renders the transport problem beyond the linear response to a solution of the continuity equation for the carrier density

$$\nabla_R(v(R, T)) = 0 \qquad (2.178)$$

with the current equation

$$v(R, T) = \mu n(R, T)F(R, T) \qquad (2.179)$$

in which the mobility depends on the driving force F via Eq. (2.176). The driving force itself was

$$F(R, T) = E(R, T) + n(R, T)^{-1} \nabla_R (u_T(R, T) n(R, T)) \qquad (2.180)$$

and is a functional of the mobility via Eq. (2.177). In linear response the driving force $\nabla_R \varepsilon_F$ was not a direct functional of the mobility. This feature introduces a higher order of complexity into the system which is especially important for a numerical solution of the problem. Many approximations have been made to reduce the very complex generalized set of semi-conductor equations displayed in Eqs. (2.104)–(2.107) to the rather simple looking Eqs. (2.176)–(2.180). We will pause here to consider some features of Eqs. (2.176)–(2.180).

The most important feature is that instead of having two more equations accounting for the energy balance there is only a modified current equation in which the mobility μ and the thermal voltage u_T have to be determined in a self-consistent way. Both are coupled via the generalized driving force F to the current equation. This automatically assures the correct behavior when equilibrium is approached. It can easily be shown that for the field free case in the presence of a density gradient the correction to the thermal voltage is $u_T \approx u_0 + 0(\tau_{\varepsilon 0} v_{sat}/L_D)$; here L_D is the diffusion length. For Si device material at room temperature we have for typical parameters $\tau_{\varepsilon 0} v_{sat} \ll L_D$. For the practical application this correction is smaller than the numerical accuracy and therefore negligible. Comparing the current equation (2.179) with the driving force equation (2.178) with the traditionally used drift–diffusion approximation we find that contributions beyond the linear response regime do not only effect the mobility but also the thermal voltage. Their influence in the mobility enters directly into the calculation of the terminal currents. An elevated carrier temperature will change the density distribution in the device and will effect the terminal currents (which are integral quantities) only little. This modified carrier distribution will, however, have an important effect on local quantities like avalanche generation and gate oxide injection, which we consider in Chapters 4 and 5. A new parameter in the theory is the energy relaxation time $\tau_{\varepsilon 0}$. In the limit $F \to \infty$ the thermal voltage is directly proportional to $\tau_{\varepsilon 0}$.

$$\lim_{F \to \infty} u_T = \frac{2}{3} \tau_{\varepsilon 0} v_{sat} F \qquad (2.181)$$

It is therefore of interest to investigate possible bounds for $\tau_{\varepsilon 0}$. This can be done by looking at the diffusion constant in high fields. The diffusion constant D, in the traditional case, is the transport coefficient that relates the diffusive current to the particle density gradient. To distinguish more clearly between drift and diffusion contributions we cast the current equation (2.179) into the following form

$$v(R, T) = -\mu n(R, T)\nabla_R[V(R, T) - u_T(R, T)]$$
$$+ \mu u_T(R, T)\nabla_R[n(R, T)] \qquad (2.182)$$

The diffusion constant $D = \mu u_T$ looks like a generalized Einstein relationship. It is easily verified that is not the case. Einstein's relationship is very closely linked to the existence of the quasi Fermi potential ε_F, the gradient of which is the driving force of the current in the linear response regime (compare Eqs. (2.17)–(2.25)). For the non-linear regime we have a driving force, but this is not the gradient of a potential field. In the limit $F \to \infty$ we get

$$\lim_{F \to \infty} D = \frac{2}{3}\tau_{\varepsilon 0} v_{sat}^2 = D_\infty \qquad (2.183)$$

From experimental material (Brunetti et al. 1981, Reggiani et al. 1986) we find that the low field diffusion constant $D_0 = \mu_0 u_0$ is an upper bound for D.

$$D_0 > D_\infty \qquad (2.184)$$

This in turn will give bounds for the energy relaxation time

$$0 < \tau_{\varepsilon 0} < \tau_{\varepsilon 0}^{max} = \frac{3}{2}\frac{\mu_0 u_0}{v_{sat}^2} \qquad (2.185)$$

From the experimental data of the diffusion constant in high fields $\tau_{\varepsilon 0}$ is related to the ratio of its low and high field values.

$$\tau_{\varepsilon 0} = \frac{D_\infty}{D_0}\tau_{\varepsilon 0}^{max} = \eta\tau_{\varepsilon 0}^{max}; \quad \eta < 1 \qquad (2.186)$$

With typical values for a silicon MOSFET device we will have: $\mu_0 = 800 \text{ cm}^2/\text{Vs}$, $v_{sat} = 10^7 \text{ cm/s}$, and $u_0 = 26 \text{ mV}$

$$\tau_{\varepsilon 0} \leq 0.3 \text{ ps} \qquad (2.187)$$

In comparison we find for the momentum relaxation time τ_v in the low field regime $F \to 0$

$$\tau_v = q^{-1}\mu_0 m \approx 0.2 \text{ ps} \qquad (2.188)$$

which reduces in the high field regime $F \approx 10^5 \text{ V/cm}$ by approximately one order of magnitude.

So far we have assumed that u_T is of weak spatial variation. Therefore we neglected contribution of order $F \cdot \nabla_R u_T/F^2$ in Eqs. (2.167) and (2.169). We will now consider the influence of these terms in our analysis and study their physical relevance. Again concentrating only on the stationary case we can derive for the mobility μ

$$\mu = \cfrac{1}{1 + \eta \cfrac{F \cdot \nabla_R u_T}{F^2}}$$

$$\cdot \cfrac{2\mu_0}{1 + \left(1 + 4\cfrac{\mu_0^2}{v_{sat}^2} F^2 \cfrac{\left[1 + \cfrac{3}{2}\cfrac{F \cdot \nabla_R u_T}{F^2}\right]\left[1 + \cfrac{F \cdot \nabla_R u_T}{F^2}\right]}{\left[1 + \eta \cfrac{F \cdot \nabla_R u_T}{F^2}\right]^2}\right)^{1/2}}$$

$$(2.189)$$

which turns into Eq. (2.176) for $\nabla_R u_T \to 0$. For the thermal voltage u_T we find

$$u_T = u_0 + \frac{2}{3}\tau_{\varepsilon 0} v_{sat}^2 \frac{1 + \eta \cfrac{F \cdot \nabla_R u_T}{F^2}}{1 + \cfrac{F \cdot \nabla_R u_T}{F^2}}\left[\cfrac{1}{\left(1 + \eta \cfrac{F \cdot \nabla_R u_T}{F^2}\right)\mu} - \cfrac{1}{\mu_0}\right]$$

$$(2.190)$$

In the velocity saturation limit $F \to \infty$ Eqs. (2.189) and (2.190) turn into

$$\lim_{F \to \infty} \mu F = \cfrac{v_{sat}}{\left(\left[1 + \cfrac{3}{2}\cfrac{F \cdot \nabla_R u_T}{F^2}\right]\left[1 + \cfrac{F \cdot \nabla_R u_T}{F^2}\right]\right)^{1/2}} \qquad (2.191)$$

$$\lim_{F \to \infty} u_T = u_0 + \frac{2}{3}\tau_{\varepsilon 0} v_{sat}\left[\left(\cfrac{1 + \cfrac{3}{2}\cfrac{F \cdot \nabla_R u_T}{F^2}}{1 + \cfrac{F \cdot \nabla_R u_T}{F^2}}\right)^{1/2} F - \cfrac{1 + \eta \cfrac{F \cdot \nabla_R u_T}{F^2}}{1 + \cfrac{F \cdot \nabla_R u_T}{F^2}}\cfrac{v_{sat}}{\mu_0}\right]$$

$$(2.192)$$

which allows for a velocity larger than v_{sat}, provided

$$-1 < \frac{3}{2}\frac{F \cdot \nabla_R u_T}{F^2} < 0 \qquad (2.193)$$

This is called velocity overshoot and is often considered to be a ballistic effect. On the other hand u_T is less influenced by the spatial gradients. We argue that velocity overshoot is an effect related to rapid spatial variations

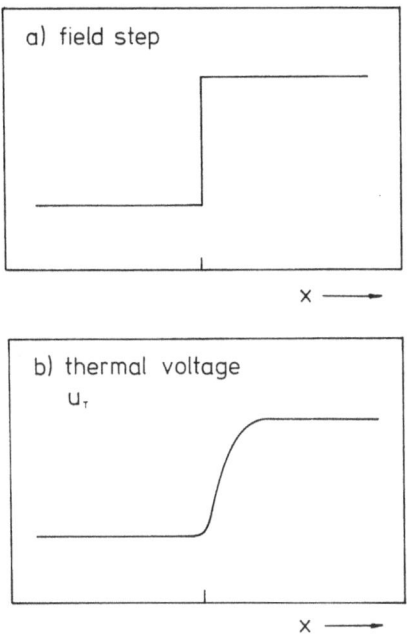

Fig. 9 The response of the thermal voltage (b) to a step in the electric field (a).

in the electric field E. The situation becomes clear if we consider the transport over a step-like field profile, Fig. 9a, connecting the fields E_1 and E_2, with $E_2 < E_1 < 0$. The charge will flow from the left to the right. To this field profile belongs the profile of the thermal voltage as shown in Fig. 9b. There is a transient length in which the carrier system has to adjust to the new stationary state. This length will be of the order of several $v_{sat}\tau_{\varepsilon 0}$. The velocity profile $v_d = \mu E$ will be calculated with the mobility from Eq. (2.167). If we neglect the spatial variation of u_T we obtain a slightly enhanced velocity compared to the stationary state. This enhancement will be increased if we take the spatial variation of u_T explicitly into account. Similar features we observe in the approximations Eqs. (2.176) and (2.189). Let us assume that both fields, E_1 and E_2, are sufficiently large so that for the stationary case we are in velocity saturation. Under these circumstances Eq. (2.176) will show no structure in the vicinity of the step. On the other hand Eq. (2.189) is even in the limit of velocity saturation sensitive to the profile of u_T. In the vicinity of the step there will be a region with a velocity larger than v_{sat}.
An important question to resolve is: when is it permitted to use Eq. (2.176)? Because we are interested in the behavior for $F \rightarrow \infty$ we will use Eq. (2.192)

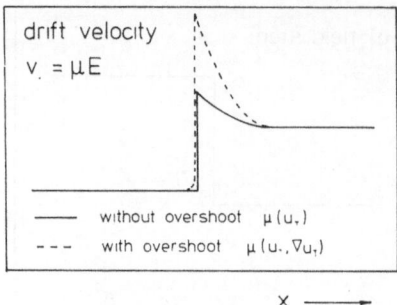

Fig. 10 The drift velocity for the field step of Fig. 9a. The solid line assumes a mobility that depends on u_T. The dashed line assumes a mobility that depends on u_T and its spatial derivatives ∇u_T

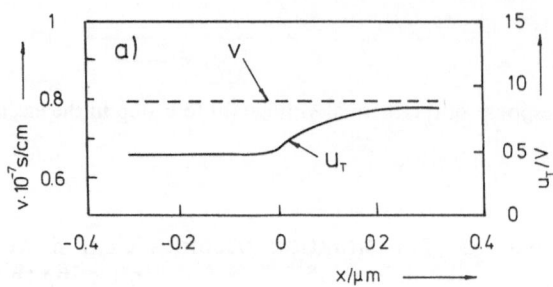

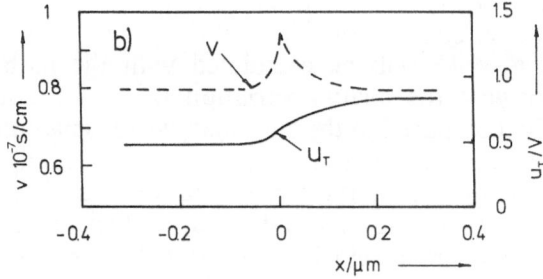

Fig. 11 The drift velocity for a field step in the saturation region. For the calculation we assumed a saturation velocity $v_{sat} = 0.8 \cdot 10^7$ cm/s and an energy relaxation time $\tau_\varepsilon = 0.4$ ps. The jump in the field is from $E(x < 0) = 3 \cdot 10^5$ V/cm to $E(x > 0) = 5 \cdot 10^5$ V/cm. (a) The mobility depends only on u_T, Eq. (2.176). (b) The mobility depends on u_T and its spatial derivatives ∇u_T, Eq. (2.189)

to estimate the influence of $F \cdot \nabla_R u_T / F^2$

$$\lim_{F \to \infty} \frac{F \cdot \nabla_R u_T}{F^2} = \frac{2}{3} \tau_{\varepsilon 0} v_{\text{sat}} \frac{F \cdot \nabla_R F}{F^2} = \lambda_{\text{crit}} \frac{F \cdot \nabla_R F}{F^2}; \ \lambda_{\text{crit}} = \frac{2}{3} \tau_{\varepsilon 0} v_{\text{sat}}$$

$$(2.194)$$

The influence of the spatial variation of u_T will be no more negligible if the generalized driving force changes considerably parallel to the current direction over the critical distance $\lambda_{\text{crit}} \approx 20$ nm. This is approximately twice the mean free path at room temperature. It is clear if variations in the fields are on the scale of the mean free path there will be not sufficient collisions to maintain a locally stationary state with the field. We would expect in this case the theory to become more non-local, which means in our case that we can no longer neglect $\nabla_R u_T$.

To utilize Eqs. (2.189) and (2.191) in practical modeling further simplification is necessary. The objective is to keep the physical features discussed above within a structure as closely related to Eqs. (2.176), (2.177), (2.179), and (2.180). Because we are mostly concerned with the velocity saturation regime we will investigate limit $F \to \infty$ to find suitable approximations. The lower bound given in Eq. (2.193) is an artifact of our approximation scheme. Nevertheless, it is also a signal to indicate the limits of the hydrodynamical model which is not fitted to describe the ballistical regime correctly. However, if the correction term $F \cdot \nabla_R u_T / F^2$ is small the hydrodynamic model is still justified. Therefore it seems consistent, within that model, to expand Eqs. (2.191) and (2.192) in terms of $F \cdot \nabla_R u_T / F^2$.

$$\mu = \frac{\mu|_{\nabla_R u_T = 0}}{1 + \frac{5}{6} \tau_{\varepsilon 0} v_{\text{sat}} e_F \cdot \nabla_R \left[\ln \left(\frac{u_T}{u_0} \right) \right]} \qquad (2.195)$$

$$u_T = u_0 + \frac{2}{3} \tau_{\varepsilon 0} v_{\text{sat}}^2 \left[\left(1 + \frac{1}{6} \tau_{\varepsilon 0} v_{\text{sat}} e_F \cdot \nabla_R \ln \left(\frac{u_T}{u_0} \right) \right) \frac{1}{\mu|_{\nabla_R u_T = 0}} \right.$$

$$\left. - \left(1 + (\eta - 1) \frac{2}{3} \tau_{\varepsilon 0} v_{\text{sat}} e_F \cdot \nabla_R \ln \left(\frac{u_T}{u_0} \right) \right) \frac{1}{\mu_0} \right] \qquad (2.196)$$

Here $\mu|_{\nabla_R u_T = 0}$ is the mobility from Eq. (2.176) which is sufficient to cover the regime between ohmic behavior and velocity saturation. In the limit $F \to \infty$ Eqs. (2.195) and (2.196) are equivalent to Eqs. (2.191) and (2.192). The unit vector e_F has the direction of F.

With Eqs. (2.195) and (2.196) our discussion of the hydrodynamic model within the one-band effective mass approximation is completed. Application of Eqs. (2.195) and (2.196) to a real MOSFET device with channel length down to 60 nm showed that the overshoot correction is less than

10% at room temperature and is therefore not significant for the terminal currents of the device (Hänsch and Jacobs 1989). Our aim for this section was to demonstrate that although the moment method for the Boltzmann equation is a very complex theory it is still possible to derive feasible results that contain the physics that is needed beyond the linear response regime. To obtain these results it is not sufficient to investigate only the kinetic terms, which come from the left hand side of Boltzmann's equation Eq. (2.92). It is absolutely mandatory to consider the collision terms as well.

2.3.3 Two-band system

Although the effective mass approximation is a very simplified model it seems nevertheless to cover most of the physics that is essential to describe charge transport in a Si device. More complex models are available to account for the influence of various deviations from the simple model such as: non-parabolicity, scattering between equivalent valleys, umklapp processes, and Brillouin zone edge effects. The price to pay is, however, the more detailed the physical model, the less complex the device it can describe. We have to remember that the physics has to be cast into a numerical code that has to return the solution of the problem the engineer is interested in. The trade off in modeling is between maintaining the essential physics and allowing for very complex device structures. In the previous sections we have developed a formulation for charge transport in non-degenerate semiconductors within the hydrodynamical approach. As the major result we were able to reduce the transport problem to one current equation for both linear and non-linear response regime. We did not consider particle exchange between different bands. In Chapter 4 impact ionization, which couples conduction and valence band, will be added. For Si material this, Shockley Read Hall, and Auger recombination are the important mechanisms that connect conduction and valence band. They enter the formulation as right hand side of the continuity equation Eq. (2.163). The analysis we did to derive the current equation Eq. (2.178) with the mobility Eq. (2.195) and the thermal voltage (2.196) is still valid. The situation is different for III–V type semiconductors. Here the conduction band consists of different non-equivalent bands which are separated by an energy that is small compared to the principal band gap E_G. As a result there is a considerable particle transfer already at small fields among the different bands in the conduction band. Because the bands differ considerably the transport properties will be different from that in Si where the one-band approximation is sufficient for the conduction band. A simple approach to take account of this complex situation is to approximate the complex band structure by a number of independent energy parabolas. Their effective mass and position in energy is chosen to reproduce the

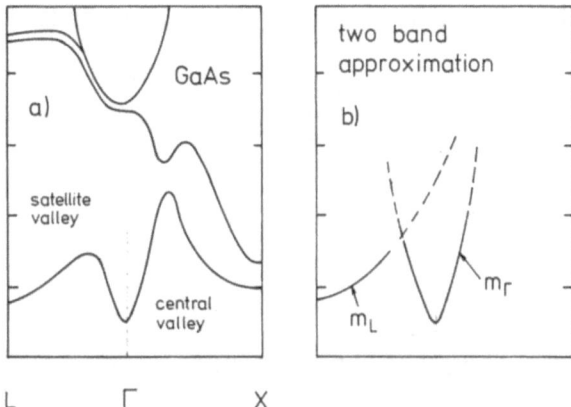

Fig. 12 (a) The band diagram of GaAs after Chelikowsky and Cohen (Chelikowsky and Cohen 1976). (b) Central (Γ) and satellite (L) valley are approximated by two bands with different effective masses

correct band structure as close as possible. In Fig. 12 we show for example the band structure of GaAs, the most important III–V compound material. An effective mass approximation is possible in the vicinity of the band minima. For a two-band approximation we have the Γ-band located at the Brillouin zone center and the L-band with a position off the center. The energy off-set $\Delta_{\Gamma L}$ is approximately 300 meV and the effective mass in the L-band is larger than that in the Γ-band. There is a third band located at the X-point, which is usually neglected.

The next step is to develop an appropriate formulation for charge transport that takes account of coupling between bands. There are two possible choices:

(i) A separate formulation for each single band. The coupling between the bands will then appear on the right hand side of the continuity equation of the single bands. The total current density is the sum of all contributions coming from the separate bands

(ii) The coupling between bands is lumped into an effective mobility μ and thermal voltage u_T. These enter an effective current equation to calculate the current densities.

The number of equations to solve in the first approach increases with the number of bands considered and therefore is larger than in the second one. If minority and majority carriers are important in the device this would create a high degree of complexity and it would be impossible to develop a code that is compatible for Si and III–V compound devices. This compatibility is possible within the second approach. An exchange of the models for

μ and u_T would be sufficient if only the material is changed and not the device type.

Following the same steps that gave us Eqs. (2.163)–(2.167) for the one-band model we derive for the two-band model a similar set of equations. In the stationary case we find

$$\nabla_R v_\Gamma = \frac{n_L - n_\Gamma^>}{\tau_p} \tag{2.197}$$

$$\nabla_R v_L = -\frac{n_L - n_\Gamma^>}{\tau_p} \tag{2.198}$$

$$v_\Gamma = n_\Gamma \mu_\Gamma E + \mu_\Gamma \nabla_R(u_{T\Gamma} n_\Gamma) = n_\Gamma \mu_\Gamma F_\Gamma \tag{2.199}$$

$$v_L = n_L \mu_L E + \mu_L \nabla_R(u_{TL} n_L) = n_L \mu_L F_L \tag{2.200}$$

$$\frac{2}{3} \nabla_R v_{\varepsilon\Gamma} = \frac{u_{T\Gamma} - u_0}{\tau_{\varepsilon 0\Gamma}} n_\Gamma + \frac{u_{TL} - u_0}{\tau_s} n_L - \frac{u_{T\Gamma} - u_0}{\tau_s} n_\Gamma^> - \frac{2}{3} E v_\Gamma \tag{2.201}$$

$$\frac{2}{3} \nabla_R v_{\varepsilon L} = \frac{u_{TL} - u_0}{\tau_{\varepsilon 0 L}} n_L - \frac{u_{TL} - u_0}{\tau_s} n_L + \frac{u_{T\Gamma} - u_0}{\tau_s} n_\Gamma^> - \frac{2}{3} E v_L \tag{2.202}$$

$$v_{\varepsilon\Gamma} = \frac{5}{2} u_{T\Gamma} v_\Gamma + \frac{5}{2} \mu_\Gamma n_\Gamma u_{T\Gamma} \nabla_R u_{T\Gamma} \tag{2.203}$$

$$v_{\varepsilon L} = \frac{5}{2} u_{TL} v_L + \frac{5}{2} \mu_L n_L u_{TL} \nabla_R u_{TL} \tag{2.204}$$

$$\mu_\Gamma = \frac{\mu_{0\Gamma}}{1 + \gamma_\Gamma \left(u_{T\Gamma} \left[1 + \dfrac{F_\Gamma \cdot \nabla_R u_{T\Gamma}}{F_\Gamma^2} \right] - u_0 \right)}; \quad \gamma_\Gamma = \frac{\mu_{0\Gamma}}{v_{\text{sat}\,\Gamma} \tau_{\varepsilon\Gamma}} \tag{2.205}$$

$$\mu_L = \frac{\mu_{0L}}{1 + \gamma_L \left(u_{TL} \left[1 + \dfrac{F_L \cdot \nabla_R u_{TL}}{F_L^2} \right] - u_0 \right)}; \quad \gamma_L = \frac{\mu_{0L}}{v_{\text{sat}\,L} \tau_{\varepsilon L}} \tag{2.206}$$

Here $n_\Gamma^>$ is the partial particle density in the Γ-band for carriers with energy larger than the off-set energy $\Delta_{\Gamma L}$. We look for a solution similar to Eq. (2.179)

$$v = n\mu E + \mu \nabla_R(u_T n) = n\mu(F)F; \quad F = E + n^{-1}\nabla_R(u_T n) \tag{2.207}$$

with $n = n_\Gamma + n_L$. The total velocity v is

$$v = v_\Gamma + v_L = n_\Gamma \mu_\Gamma F_\Gamma + n_L \mu_L F_L = n\mu F \tag{2.208}$$

and the total thermal voltage is,

$$u_T = \frac{n_\Gamma}{n} u_{T\Gamma} + \frac{n_L}{n} u_{TL} = \Lambda u_{T\Gamma} + \lambda u_{TL}; \qquad \Lambda = \frac{n_\Gamma}{n} \qquad \lambda = \frac{n_L}{n}$$

(2.209)

which are derived by adding the velocities and energy densities of both bands. To define an effective mobility form Eq. (2.208) we have to introduce the hypothesis that through the coupling of the bands both will respond to the same driving force.

$$F_\Gamma, F_L \to F$$

(2.210)

$$\mu = \Lambda \mu_\Gamma + \lambda \mu_L$$

(2.211)

Because μ_Γ and μ_L are functions of u_Γ and u_L, respectively, we are finished if we can relate these thermal voltages to the driving force F. The simplest way to accomplish this task is to utilize Eq. (2.177) from the one-band model. In this case we do not account for particle transfer between the bands explicitly. Only indirectly it is contained in the formulation by assuming the common driving force

$$u_{T\Gamma} = u_0 + \frac{2}{3} \tau_{\varepsilon 0 \Gamma} v_{\text{sat}\Gamma}^2 \frac{\mu_{\Gamma 0} - \mu_\Gamma}{\mu_{\Gamma 0} \mu_\Gamma}$$

(2.212)

$$u_{TL} = u_0 + \frac{2}{3} \tau_{\varepsilon 0 L} v_{\text{sat}L}^2 \frac{\mu_{L0} - \mu_L}{\mu_{L0} \mu_L}$$

(2.213)

We did not yet utilize the information contained in the continuity equations Eqs. (2.197), (2.198), (2.201), and (2.202) for particle and energy conservation. If we use Eqs. (2.212) and (2.213) we have to know the parameters $\mu_{0\Gamma}$, μ_{0L}, $v_{\text{sat}\Gamma}$, $v_{\text{sat}L}$, $\tau_{\varepsilon\Gamma}$, and $\tau_{\varepsilon L}$ which are required for each single band. This has to be contrasted with the number of parameters that can be measured: low field mobility $\mu_0 \approx \mu_{0\Gamma}$, energy relaxation time $\tau_{\varepsilon 0} \approx \tau_{\varepsilon 0\Gamma}$ from the curvature of u_T in the warm carrier regime, saturation velocity $v_{\text{sat}} = \lim \mu F$ in the high field region, and μ_{0L} can be measured as low frequency mobility (Ohmi 1967). Therefore three of the independent band parameters are directly accessible the rest is not. The energy continuity equations Eqs. (2.201) and (2.202) have to be used to relate them to the measurable quantities. As done before we will again consider only leading contributions in F. Combining Eqs. (2.201)–(2.206) we obtain the coupled set of equations for the thermal voltages $u_{T\Gamma}$ and u_{TL}

$$\frac{1}{\tau_{\varepsilon 0 \Gamma}} \Omega_\Gamma + \beta \frac{1}{\tau_s} \Omega_L - \alpha^> \frac{1}{\tau_s} \Omega_\Gamma = \frac{2}{3} \frac{\mu_{0\Gamma} F^2}{1 + \gamma_\Gamma \Omega_\Gamma}$$

(2.214)

$$\frac{1}{\tau_{\varepsilon 0 L}} \Omega_L + \frac{\alpha^>}{\beta \tau_s} \Omega_\Gamma - \frac{1}{\tau_s} \Omega_L = \frac{2}{3} \frac{\mu_{0L} F^2}{1 + \gamma_L \Omega_L}$$

(2.215)

Here we have used the abbreviations $\Omega_\Gamma = u_{T\Gamma} - u_0$, $\Omega_L = u_{TL} - u_0$, $\beta = n_L/n_\Gamma$, and $\alpha^> = n_\Gamma^>/n_\Gamma$. The solution of these equations will provide an alternative way to determine $u_{T\Gamma}$ and u_{TL}, including the effect of energy exchange on the thermal voltages. Unfortunately this will lead to an algebraic equation of fourth order the solution of which is very complex. Instead we will use Eqs. (2.214) and (2.215) to find the relationship between the independent band parameters in Eqs. (2.212) and (2.213) and the measurable ones. To this end we consider low and high field limits of Eqs. (2.212) and (2.213) in connection with Eqs. (2.214) and (2.215). This provides us with a set of parameters that is consistent with the energy continuity equation and allows us to maintain the simple formulation for the calculation of the thermal voltages $u_{T\Gamma}$ and u_{TL}. In the low field limit we have

$$\lim_{F \to 0} \Omega_\Gamma = \frac{2}{3} \tau_{\varepsilon 0 \Gamma} \mu_{0\Gamma} F^2 \tag{2.216}$$

$$\lim_{F \to 0} \Omega_L = \frac{2}{3} \tau_{\varepsilon 0 L} \mu_{0L} F^2 \tag{2.217}$$

which gives

$$\tau_{\varepsilon 0 L} = \frac{\alpha_0^> \mu_{0\Gamma}}{\beta_0 \mu_{0L}} \tau_{\varepsilon 0 \Gamma} \tag{2.218}$$

where the subscript $_0$ on $\alpha^>$ and β refers to their equilibrium values. In the high field limit we obtain

$$\lim_{F \to \infty} \Omega_\Gamma = \frac{2}{3} \tau_{\varepsilon 0 \Gamma} v_{\mathrm{sat}\Gamma} F^2 \tag{2.219}$$

$$\lim_{F \to \infty} \Omega_L = \frac{2}{3} \tau_{\varepsilon 0 L} v_{\mathrm{sat}L} F^2 \tag{2.220}$$

Inserting these equations in Eqs. (2.214) or (2.215) we find

$$v_{\mathrm{sat}\Gamma} = \frac{\beta_\infty \alpha_0^> \mu_{0\Gamma}}{\alpha_\infty^> \beta_0 \mu_{0L}} v_{\mathrm{sat}L} \tag{2.221}$$

Here the subscript $_\infty$ refers to the high field limit. The measurable saturation velocity v_{sat} we obtain in the high field limit of Eq. (2.208) with the mobility Eq. (2.211)

$$\lim_{F \to \infty} v = v_{\mathrm{sat}} = \Lambda_\infty v_{\mathrm{sat}\Gamma} + \lambda_\infty v_{\mathrm{sat}L} \tag{2.222}$$

We can now relate the not measurable quantities $v_{\text{sat}\,\Gamma}$, $v_{\text{sat}\,L}$, and $\tau_{\varepsilon 0 L}$ to the measurable ones

$$\tau_{\varepsilon 0 L} = \frac{\alpha_0^> \mu_{0\,\Gamma}}{\beta_0 \mu_{0 L}} \tau_{\varepsilon 0} \tag{2.223}$$

$$v_{\text{sat}\,\Gamma} = \frac{\dfrac{1}{\Lambda_\infty}}{1 + \dfrac{\lambda_\infty \alpha_\infty^> \beta_0 \mu_{0 L}}{\Lambda_\infty \beta_\infty \alpha_0^> \mu_{0\,\Gamma}}} v_{\text{sat}} \tag{2.224}$$

$$v_{\text{sat}\,L} = \frac{\dfrac{\alpha_\infty^> \beta_0 \mu_{0 L}}{\Lambda_\infty \beta_\infty \alpha_0^> \mu_{0\,\Gamma}}}{1 + \dfrac{\lambda_\infty \alpha_\infty^> \beta_0 \mu_{0 L}}{\Lambda_\infty \beta_\infty \alpha_0^> \mu_{0\,\Gamma}}} v_{\text{sat}} \tag{2.225}$$

After having determined the independent band parameters in a consistent way the next step is now to find the expressions for $\alpha^>$, and β because $\Lambda = (1 + \beta)^{-1}$ and $\lambda = \beta(1 + \beta)^{-1}$. To do this we have to consider in more detail the transfer of particles from one band to the other. The particle exchange between Γ- and L-band will be described with the following rate equations (Baranger 1986)

$$-q F v_\Gamma(\varepsilon) \frac{\partial}{\partial \varepsilon} f_\Gamma(\varepsilon) = -\frac{f_\Gamma(\varepsilon) - a f_{\Gamma 0}(\varepsilon)}{\tau_{\Gamma\Gamma}(\varepsilon)}$$

$$-\frac{f_\Gamma(\varepsilon) - b f_{\Gamma 0}(\varepsilon)}{\tau_{\Gamma L}(\varepsilon)} \theta(\varepsilon - \Delta_{\Gamma L}) \tag{2.226}$$

$$-q F v_L(\varepsilon) \frac{\partial}{\partial \varepsilon} f_L(\varepsilon) = -\frac{f_L(\varepsilon) - c f_{L 0}(\varepsilon)}{\tau_{LL}(\varepsilon)}$$

$$-\frac{f_L(\varepsilon) - d f_{L 0}(\varepsilon)}{\tau_{L\Gamma}(\varepsilon)}; \quad \varepsilon > \Delta_{\Gamma L} \tag{2.227}$$

The scattering term has one intra-band term $(\tau_{\Gamma\Gamma}, \tau_{LL})$ and one inter-band contribution $(\tau_{\Gamma L}, \tau_{L\Gamma})$ that accounts for $\Gamma \to L$ and $L \to \Gamma$ transitions, respectively. The energy-dependent velocity is $v(\varepsilon) = \sqrt{(2m\varepsilon)}$ and F is field that drives the system off equilibrium. The equations are coupled through the requirements of particle conservation. Intra-band scattering does not change the number of particles in one band. The inter-band scattering terms in both equations have to assure that the net transfer is zero otherwise the number of particles in the system would change, which is impossible. The coupling enters through the as yet undetermined energy independent a, b, c, d. As a first simplification in Eqs. (2.226) and (2.227) we will replace the energy-dependent scattering times by appropriately averaged ones.

$$\langle \tau_{ij}(\varepsilon)^{-1} \rangle_k = \tau_{ij}(u_{Tk}/q)^{-1}; \quad i, j, k = \Gamma, L \tag{2.228}$$

The energy argument is replaced by the thermal voltage of the band in which the average is performed. With this manipulation Eqs. (2.226) and (2.227) can be solved analytically. If we normalize all energies to $k_B T$, and introduce the normalized field $F_i = -q F v_{\text{th}, i} \tau_{ii}$, $i = \Gamma, L$, with $v_{\text{th}, i}$ as the thermal velocity in the ith band, we have to solve the equations

$$F_\Gamma \sqrt{\varepsilon} \frac{\partial}{\partial \varepsilon} f_\Gamma(\varepsilon) = -f_\Gamma(\varepsilon) + a f_{\Gamma 0}(\varepsilon); \quad \varepsilon < \Delta_{\Gamma L} \tag{2.229}$$

$$F_\Gamma^* \sqrt{\varepsilon} \frac{\partial}{\partial \varepsilon} f_\Gamma(\varepsilon) = -f_\Gamma(\varepsilon) + a^* f_{\Gamma 0}(\varepsilon); \quad \varepsilon > \Delta_{\Gamma L} \tag{2.230}$$

$$F_L^* \sqrt{\varepsilon} \frac{\partial}{\partial \varepsilon} f_L(\varepsilon) = -f_L(\varepsilon) + c^* f_{L 0}(\varepsilon); \quad \varepsilon > \Delta_{\Gamma L} \tag{2.231}$$

with

$$F_i^* = F_i \frac{\tau_{i, \text{eff}}}{\tau_{ii}}; \quad i = \Gamma, L \tag{2.232}$$

$$\tau_{\Gamma, \text{eff}} = \frac{\tau_{\Gamma\Gamma} \tau_{\Gamma L}}{\tau_{\Gamma\Gamma} + \tau_{\Gamma L}}; \quad \tau_{L, \text{eff}} = \frac{\tau_{LL} \tau_{L\Gamma}}{\tau_{LL} + \tau_{L\Gamma}} \tag{2.233}$$

$$a^* = \frac{\tau_{\Gamma, \text{eff}}}{\tau_{\Gamma\Gamma}} \left(a + b \frac{\tau_{\Gamma\Gamma}}{\tau_{\Gamma L}} \right); \quad c^* = \frac{\tau_{L, \text{eff}}}{\tau_{LL}} \left(c + d \frac{\tau_{LL}}{\tau_{L\Gamma}} \right) \tag{2.234}$$

The solutions of Eqs. (2.229)–(2.231) are

$$f_\Gamma(\varepsilon) = \sqrt{\pi} \frac{a}{F_\Gamma} \exp(-\varepsilon) \, \text{Gerfc} \left(\frac{1}{F_\Gamma} + \sqrt{\varepsilon} \right); \quad \varepsilon < \Delta_{\Gamma L} \tag{2.235}$$

$$f_\Gamma(\varepsilon) = \sqrt{\pi} \frac{a^*}{F_\Gamma^*} \exp(-\varepsilon) \, \text{Gerfc} \left(\frac{1}{F_\Gamma^*} + \sqrt{\varepsilon} \right); \quad \varepsilon > \Delta_{\Gamma L} \tag{2.236}$$

$$f_L(\varepsilon) = \sqrt{\pi} \frac{c^*}{F_L^*} \exp(-\varepsilon) \, \text{Gerfc} \left(\frac{1}{F_L^*} + \sqrt{\varepsilon} \right); \quad \varepsilon > \Delta_{\Gamma L} \tag{2.237}$$

with

$$\text{Gerfc}(x) = \exp(x^2) \, \text{erfc}(x) \tag{2.238}$$

Here erfc is the complementary error function. The particle densities are now given by

$$n_\Gamma^< = a n_{\Gamma 0} T_{\Gamma 1}; \quad T_{\Gamma 1} = \frac{2}{F_\Gamma} \int_0^{\Delta_{\Gamma L}} d\varepsilon \sqrt{\varepsilon} \exp(-\varepsilon) \mathrm{Gerfc}\left(\frac{1}{F_\Gamma} + \sqrt{\varepsilon}\right)$$

(2.239)

$$n_\Gamma^> = a^* n_{\Gamma 0} T_{\Gamma 2}; \quad T_{\Gamma 2} = \frac{2}{F_\Gamma^*} \int_{\Delta_{\Gamma L}}^\infty d\varepsilon \sqrt{\varepsilon} \exp(-\varepsilon) \mathrm{Gerfc}\left(\frac{1}{F_\Gamma^*} + \sqrt{\varepsilon}\right)$$

(2.240)

$$n_L = c^* n_{L0} T_L; \quad T_L = \frac{2}{F_L^*} \int_0^\infty d\varepsilon \sqrt{\varepsilon} \exp\left(-(\varepsilon + \Delta_{\Gamma L})\right)$$

$$\cdot \mathrm{Gerfc}\left(\frac{1}{F_L^*} + \sqrt{(\varepsilon + \Delta_{\Gamma L})}\right)$$

(2.241)

For the particle density in the Γ-band we have

$$n_\Gamma = n_\Gamma^< + n_\Gamma^>$$

(2.242)

and particle conservation requires

$$n_\Gamma + n_L = n$$

(2.243)

We now have to determine the pre-factors a, b, c, and d. The coefficients a and c are related to the intra-band scattering properties. As mentioned above this scattering is particle conserving, in the corresponding band, therefore we have

$$a = \frac{\int d\varepsilon \sqrt{\varepsilon} f_\Gamma(\varepsilon)}{\int d\varepsilon \sqrt{\varepsilon} f_{\Gamma 0}(\varepsilon)} = \frac{n_\Gamma}{n_{\Gamma 0}}$$

(2.244)

$$c = \frac{\int d\varepsilon \sqrt{(\varepsilon - \Delta_{\Gamma L})} f_L(\varepsilon)}{\int d\varepsilon \sqrt{(\varepsilon - \Delta_{\Gamma L})} f_{L0}(\varepsilon)} = \frac{n_L}{n_{L0}}$$

(2.245)

Combining Eqs. (2.239) and (2.244) we find

$$\frac{n_\Gamma^>}{n_\Gamma} = 1 - T_{\Gamma 1}$$

(2.246)

For the coefficients b and d we obtain the following relationships

$$b n_{\Gamma 0} = n_\Gamma \left(\frac{1 - T_{\Gamma 1}}{T_{\Gamma 2}} \frac{\tau_{\Gamma L}}{\tau_{\Gamma \mathrm{eff}}} - \frac{\tau_{\Gamma L}}{\tau_{\Gamma \Gamma}}\right)$$

(2.247)

$$d n_{L0} = n_L \left(\frac{1}{T_L} \frac{\tau_{L\Gamma}}{\tau_{L\mathrm{eff}}} - \frac{\tau_{L\Gamma}}{\tau_{LL}}\right)$$

(2.248)

In addition to Eq. (2.243) we need another equation to determine their value with respect to the transfer integrals T_i and the total particle density. This second equation comes from the requirement that inter-band scattering has to be consistent with particle conservation. In the literature it is suggested that it can be assured by subjecting the inter-band scattering contribution to the following boundary condition (Baranger 1986)

$$\frac{n_\Gamma^> - b n_{\Gamma 0}^>}{\tau_{\Gamma L}} + \frac{n_L - d n_{L0}}{\tau_{L\Gamma}} = 0 \tag{2.249}$$

A careful analysis of the equilibrium limit $F \to 0$ gives the unphysical result

$$\lim_{F \to 0} \frac{n_\Gamma}{n_L} = -\left(\frac{m_\Gamma}{m_L}\right)^{1/2} \tag{2.250}$$

The reason for this is to be sought in the rate equations Eqs. (2.226) and (2.227). The relaxation time like scattering rates cannot distinguish between particles arriving and particles leaving the band. In and out scattering terms are represented in one effective relaxation type expression. In the strong coupling case inter-band scattering is much faster than intra-band scattering and as a result stationary state will evolve: the number of particles leaving one band has to be equal to the number of particles arriving in the other independent from the intra-band scattering. This is a generalization of the detailed balance at equilibrium.

$$\frac{n_\Gamma^>}{\tau_{\Gamma L}} - \frac{n_L}{\tau_{L\Gamma}} = 0 \tag{2.251}$$

With Eqs. (2.243) and (2.246)–(2.248) we are now able to calculate b and d

$$b = \frac{n}{n_{\Gamma 0}} \left(\frac{1 - T_{\Gamma 1}}{T_{\Gamma 2}} \frac{\tau_{\Gamma L}}{\tau_{\Gamma \text{eff}}} - \frac{\tau_{\Gamma L}}{\tau_{\Gamma \Gamma}}\right) \left[(1 - T_{\Gamma 1}) \frac{\tau_{L\Gamma}}{\tau_{\Gamma L}} + 1\right]^{-1} \tag{2.252}$$

$$d = \frac{n}{n_{L0}} \left(\frac{1}{T_L} \frac{\tau_{L\Gamma}}{\tau_{L\text{eff}}} - \frac{\tau_{L\Gamma}}{\tau_{LL}}\right) \frac{\tau_{L\Gamma}}{\tau_{\Gamma L}} (1 - T_{\Gamma 1}) \left[(1 - T_{\Gamma 1}) \frac{\tau_{L\Gamma}}{\tau_{\Gamma L}} + 1\right]^{-1} \tag{2.253}$$

which gives for n_Γ and n_L

$$n_\Gamma = n \left[(1 - T_{\Gamma 1}) \frac{\tau_{L\Gamma}}{\tau_{\Gamma L}} + 1\right]^{-1} \tag{2.254}$$

$$n_L = n \frac{\tau_{L\Gamma}}{\tau_{\Gamma L}} (1 - T_{\Gamma 1}) \left[(1 - T_{\Gamma 1}) \frac{\tau_{L\Gamma}}{\tau_{\Gamma L}} + 1\right]^{-1} \tag{2.255}$$

$$n_\Gamma^> = n_\Gamma (1 - T_{\Gamma 1}) \tag{2.256}$$

We added here Eq. (2.246) for completion. A surprising result is that only one transfer integral $T_{\Gamma 1}$ is needed to calculate the coefficients $\alpha^>$, and β. It is easily verified that in the equilibriums limit $F \to 0$ Eqs. (2.254)–(2.256) assume the correct value. More interesting, however, is the high field limit $F \to \infty$ in the Γ-band

$$\lim_{F \to \infty} n_\Gamma = n \left(\left. \frac{\tau_{L\Gamma}}{\tau_{\Gamma L}} \right|_\infty + 1 \right)^{-1} \tag{2.257}$$

As we will find below the ratio $\tau_{L\Gamma}/\tau_{\Gamma L}$ is approximately of order ten to several hundred, depending on the model for the scattering rates. Therefore the number of particles in the Γ-band is only a small fraction of the total number of particles. It is clear that transport is now dominated by the L-type band.

We now have to specify the scattering rates. For the calculation we need $\tau_{\Gamma\Gamma}$ and the ratio $\tau_{L\Gamma}/\tau_{\Gamma L}$. The energy dependent scattering rates we take from Conwell (1967). The intra-band scattering is dominated by polar optical phonons with energy ω_{opt}

$$\frac{1}{\tau_{\Gamma\Gamma}(\varepsilon)} = C \cdot \coth(\omega_{\mathrm{opt}}/k_B T)/\sqrt{\varepsilon} \tag{2.258}$$

Here the constant C is related to the scattering matrix element and will serve as a fitting parameter. For the $\Gamma \to L$ and $L \to \Gamma$ transitions we have, respectively

$$\frac{1}{\tau_{\Gamma L}(\varepsilon)} = \frac{1}{\tau_T} \frac{(\varepsilon + \omega_{\mathrm{opt}} - \Delta_{\Gamma L})^{1/2} + \exp(\omega_{\mathrm{opt}}/k_B T)(\varepsilon - \omega_{\mathrm{opt}} - \Delta_{\Gamma L})^{1/2}}{\sqrt{\Delta_{\Gamma L}}} \tag{2.259}$$

$$\frac{1}{\tau_{L\Gamma}(\varepsilon)} = \left(\frac{m_\Gamma}{m_L} \right)^{3/2} \frac{1}{\tau_T} \frac{(\varepsilon + \omega_{\mathrm{opt}})^{1/2} + \exp(\omega_{\mathrm{opt}}/k_B T)(\varepsilon - \omega_{\mathrm{opt}})^{1/2}}{\sqrt{\Delta_{\Gamma L}}}$$

$$\varepsilon > \Delta_{\Gamma L} \tag{2.260}$$

Here τ_T contains the exchange matrix element between L and Γ band. With Eqs. (2.258)–(2.259) a suitable average has to be calculated. For this purpose we choose an exponential that is scaled with the carrier temperature $k_B T_{ci} = u_{Ti}/q$, $i = \Gamma, L$ in the corresponding band to have a rough approximation of the tails of the distribution function. The average Eq. (2.228) is calculated by virtue of

$$\langle \ldots \rangle_k = \frac{\displaystyle\int_{\varepsilon_1}^{\varepsilon_2} d\varepsilon \ldots \exp(-\varepsilon/k_B T_{ck})}{\displaystyle\int_{\varepsilon_1}^{\varepsilon_2} d\varepsilon \exp(-\varepsilon/k_B T_{ck})} \tag{2.261}$$

Here ε_1 and ε_2 are the lower and upper bound of the region the average is performed over. The average for the intra-band scattering rate is

$$\frac{1}{\tau_{\Gamma\Gamma}} = \frac{1}{\tau_{\Gamma\Gamma 0}}\left(\frac{k_B T}{k_B T_{c\Gamma}}\right)^{1/2} \coth(\omega_{opt}/k_B T) \tag{2.262}$$

and for the ratio of the inter-band scattering rates we find

$$\frac{\tau_{L\Gamma}}{\tau_{\Gamma L}} = \Gamma(3/2)\left(\frac{k_B T_{c\Gamma}}{k_B T_{cL}}\right)^{1/2}\left(\frac{m_L}{m_\Gamma}\right)^{3/2}\exp(-\Delta_{\Gamma L}/k_B T_{cL}).$$

$$\cdot\frac{1 + \exp(\omega_{opt}/k_B T)}{\Gamma\left(3/2, \dfrac{\Delta_{\Gamma L} - \omega_{opt}}{k_B T_{cL}}\right) + \exp(\omega_{opt}/k_B T)\Gamma\left(3/2, \dfrac{\Delta_{\Gamma L} + \omega_{opt}}{k_B T_{cL}}\right)}$$

$$\tag{2.263}$$

The energy of the optical phonon is much smaller than the band off-set, therefore we can neglect the phonon energy in the argument of the incomplete Γ-function. This simplifies Eq. (2.263) further

$$\frac{\tau_{L\Gamma}}{\tau_{\Gamma L}} = \Gamma(3/2)\Gamma\left(3/2, \frac{\Delta_{\Gamma L}}{k_B T_{cL}}\right)^{-1}\left(\frac{k_B T_{c\Gamma}}{k_B T_{cL}}\right)^{1/2}\left(\frac{m_L}{m_\Gamma}\right)^{3/2}$$

$$\cdot\exp(-\Delta_{\Gamma L}/k_B T_{cL}) \tag{2.264}$$

We are now able to calculate the high field limit required in Eq. (2.257). With Eqs. (2.219) and (2.220)

$$\lim_{F\to\infty}\frac{\tau_{L\Gamma}}{\tau_{\Gamma L}} = \left(\frac{\tau_{\varepsilon 0\Gamma}v_{sat\Gamma}}{\tau_{\varepsilon 0L}v_{satL}}\right)^{1/2}\left(\frac{m_L}{m_\Gamma}\right)^{3/2} = \left(\frac{n_L}{n_\Gamma^>}\right)^{1/2}\Bigg|_{\infty}\left(\frac{m_L}{m_\Gamma}\right)^{3/2} \tag{2.265}$$

If we use the generalized detailed balance equation Eq. (2.251) we can eliminate the ratio of the densities and obtain

$$\lim_{F\to\infty}\frac{\tau_{L\Gamma}}{\tau_{\Gamma L}} = \left(\frac{m_L}{m_\Gamma}\right)^3 \approx 700 \tag{2.266}$$

If we had not taken the energy dependence of the scattering rates into account the ratio of the inter-band transitions would have been fixed by Eq. (2.251). Independent of the field we would have

$$\frac{\tau_{L\Gamma}}{\tau_{\Gamma L}} = \frac{n_{L0}}{n_{\Gamma 0}^>} \approx 7 \tag{2.267}$$

To calculate Eqs. (2.266) and (2.267) we have used $m_L = 0.55 m_0$, $m_\Gamma = 0.0632 m_0$, and an energy off-set $\Delta_{\Gamma L} = 284$ meV.
Using Eqs. (2.209) and (2.211) for the thermal voltage and the drift velocity, the transfer integral from Eq. (2.239), and the scattering rates from Eqs. (2.262), (2.263), and (2.264) we can calculate some interesting features for the

two-band model, such as drift velocity versus field, thermal voltage versus field, and the various particle ratios versus field. The parameters we used in the calculation are:

$\tau_{\Gamma\Gamma} = 0.1$ ps, determines the onset of scattering from the Γ-band into the L-band. From the interaction matrix element an estimated order of magnitude is 2 ps $< \tau_{\Gamma\Gamma} < 0.05$ ps.

$\mu_{\Gamma 0}$, is the low field mobility. Its doping dependence is accounted for by using the expression found in the work of Hilsum (1974), for instance, with its minimal value for pure GaAs $\mu_{\Gamma 0} = 10^4$ cm^2/Vs.

μ_{L0}, is the mobility of the L-band electrons near equilibrium. A value of $\mu_{L0} = \mu_{\Gamma 0}/3$ was chosen in accordance with available experimental data (Ohmi 1967, Blakemore 1982).

$\tau_{\varepsilon} = 4$ ps, low field energy relaxation time. Its value was adjusted to fit Monte Carlo results for homogeneous field in the low field region (Maloney and Frey 1977)

$v_{sat} = 10^7$ cm/s, high field saturation velocity. With increasing field experimental results show a decreasing drift velocity. In contrast to the Si-case higher bands become more and more important in the high field region. In the two band model only Γ and L band are accounted for. Therefore we assign a value to the saturation velocity that at $F = 20$ kV/cm

The energy off-set $\Delta_{\Gamma L} = 284$ meV and the effective masses $m_\Gamma = 0.0632 m_0$ and $m_L = 0.55 m_0$ are taken from literature (Blakemore 1982).

Figure 13 shows how the energy dependence of the scattering rates influences the particle ratios in the different valleys and in Fig. 14 the resulting

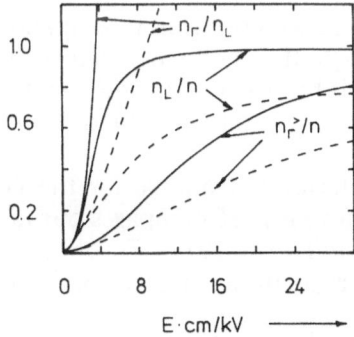

Fig. 13 The influence of the scattering rates on the particle distribution in the different bands. Dashed lines—constant scattering rates $\tau_{\Gamma\Gamma} = \tau_{\Gamma\Gamma 0} = 0.1$ ps and $\tau_{L\Gamma}/\tau_{\Gamma L}$ according to Eq. (2.267). Solid lines—energy-dependent scattering rates according to Eqs (2.262) and (2.263)

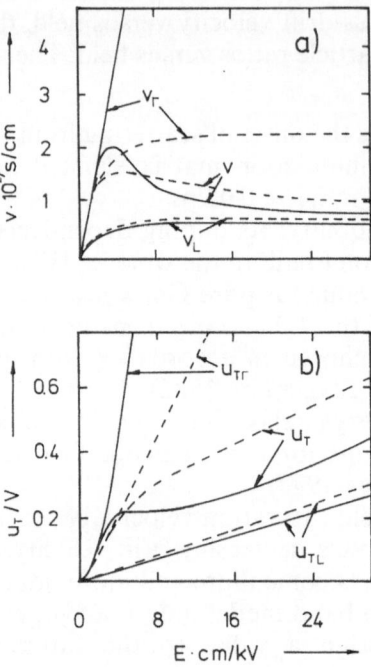

Fig. 14 The influence of the scattering rates on drift velocity (a) and thermal voltage (b). Dashed lines—constant scattering rates. Solid lines—energy-dependent scattering rates

total drift velocity and thermal voltage are displayed. The results for constant scattering rates (broken lines) were obtained by replacing the carrier temperatures in Eqs. (2.262) and (2.263) by the lattice temperature. The particle ratio $n_L/n = 1 - n_\Gamma/n$ very clearly shows how important it is to account for the energy dependence of the scattering rates. This was already estimated in Eq. (2.257) as we used Eqs. (2.266) and (2.267) for the transfer scattering rates. Neglect of the energy dependence gives a particle transfer between the bands that is too weak to produce sufficient structure in v_d and u_T. Figure 15 compares our model for the drift velocity versus field with experimental data.

Little experimental material is available for the field dependent thermal voltage in GaAs. Therefore Fig. 16 compares our model with Monte Carlo calculations (Maloney and Frey 1977). To obtain the best fit for the low field region the energy relaxation time $\tau_{\varepsilon 0}$ was adjusted to have the value 4 ps.

Figure 17 shows the lattice temperature dependence of the particle ratios, the drift velocity, and the thermal voltage. The particle ratios are only very weakly dependent on the lattice temperature for the considered range. The reason for this is that the electric field is already a strong perturbation so

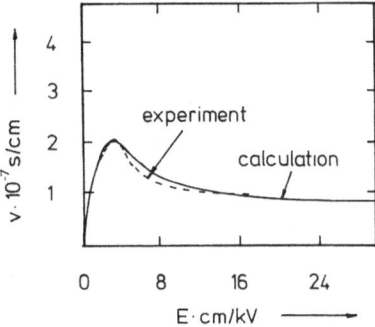

Fig. 15 Comparison of calculated drift velocity (solid lines) with measured data (dashed lines) compiled from Ruch and Kino (Ruch and Kino 1968) and Houston and Evans (Houston and Evans 1977)

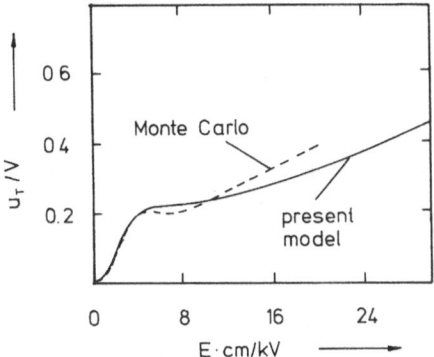

Fig. 16 Comparison of calculated thermal voltage (solid line) with results from Monte Carlo calculations (dashed line) by Maloney and Frey (Maloney and Frey 1977). The energy relaxation time $\tau_\varepsilon = 4$ ps was adjusted to fit the low field regime

that it has the dominant influence. The main influence v_d and u_T comes from the temperature dependence of the low field mobilities and the saturation velocity for which we used the results of Nichols et al. (1980) and Blakemore (1982), respectively

$$\mu_{\Gamma 0}(T) = \mu_{\Gamma 0, T = 300} \frac{300\,K}{T} \qquad\qquad (2.268)$$

$$\mu_{L0}(T) = \mu_{L0, T = 300} \left(\frac{300\,K}{T}\right)^2 \qquad\qquad (2.269)$$

$$v_{\text{sat}} = (1.45 - 0.0015 \cdot T) \cdot 10^7 \text{ cm/s} \qquad\qquad (2.270)$$

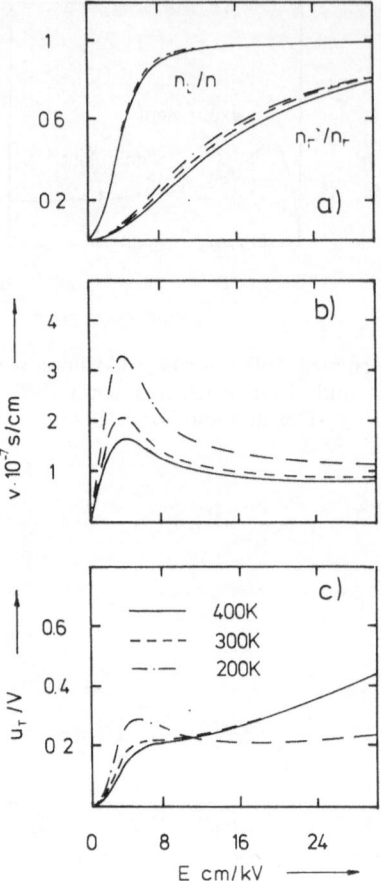

Fig. 17 Comparison of the temperature dependence of (a) particle distribution among valleys, (b) drift velocity, and (c) thermal voltage. The temperatures are 400 K (solid line), 300 K (dashed line), and 200 K (dashed–dotted line)

Figure 18 shows how the drift velocity and the thermal voltage varies with doping concentration. For the low field mobility we used the doping dependence according to Hilsum (1974)

$$\mu_{\Gamma 0}(N_D) = \frac{\mu_{\Gamma 0,\,\text{intr}}}{1 + (N_D/10^{17}\ \text{cm}^3)} \tag{2.271}$$

For the L-band a similar expression was used. Equation (2.271) does not account for a saturation at high doping concentrations. For the doping levels shown in Fig. 18 this saturation is of minor importance. An expression that accounts for that saturation is found in the work by Nakagawa (1986).

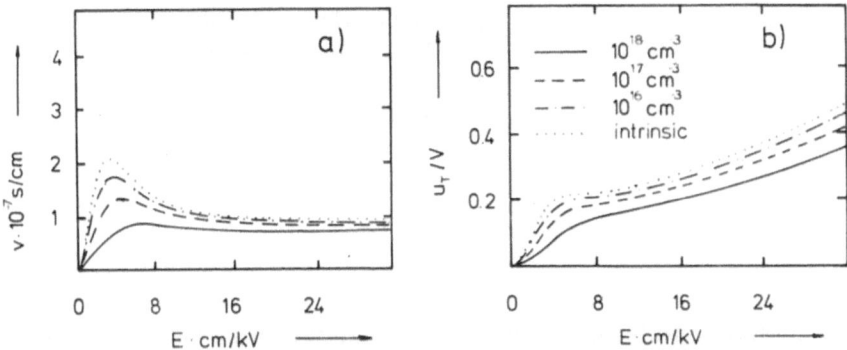

Fig. 18 Doping dependence of drift velocity (a) and thermal voltage (b). The doping concentrations are 10^{18} cm^{-3} (solid line), 10^{17} cm^{-3} (dashed line), 10^{16} cm^{-3} (dashed–dotted line), and intrinsic material

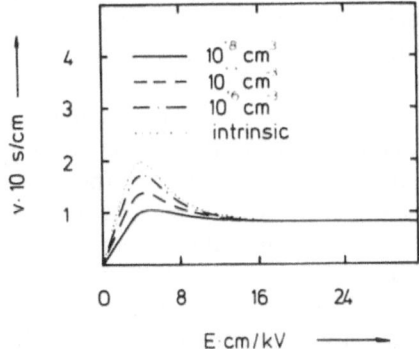

Fig. 19 Doping dependence of drift velocity extracted from experimental material after Nakagawa (Nakagawa 1986). The doping concentrations are 10^{18} cm^{-3} (solid line), 10^{17} cm^{-3} (dashed line), 10^{16} cm^{-3} (dashed–dotted line), and intrinsic material

Experimental results for the doping dependence of the field dependent drift velocity (Nakagawa 1986) are shown in Fig. 19.

We observe a similar trend in the reduction of the velocity peak with doping concentration which is located at the critical field $E_\Delta \approx 3.8$ kV/cm. In our model this critical field is determined by the band off set $\Delta_{\Gamma L}$ between central and satellite valley and the intra-valley relaxation rate $\tau_{\Gamma\Gamma}$. For higher doping levels the saturation velocity v_{sat} is reached at much larger fields than that shown in Fig. 19. This is directly related to the reduced heating of the carriers at high doping level as shown in Fig. 18b. For higher doping level the drift velocity drops after its peak value at E_Δ to a somewhat

lower value than v_{sat} and then approaches the saturation velocity asymp-
totically in the limit of large fields. For moderate field strength the drift
velocity assumes a value that is about 20% below its asymptotic value v_{sat}.
This seems in agreement with the data provided in Fig. 19.

2.4 Summary

The foundation of the drift–diffusion approximation and its possible extensions were discussed in this chapter. Because an extensive amount of
material was covered, a brief summary of the main results will be helpful.
The investigation of the linear-response regime in the first part and the far-
off equilibrium case treated in the second part have the Boltzmann equation as a common root. We learned that the scattering term of the
Boltzmann equation is essential for the analysis in both cases.
For weakly perturbed systems, we derived the general form of the
drift–diffusion equation for the current density, Eq. (2.22), with the generalized Einstein relationship Eq. (2.23). This current density is an appropriate equation in highly doped semiconductors, which are realized in modern
bipolar devices, for example. Here, the effects of degeneracy are of some
importance. The next step was to derive the low-field mobility. This
brought us to the relaxation-time approximation of the Boltzmann equation, Eq. (2.31). This approximation allows us to characterize the scattering
term of the Boltzmann equation by a single time constant: the transport
lifetime τ_{tr}, Eq. (2.30). We found that this time constant has to be distinguished from the collision or optical lifetime τ_c that does not suppress
small-angle scattering. In the discussion of the validity of the relaxation-
time approximation, we found a limiting time scale, Equation (2.91), in
which this approximation produces physically meaningful results. This
restriction was due to the absence of any energy dissipation within that
level of approximation. Within the relaxation-time approximation, however, a very detailed analysis of the low-field mobility as limited by
acoustical phonons and randomly-distributed impurities was performed.
The commonly used expression for the ohmic mobility in bulk material, Eq.
(2.74), including saturation at high doping concentrations, was justified
from first principles.
For the nonlinear response regime, we derived a generalized set of semiconductor equations, (2.104)–(2.107), that included the energy balance of
the system of carriers. The problem with these equations was that they do
not comprise a closed set for the relevant quantities: density n, current
density j, energy density $\mathscr{E}$, and energy current density v_ε. A solution cannot
be found without further manipulation. Closure of this set of equations

involves truncation of both: a higher moment of the distribution function has to be approximated on the left-hand side and the scattering term has to be considered on the right-hand side. For this closure, we discussed the drifted Maxwellian and a generalized expansion method as approximations for the distribution function. A main result of this discussion was finding different ways of defining a carrier temperature, Eqs. (2.149) and (2.150). We again gave a very detailed analysis of the scattering terms. This brought us to an expression for the mobility that depends on the thermal voltage and contains a velocity overshoot caused by a large gradient in the thermal voltage u_T, Eq. (2.167). The general set of equations (2.163)–(2.167) allows a reduction to a modified drift–diffusion current equation, Eq. (2.182). This current-density equation is tightly coupled to mobility and thermal voltage models, Eqs. (2.176) and (2.177).

A straightforward generalization of the one-band model to a two-band model that accounts for particle exchange was given in section 2.3.4. The important result is that it is again possible to reduce the complex transport problem to one effective drift–diffusion type equation for the current density with, of course, suitable models for the mobility and thermal voltage, Eqs. (2.211) and (2.209), respectively. The particle transfer between the two bands in the effective-mass approximation is described by only one transfer integral, Eq. (2.239). The approximation derived here is consistent with particle conservation in the total system. As the numerical examples show, it is important to include energy-dependent scattering rates.

The results derived in this chapter are very important for practical modeling applications. We were able to demonstrate that for a wide range of applications it is sufficient to use a modified drift–diffusion approximation for the current density in the conduction band of a semiconductor. A suitable choice of mobility model and thermal voltage allows a large variety of different systems to be covered without adding extra equations.

References

Baranger H. U. (1986) Ballistic Electrons in a Submicron Semiconducting Structure: A Boltzmann Equation Approach. Thesis. Cornell University, Ithaca, New York.

Blakemore J. (1982): J. Appl. Phys. *53*, R123.

Blotekjaer K., Lunde E. B. (1969): Physica Status Solidi *35*, 581.

Blotekjaer K. (1970): IEEE Trans. Electron. Devices *ED17*, 38.

Brunetti R., Jacoboni C., Nava F., Reggiani L., Bosman G., Zijlstra R. J. J. (1981): J. Appl. Phys. *52*, 6713.

Chelikowsky J. R., Cohen M. L. (1976): Phys. Rev. *B14*, 556.

Conwell E. M. (1967): High Field Transport in Semiconductors. In: Solid State Physics Suppl. 9 (eds. F. Seitz, D. Thornbull, H. Ehrenreich). Academic Press, New York.

Cook R. K., Frey J. (1982): COMPEL *1*, 65.

Fischetti M. V., Laux S. E. (1989): Phys. Rev. *b38*, 9721.

Forghieri A., Guerrieri R., Camoloni P., Gnudi A. Rudan M., Baccarani G. (1986): NUPAD Symposium, Santa Clara.

Fukuma M., Übbing R. H. (1984): IEDM Digest 621.

Hänsch W., Jacobs H. (1989): IEEE Electr. Device Lett. *EDL7*, 285.

Hänsch W., Miura-Mattausch M. (1986): J. Appl. Phys. *60*, 650.

Hänsch W., Schmeisser C. (1989): J. Appl. Math. Phys. (ZAMP) *40*, 440.

Hänsch W., Selberherr S. (1986): IEEE Trans. Electr. Devices *ED34*, 1074.

Hilsum C. (1974): Electron. Lett. *10*, 259.

Houston P. A., Evans A. G. (1977): Solid State Electr. *20*, 197.

Jacoboni C., Lugli P. (1989): The Monte Carlo Method for Semiconductor Device Simulation. In: Computational Microelectronics (ed. S. Selberherr). Springer, Wien.

Jaggi R. (1969): Helv. Phys. Acta *42*, 941.

Madelung O. (1978): Introduction to Solid State Theory. Springer Series in Solid State Science Vol. 2. Springer, Berlin.

Mahan G. D. (1981): Many-Particle Physics. Plenum, New York.

Maloney T. J., Frey J. (1977): J. Appl. Phys. *48*, 781.

Nakagawa A. (1986): IEEE Trans. Electr. Devices *ED33*, 167.

Nichols K. H., Yee C. M., Wolfe C. M. (1980): Solid State Electr. *23*, 109.

Ohmi T. (1967): Proc. IEEE, October 1967.

Onsager L. (1931a): Phys. Rev. *37*, 405.

Onsager L. (1931b): Phys. Rev. *38*, 2265.

Reggiani L., Brunetti R., Normantes E. (1986): J. Appl. Phys. *59*, 1212.

Ruch J., Kino G. (1968): Phys. Rev. *174*, 921.

Selberherr S. (1984): Analysis and Simulation of Semiconductor Devices. Springer, Wien.

Stratton R. (1957): Proc. R. Soc. *A242*, 355.

Thoma R., Emunds A., Meinerzhagen B., Pfeifer H., Engl W. L. (1989): IEDM 89 Technical Digest 139.

Wingreen N. S., Stanton C. B., Wilkins W. (1986): Phys. Rev. Lett. *57*, 1084.

Yourgau W., Van Der Merwe A., Raw G. (1982): Treatise on Irreversible and Statistical Thermophysics. Dover, New York.

Carrier Transport in an Inversion Channel 3

3.1 Introduction

In the previous two chapters, we directed our attention to basic questions of carrier transport in bulk material. We derived the Boltzmann equation in Chapter 1 and in Chapter 2 studied its consequences for low- and high-field transport in an electron-impurity-phonon system. We did not consider the influence of a finite domain containing interfaces on the transport properties. Interfaces are essential in device operation. Every contact is related to an interface problem. Neither ohmic nor Schottky contacts have hitherto been described self-consistently within the framework of the drift–diffusion approximation. Modeling efforts are based on heuristic arguments and more or less do the job. We will not go into the details of how to formulate a suitable theory for current-carrying contacts. In Chapter 1, we mentioned that a self-consistent incorporation of tunneling in the transport problem is the key feature of such a theory and that Eqs. (1.151) or (1.163) are a possible starting point. Tunneling is a quantum-mechanical phenomenon and is therefore not included in the drift–diffusion approximation.

Another device-related phenomenon that exhibits quantum-mechanical features is carrier transport in inversion layers. In a MOS device, the gate electrode is separated from the semiconductor material by a thin insulating layer. An appropriate bias at the gate contact controls the transverse field and thus the position of the band edges in the silicon. Inversion channels are created by extreme band bending near the surface so that an accumulation of minority carriers is achieved in a very small layer of a few nanometers thickness at the surface (Sze 1981). The corresponding diagram for the conduction band edge for p-material is shown in Fig. 20 for the case of equilibrium, where the Fermi level in the silicon is constant.

The band edge confines the carriers in a potential well perpendicular to the surface. If the typical width of this potential well is of the order of the de Broglie wavelength of the electron, which is $\lambda \approx \sqrt{(\hbar^2/2mk_B T)} \approx 1$ nm 2 nm at room temperature, we have to account for quantum-mechanical

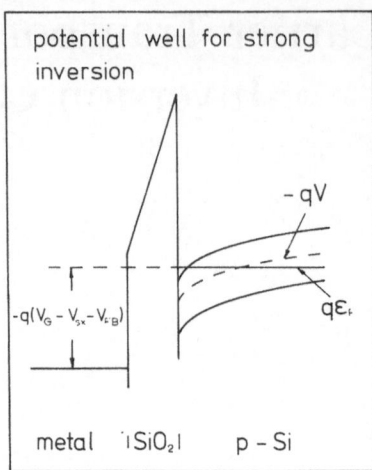

Fig. 20 Electric potential for a MOSFET in strong inversion. V_G, V_{sx}, and V_{FB} are gate voltage, substrate bias and flat band voltage, respectively, and ε_F is the quasi Fermi energy

effects. Pioneering work was done here by Stern and Howard (1967), who showed that while the motion of carriers is restricted parallel to the surface, perpendicular to it a discrete set of eigenstates exists. Many others refined this analysis (Stern 1972, Hsing et al. 1979, Übensee et al. 1985). It is by now certain that even at room temperature, the separation of the lowest eigenvalues is greater than the thermal energy $k_B T$. The puzzling question is: why do we not observe these quantized states in the terminal currents of a MOSFET at room temperature? The answer may be: why should we? All publications dealing with a rigorous approach to carrier transport in inversion layers stop after solving the first part of the problem: the coupled set of Poisson's and Schrödinger's equation to find its eigenvalues. This is not sufficient for the transport problem! This problem requires the solution of the density matrix equation that also contains the scattering in the system. For the eigenvalue problem, carrier scattering represents only a small correction and can be neglected. But it is essential for the transport problem because it yields the mobility. A self-consistent quantum-mechanical solution for carrier transport in an inversion channel at elevated temperatures that includes all eigenstates of the potential well is not available at present. We can therefore speculate that temperature and scattering effects together are responsible for the absence of quantum effects in the terminal currents.

The mobility is a key quantity for the purpose of modeling. We devoted a large proportion of the previous chapter to understand its low- and high-field behavior in bulk material from first principles. The mobility in an inversion channel is more difficult to handle. As we mentioned above, it would be the result of a fully self-consistent quantum-mechanical treatment of carrier transport in an inversion channel. Again, the modeling community cannot wait until this problem is solved. Different models are proposed which are based on a careful inspection of the experimental material (Yamaguchi, 1979, 1983, Cooper and Nelson 1981, Schwarz and Russek 1983, Hiroki *et al.* 1987, Selberherr *et al.* 1990). The most important of all these observations is a reduction of the channel mobility with increasing surface field. The experimental material only provides an effective channel mobility μ_{eff} which is related to the measured terminal current I_D and the measured inversion layer carrier densities n_{inv} (from capacity measurements) for small drain voltages V_D.

$$\mu_{\text{eff}} = \lim_{V_D \to 0} \frac{I_D}{V_D \, q n_{\text{inv}}} \tag{3.1}$$

The microscopic details are not available. Very useful models have been proposed by parameterizing a channel mobility by effective fields E_{eff} which are also insensitive to the microscopic details in the inversion channel (Cooper and Nelson 1981, Schwarz and Russek 1983, Walker and Woerlee 1988, Takagi *et al.* 1988). The problem with this parameterization is that it is not always clear how this effective field can be extracted from the internal device-specific field distributions. Different groups favor different approaches. The physics content is that the channel mobility is obviously somehow not simply a function of the local electric field component perpendicular to the interface.

In the following sections, we will study the carrier transport in an inversion channel starting from first principles. Although our analysis will start by using the Wigner–Boltzmann equation, Eq. (1.163), for the situation of an inversion channel, we will not try the fully self-consistent quantum-mechanical solution. By calculating the limit $\hbar \to 0$, we can obtain a link between the classical and quantum-mechanical situation. A solution for the density distribution and the channel mobility is found in a closed analytical form. It is suitable for incorporation within the drift diffusion equation and will contain the essential features of the correct quantum-mechanical solution. We will then compare this solution, in the case of equilibrium, with the exactly solvable quantum-mechanical problem.

3.2 The Classical Limit $\hbar \to 0$

As device dimensions become smaller, and the future promises device features of atomic dimensions, quantum transport is a challenge for device modeling. The quantum transport problem separates into two subproblems.

(1) The quantum-mechanical eigenvalue problem has to be solved for the structure of interest. Technically this means solving the eigenvalue problem of a coupled Schrödinger–Poisson system which provides the eigenstates in a self-consistently calculated electric field. The Schrödinger–Poisson system is equivalent to the Hartree approximation of an interacting electron gas and is a one-body problem. The density matrix reduces to the one-particle occupation number. If the quantum mechanical one-particle states φ_n of the system are known, the particle density distribution is calculated by

$$n(R) = \sum_n n_F(\varepsilon_n)|\varphi_n(R)|^2 \tag{3.2}$$

where n_F is the Fermi distribution function.

(2) To solve the transport problem, we have to find a solution for the density matrix ρ for the non-equilibrium case. If it is available, the particle and current density are given by

$$n(R, T) = \mathrm{Tr}(\rho) \tag{3.3}$$

$$j(R, T) = q\,\mathrm{Tr}(\rho v) \tag{3.4}$$

The trace is calculated on a conveniently chosen basis, for instance, the previously determined eigenstates can serve as a proper basic set. If the deviations are not too far from equilibrium, this would be sufficient. For large deviations from equilibrium, the solution of the Poisson equation might change dramatically, so that the eigenstates self-consistently determined under (1) are a bad choice.

Fortunately, if we neglect the exchange interaction (Pauli's principle) and higher-order correlation effects in an electron system, the whole problem is equivalent to a one-body problem where the Poisson equation takes care of the repulsive electron–electron interaction (Hartree approximation). However, it may be necessary to include the exchange and correlation effects in the eigenvalue problem as well. A degree of complexity is then reached so that even the first step is very elaborate (Laux and Warren 1986) and the second nearly impossible. However, the situation is less desperate for very special situations: a complete theory exists for the case of small deviations from equilibrium in a homogeneous field (Mahan 1981). In this case, the transport coefficients are independent of the driving fields and can therefore be directly calculated with equilibrium quantities. The powerful instrument of equilibrium many-body theory can now be utilized. These approaches are listed in the literature under linear-response theory or Kubo's formula.

Their main feature is that they do not calculate the density matrix explicitly but rather evaluate the expectation values for n and j directly. The problem in such an approach is to find approximations of n and j that are consistent with the continuity equation. The linear-response theory provides such approximation schemes. A direct calculation of the non-equilibrium density matrix solves its equation of motion. The Wigner function method which we introduced in Chapter 1 is an approximation scheme to obtain the relevant information for the non-equilibrium density matrix (Kadanoff and Baym 1962).

For the case of linear response and a constant electric field, it has been shown that the linearized generalized Boltzmann equation (compare discussion of Eq. (1.162)) including quantum corrections is exactly equivalent to Kubo's formula for the conductivity of a metal (Hänsch and Mahan 1983). For the case of a semiconductor, we do not in general have a constant field. We must allow for space charges and thus for spatial field variations. In this case, we can consider Eq. (1.163) as a suitable transport equation. It is correct as it stands if the scattering in the system is sufficiently weak, so that the spread of the spectral function is small, and if the appropriate zero-order eigenstates are free particles. A generalization of this equation that includes bound states in the potential field, for example, can be found by reconsidering the generalized Boltzmann equation (1.150). In the quasi-particle limit, we can neglect the contributions from a finite spread of the spectral function. We can therefore omit the last three terms in Eq. (1.150)

$$(g_0^{r-1} G^< - G^< g_0^{a-1}) = \{G^> \Sigma^< - \Sigma^> G^<\} \tag{3.5}$$

This is the Wigner–Boltzmann equation for quantum systems with weak interactions. The left-hand side is, according to Eq. (1.153),

$$g_0^{r-1} G^< - G^< g_0^{a-1} \to i \frac{\partial}{\partial T} G^<(R, T, k, \omega) + i \frac{k}{m} \nabla_R G^<(R, T, k, \omega)$$

$$+ q \left(V \left(R + i\frac{1}{2}\nabla_k T - i\frac{1}{2}\frac{\partial}{\partial \omega} \right) - V \left(R - i\frac{1}{2}\nabla_k, T + i\frac{1}{2}\frac{\partial}{\partial \omega} \right) \right)$$

$$\cdot G^< f(R, T, k, \omega) \tag{3.6}$$

For the scattering term or the right-hand side of Eq. (3.5), we utilize Eq. (1.159) and follow the same steps that brought us to Eq. (1.163). However, we will treat the spatial dependence rigorously and assume a slow variation in time. In this limit, we can again perform a Taylor expansion in the relative time coordinates and keep the leading term (see also Eqs. (1.159)–(1.162)). After Fourier transformation of the complete right-hand side with respect to the Wigner coordinates r and t we obtain,

$$\{G^> \Sigma^< - \Sigma^> G^<\} \rightarrow \int dr \exp(-ikr)$$

$$- \left[\int dr' \int \frac{dk'}{(2\pi)^3} \exp(ik'r')G^> (R + \tfrac{1}{2}(r - r'), T, k', \omega) \right.$$

$$\cdot \Sigma^< (R - \tfrac{1}{2}r', T, r - r', \omega) \tag{3.7}$$

$$- \int dr' \int \frac{dk'}{(2\pi)^3} \exp(ik'(r - r'))\Sigma^> (R + \tfrac{1}{2}(r - r'), T, r', \omega)$$

$$\left. \cdot G^< (R - \tfrac{1}{2}r', T, k'\omega) \right]$$

We have not yet assumed any restrictions on the spatial variation of the potential, the Green's functions $G^{</>}$, and the self-energies $\Sigma^{</>}$. Therefore Eqs. (3.6) and (3.7) contain the complete spatial information of the system. For the Green's functions $G^{</>} (R, T, k, \omega)$, which are closely related to the distribution function, we obtain the following according to Eqs. (1.94) and (1.95)

$$G^< (R, T, k, \omega) = i \int dr \int_{-\infty}^{\infty} d\tau \exp(-ikr + i\omega\tau)$$

$$\cdot \langle \Psi^+ (R - \tfrac{1}{2}r, T - \tfrac{1}{2}\tau)\Psi(R + \tfrac{1}{2}r, T + \tfrac{1}{2}\tau) \rangle \tag{3.8}$$

$$G^> (R, T, k, \omega) = - i \int dr \int_{-\infty}^{\infty} d\tau \exp(-ikr + i\omega\tau)$$

$$\cdot \langle \Psi (R + \tfrac{1}{2}r, T + \tfrac{1}{2}\tau)\Psi^+ (R - \tfrac{1}{2}r, T - \tfrac{1}{2}\tau) \rangle$$

with the field operators Ψ given by Eq. (1.6)

$$\Psi(R, T) = \sum_i \varphi_i(R)a_i(T)$$

$$\Psi^+ (R, T) = \sum_i \varphi_i^*(R)a_i^+ (T) \tag{3.9}$$

In the previous analysis, we implicitly assumed the wave functions to be plane waves. This is by no means necessary, although it simplifies the algebra considerably. The summation index in Eq. (3.8) runs over an arbitrary complete set of eigenstates. In the case of bulk material, plane waves are a good zero-order approximation and equivalent to the effective mass approximation, although Bloch waves would have been a better choice due to the periodic crystal structure. In the following, we will disregard band-structure effects again. In principle, they can be considered with the same approach outlined below.

Our goal for this section is to derive a transport equation that can be used if the system contains bound states. We will then take the limit $\hbar \to 0$ to obtain the correct match to the classical results which we discussed in Chapter 2. We will not investigate this problem in the most general way, but rather restrict our discussion to the most relevant situation where charge transport appears in a lateral field while the perpendicular field confines the particles in a potential well so that bound states exist. This situation is satisfied in the inversion channel of a MOSFET and also approximately in hetero-junction devices, for instance the MODFET, with charge transport dominantly parallel to the layered structure.

In equilibrium, the lateral field is zero and the potential depends only on R_t, the coordinate transverse to the direction of charge transport. The eigenvalue problem to be solved is the following Schrödinger equation

$$\left\{ -\frac{1}{2m}\frac{\partial^2}{\partial R_t^2} - qV(R_t) \right\} \varphi_n(R_t) = \varepsilon_n \varphi_n(R_t) \tag{3.10}$$

together with Poisson's equation to determine the self-consistent potential $V(R_t)$. If we now apply a small lateral field, current flow is possible. We will assume that the lateral field is sufficiently weak so that its effect on the eigenvalues ε_n and eigenfunctions $\varphi_n(R_t)$ is negligible. In this case, the lateral and transverse directions decouple completely and we can consider a potential of the form

$$V(R, T) \to V_1(R_1, T) + V_t(R_t, T) \tag{3.11}$$

For the wave functions we then have

$$\varphi_i(R) \to \exp(ik_1 R_1)\varphi_n(R_t) \tag{3.12}$$

Inserting Eqs.. (3.9) with the wave functions from Eq. (3.12) into Eq. (3.9) we obtain the following for the Green's functions

$$G^<(R, T, k, \omega) = i\sum_{nn'} \int dr_t \exp(-ik_t r_t)\varphi_{n'}^*(R_t - \tfrac{1}{2}r_t)\varphi_n(R_t + \tfrac{1}{2}r_t)$$
$$\cdot F_{nn'}^<(r_1, T, k_1, \omega) \tag{3.13}$$

$$G^>(R, T, k, \omega) = -i\sum_{nn'} \int dr_t \exp(-ik_t r_t)\varphi_{n'}^*(R_t - \tfrac{1}{2}r_t)\varphi_n(R_t + \tfrac{1}{2}r_t)$$
$$\cdot F_{nn'}^>(r_1, T, k_1, \omega) \tag{3.14}$$

The auxiliary functions $F^{</>}$ correspond to the $G^{</>}$ for the reduced lateral transport problem and are given by

$$F_{nn'}^{<}(R_1, T, k_1, \omega) = \sum_{Q_1} \int d\tau \exp(-iR_1 Q_1 + i\omega\tau)$$

$$\cdot \langle a_{k_1+1/2Q_1n'}^{+}(T - \tfrac{1}{2}\tau) a_{k_1-1/2Q_1n}(T + \tfrac{1}{2}\tau)\rangle \tag{3.15}$$

$$F_{nn'}^{>}(R_1, T, k_1, \omega) = \sum_{Q_1} \int d\tau \exp(-iR_1 Q_1 + i\omega\tau)$$

$$\cdot \langle a_{k_1+1/2Q_1n}(T + \tfrac{1}{2}\tau) a_{k_1-1/2Q_1n'}^{+}(T - \tfrac{1}{2}\tau)\rangle \tag{3.16}$$

The matrix F has the property

$$[F^{</>} nn']^{+} = [F^{</>} n'n] \tag{3.17}$$

which means that its diagonal elements have real values. We will now show that $F^{<}$ is sufficient to obtain carrier and current densities. To this end, we have to remember that the distribution function $f(R, T, k)$ was related to the Green's function $G^{<}(R, T, k, \omega)$ by a simple ω integration,

$$f(R, T, k) = -i \int \frac{d\omega}{2\pi} G^{<}(R, T, k, \omega) \tag{3.18}$$

Carrier and current density can be expressed as

$$n(R, T) = \sum_k f(R, T, k)$$

$$= \sum_{\substack{nn' \\ k_1}} \varphi_{n'}^{*}(R_t)\varphi_n(R_t) \int \frac{d\omega}{2\pi} F_{nn'}^{<}(r_1, T, k_1, \omega) \tag{3.19}$$

$$j(R, T) = q \sum_k \frac{k}{m} f(R, T, k)$$

$$= \sum_{\substack{nn' \\ k_1}} \frac{k_1}{m} \varphi_{n'}^{*}(R_t)\varphi_n(R_t) \int \frac{d\omega}{2\pi} F_{nn'}^{<}(r_1, T, k_1, \omega)$$

$$+ iq\frac{1}{2m} \sum_{\substack{nn' \\ k_1}} \left\{ \frac{\partial}{\partial R_t} \varphi_{n'}^{*}(R_t)\varphi_n(R_t) - \varphi_{n'}^{*}(R_t)\frac{\partial}{\partial R_t} \varphi_n(R_t) \right\} \tag{3.20}$$

$$\cdot \int \frac{d\omega}{2\pi} F_{nn'}^{<}(r_1, T, k_1, \omega)$$

We observe that the current density has two contributions. The first comes from the lateral motion of the carriers, as expected. The second comes from the transversal motion. Bound states in one dimension have wave functions with real values, and would therefore not contribute to a transversal current flow if $F_{nn'}^<$ is diagonal and therefore real-valued. This is in general not the case. We will show below that it is essential to keep the non-diagonal parts of $F_{nn'}^<$ to obtain the correct classical limit for its equation of motion. Nevertheless, it is probably reasonable to assume a diagonal $F_{nn'}^<$ for the lowest bound states that have the greatest energy separation. In equilibrium it is easy to verify that $F^<$ is diagonal in n and that only $Q = 0$ contributes to the Q-summation. We then get

$$F_{nn'}^<(R_1, T, k_1, \omega) = \int d\tau \exp(i\omega\tau) \exp(-i(\varepsilon_{k_1} + \varepsilon_n)\tau) n_F(\varepsilon_{k_1} + \varepsilon_n)\delta_{nn'}$$

(3.21)

which gives for the particle density the well-known expression

$$n(R) = \sum_{\substack{n \\ k_1}} |\varphi_n(R_t)|^2 n_F(\varepsilon_{k_1} + \varepsilon_n)$$

(3.22)

Because $F^<$ is the generalized distribution function, we have to find its Wigner–Boltzmann equation. We will separately investigate the kinetic term, Eq. (3.6), and the scattering term, Eq. (3.7) which comprise the left-hand and the right-hand sides, respectively, of this equation. For the first term of Eq. (3.6), we obtain, with Eq. (3.11) after performing a summation over k_t

$$i\sum_{k_t} \frac{\partial}{\partial T} G^<(R, T, k, \omega) = \sum_{nn'} \varphi_{n'}^*(R_t)\varphi_n(R_t) \frac{\partial}{\partial T} F_{nn'}^<(R_1, T, k_1, \omega)$$

(3.23)

For the second term we get

$$i\sum_{k_t} \frac{k}{m} \nabla_R G^<(R, T, k, \omega) = \sum_{nn'} \varphi_{n'}^*(R_t)\varphi_n(R_t) \frac{k_1}{m} \nabla_{R_1} F_{nn'}^<(R_1, T, k_1, \omega)$$

$$+ i\frac{1}{2m} \sum_{nn'} [\nabla_{R_t}^2 \varphi_{n'}^*(R_t)\varphi_n(R_t)$$

$$- \varphi_{n'}^*(R_t)\nabla_{R_t}^2 \varphi_n(R_t)]$$

$$\cdot F_{nn'}^<(R_1, T, k_1, \omega)$$

$$= \sum_{nn'} \varphi_{n'}^*(R_t)\varphi_n(R_t) \frac{k_1}{m} \nabla_{R_1} F_{nn'}^<(R_1, T, k_1, \omega)$$

$$- i\sum_{nn'} \varphi_{n'}^*(R_t)\varphi_n(R_t)[\varepsilon_{n'} - \varepsilon_n] F_{nn'}^<(R_1, T, k_1, \omega)$$

(3.24)

Here we used the Schrödinger equation (3.10) to express the second-order derivatives of the wave functions φ_n. Finally, we obtain for the last term of Eq. (3.6)

$$q\sum_{k_t}\left(V\left(R + i\frac{1}{2}\nabla_k T - i\frac{1}{2}\frac{\partial}{\partial\omega}\right)\right.$$

$$\left. - V\left(R - i\frac{1}{2}\nabla_k, T + i\frac{1}{2}\frac{\partial}{\partial\omega}\right)\right)G_{nn'}^{<}(R, T, k, \omega)$$

$$= -iq\sum_{nn'}\varphi_{n'}^{*}(R_t)\varphi_n(R_t)\left(V\left(R_1 + i\frac{1}{2}\nabla_{k_1}, T - i\frac{1}{2}\frac{\partial}{\partial\omega}\right)\right.$$

$$\left. - V\left(R_1 - i\frac{1}{2}\nabla_{k_1}, T + i\frac{1}{2}\frac{\partial}{\partial\omega}\right)\right)$$

$$\cdot F_{nn'}^{<}(R_1, T, k_1, \omega) \tag{3.25}$$

where we utilized the circumstance that transverse and lateral coordinates decouple in the potential, Eq. (3.11). Because we will not consider any quantum size effects in the lateral direction, we can take the limit $\hbar \to 0$ in Eq. (3.25) for the potential difference. This also corresponds to the assumption that the lateral spatial variation of the electric field is small. Collecting all contributions from Eqs. (3.23) to (3.25), we obtain the following for the left-hand side of the generalized Wigner–Boltzmann equation

$$\sum_{nn'}\varphi_{n'}^{*}(R_t)\varphi_n(R_t)\left[\frac{\partial}{\partial T} + \frac{k_1}{m}\nabla_{R_1} - qE_1(R_1, T) - i(\varepsilon_{n'} - \varepsilon_n)\right]$$

$$\cdot F_{nn'}^{<}(R_1, T, k_1, \omega) \tag{3.26}$$

We will now turn to the scattering term, Eq. (3.7). The variables in the integral refer to Wigner coordinates which are center-of-mass and relative coordinates defined by Eqs. (1.119) and (1.120). The scattering term assumes a particularly simple form if we expand the self-energies in the physical coordinates x and x' with respect to the complete basic set φ_n.

$$\Sigma^{</>}(xt, x't') = \sum_{nn'}\sigma_{nn'}^{</>}(x_1 t, x_1' t')\varphi_n(x_t)\varphi_{n'}^{*}(x_t') \tag{3.27}$$

Inserting Eqs. (3.11) and (3.27) (but now in relative coordinates!) into Eq. (3.7), the scattering terms can be expressed as

$$i\sum_{nn'}\varphi_{n'}^{*}(R_t)\varphi_n(R_t)\sum_{l}[F_{n1}^{>}(R_1, T, k_1, \omega)\sigma_{1n'}^{<}(R_1, T, k_1, \omega)$$

$$+ \sigma_{n1}^{>}(R_1, T, k_1, \omega)F_{1n'}^{<}(R_1, T, k_1, \omega)] \tag{3.28}$$

Equations (3.26) and (3.28) together give the appropriate Wigner–Boltzmann equation for charge transport in inversion channels. If we consider only the stationary $(\partial/\partial T)V = 0$ case, it reads

$$\frac{k_1}{m} \nabla_{R1} F_{nn'}^{<}(R_1, k_1, \omega) - qE_1(R_1)\nabla_{k1} F_{nn'}^{<}(R_1, k_1, \omega)$$

$$- i(\varepsilon_{n'} - \varepsilon_n) F_{nn'}^{<}(R_1, k_1, \omega)$$

$$= i \sum_l [F_{n1}^{>}(R_1, k_1, \omega)\sigma_{1n'}^{<}(R_1, k_1, \omega)$$

$$+ \sigma_{n1}^{>}(R_1, k_1, \omega)F_{1n'}^{<}(R_1, k_1, \omega)] \tag{3.29}$$

Equation (3.29) shows a certain resemblance to the ordinary Boltzmann equation (2.1). We observe classical behaviour in the lateral direction, which is hardly surprising as this was presumed from the beginning. The bound states appear in the scattering term, which is also nothing unusual as the interaction with optical phonons, for instance, can cause transitions from one bound state to another. The remarkable thing is that the scattering term is no longer a simple product but a matrix multiplication which runs over the complete spectrum of the Schrödinger equation (3.10). If the energetic separation of the eigenstates ε_n is large compared with a characteristic phonon energy, the coupling will be weak and Eq. (3.29) will indeed reduce to the Boltzmann transport equation for a two-dimensional electron gas. In this case, we consider the limit $n = n'$ in Eq. (3.29) and keep only the dominant intra-subband scattering term $n = 1 = n'$ in the summation on the right-hand side.

$$\frac{k_1}{m} \nabla_{R1} F_{nn}^{<}(R_1, k_1, \omega) - qE_1(R_1)\nabla_{k1} F_{nn}^{<}(R_1, k_1, \omega)$$

$$= i[(1 - F_{nn}^{<}(R_1, k_1, \omega))\sigma_{nn}^{<}(R_1, k_1, \omega)$$

$$+ \sigma_{nn}^{>}(R_1, k_1, \omega)F_{nn}^{<}(R_1, k_1, \omega)] \tag{3.30}$$

Here we have used the identity $F^{<} = 1 - F^{>}$, which is easily proven by utilizing Eqs. (3.15) and (3.16). This transport equation is valid, for instance at low temperatures where the high-energy phonons are frozen out and transitions between the lowest subbands are unlikely. This situation is unfortunately not realized in device operation at room temperature. As we will see below for a typical MOSFET, the energy separation of the lowest bound states is approximately 100 meV, which is not large compared with the optical phonon energy, which is of the order of 60 meV. As a result, we cannot neglect the non-diagonal parts in $F^{<}$ if we want to describe the MOSFET realistically. The situation becomes even more dramatic if we have to consider the influence of the first few bound states, which is necessary at room temperature to obtain the correct carrier density near the surface. The energy separation $\Delta\varepsilon \sim n^{-1/3}$ and therefore the coupling of the bands then becomes more important. For the eigenstates with very high energies, we have to approach the classical limit $\hbar \to 0$. We will now show that this limit can be obtained correctly only if we maintain the non-diagonal terms in Eq. (3.29). In the stationary case, the third term on the left-hand side of Eq. (3.29) is nothing other than the time derivative of the density operator $F^<$, which can be expressed by the following commutator relation

$$- i(\varepsilon_{n'} - \varepsilon_n) F^<_{nn'} = \frac{d}{d\tau} F^<_{nn'} = \frac{i}{\hbar} [h, F]_{nn'} \to \{h, f\}_{pq}$$

$$= \frac{\partial}{\partial p} h \frac{\partial}{\partial q} f - \frac{\partial}{\partial q} h \frac{\partial}{\partial p} f$$

$$= \frac{k_t}{m} \nabla_{R_t} f - q E(R_t) \nabla_{k_t} f \qquad (3.31)$$

Here we have taken the classical limit $\hbar \to 0$ by replacing the quantum mechanical commutator relationship by the corresponding Poisson bracket. This transition also means replacing the quantum labels by the classical canonical variables q and p, $nn' \to qp$. Replacing the third term on the left-hand side of equation (3.29) by the right-hand side of Eq. (3.31) and interchanging $nn' \to k_t R_t$ we discover the conventional Boltzmann equation. We have shown that the classical Boltzmann equation is indeed the correct limit of the Wigner–Boltzmann equation (3.29). The classical limit is

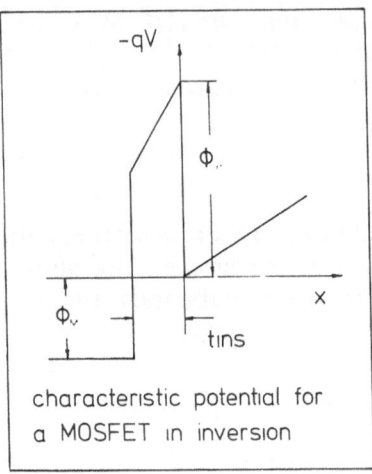

characteristic potential for
a MOSFET in inversion

Fig. 21 Approximation of the electric potential for a MOSFET in strong inversion by piece-wise linear functions. This approximation holds in the close vicinity of the Si/SiO$_2$ interface. t_{ins} is the insulator thickness, Φ_B is the barrier height between the conduction band edge of Si and SiO$_2$, and Φ_M is the potential difference of the Si surface potential and the intrinsic Fermi level of the gate metal

justified if the characteristic potential width $a \gg a_0(Ry/\varepsilon_n)^{1/2}$ for the corresponding eigenstate ε_n.

After having found the correct classical limit for the distribution function, the same task remains to be performed for the wave functions φ_n. We have to specify the potential in the Schrödinger equation (3.10). For the MOSFET, it is sufficient to consider a piecewise linear potential with discontinuities at $x = R_t = 0$ (Si/SiO$_2$ interface) and $x = -t_{ins}$, when t_{ins} is the insulator thickness. We show this potential in Fig. 21.

$$-qV(x) = \begin{matrix} F_s x, & x > 0 \\ \Phi_B + F_i x, & -t_{ins} < x < 0 \\ \Phi_M \end{matrix}$$

(3.32)

The solution of the Schrödinger equation with the potential Eq. (3.32) is found to be the following for positive x

$$\varphi_s(x) = \alpha_s Ai\left(\left[\frac{2m}{\hbar^2}\right]^{1/3} \frac{1}{F_s^{2/3}}(F_s x - \varepsilon)\right)$$

(3.33)

and

$$\varphi_i(x) = \alpha_i Ai\left(\left[\frac{2m}{\hbar^2}\right]^{1/3} \frac{1}{F_i^{2/3}}(F_i x + \Phi_B - \varepsilon)\right) + \beta_i Bi\left(\left[\frac{2m}{\hbar^2}\right]^{1/3}\right.$$
$$\left. \cdot \frac{1}{F_i^{2/3}}(F_i x + \Phi_B - \varepsilon)\right)$$

(3.34)

for $-t_{ins} < x < 0$. For $x < -t_{ins}$, we find plane wave solutions.

$$\varphi_M(x) = \alpha_M \exp\left(-i\left(\frac{2m}{\hbar^2}\right)^{1/2}(\varepsilon - \Phi_M)^{1/2} x\right)$$
$$+ \beta_M \exp\left(i\left(\frac{2m}{\hbar^2}\right)^{1/2}(\varepsilon - \Phi_M)^{1/2} x\right)$$

(3.35)

Here Ai and Bi are the Airy function of the first and second kinds and α_s, α_i, β_i, α_M, and β_M are coefficients to be determined by the boundary conditions at $x = 0$ and $x = -t_{ins}$ and normalization of the wave function. For positive x, the coefficient of the Airy function Bi is zero because it is divergent in the limit $x \rightarrow \infty$. The boundary condition at $x = 0$ is

$$\varphi_s(0) = \varphi_i(0)$$
$$\frac{\partial}{\partial x}\varphi_s(0) = \frac{\partial}{\partial x}\varphi_i(0)$$

(3.36)

which gives the following with Eqs. (3.33) and (3.34)

$$\alpha_s Ai\left(-\left[\frac{2m}{\hbar^2}\right]^{1/3}\frac{1}{F_s^{2/3}}\varepsilon\right) =$$

$$= \alpha_i Ai\left(\left[\frac{2m}{\hbar^2}\right]^{1/3}\frac{1}{F_i^{2/3}}(\Phi_B - \varepsilon)\right) + \beta_i Bi\left(\left[\frac{2m}{\hbar^2}\right]^{1/3}\frac{1}{F_i^{2/3}}(\Phi_B - \varepsilon)\right)$$

$$\tag{3.37}$$

$$F_s^{1/3}\alpha_s Ai'\left(-\left[\frac{2m}{\hbar^2}\right]^{1/3}\frac{1}{F_s^{2/3}}\varepsilon\right) =$$

$$= F_i^{1/3}\alpha_i Ai'\left(\left[\frac{2m}{\hbar^2}\right]^{1/3}\frac{1}{F_i^{2/3}}(\Phi_B - \varepsilon)\right)$$

$$+ F^{-1/3}\beta_i Bi'\left(\left[\frac{2m}{\hbar^2}\right]^{1/3}\frac{1}{F_i^{2/3}}(\Phi_B - \varepsilon)\right) \tag{3.38}$$

Here the apostrophe means the derivative of the Airy functions Ai and Bi with respect to the argument. From Eqs. (3.37) and (3.38), we can express the coefficients α_i and β_i in terms of α_s. The characteristic determinant of this set of equations is the Wronskian of the two independent Airy functions Ai and Bi. Its value is π^{-1}.

$$\alpha_i = \alpha_s \pi \left[Ai\left(-\left[\frac{2m}{\hbar^2}\right]^{1/3}\frac{1}{F_i^{2/3}}\varepsilon\right) Bi'\left(\left[\frac{2m}{\hbar^2}\right]^{1/3}\frac{1}{F_i^{2/3}}(\Phi_B - \varepsilon)\right)\right.$$

$$-\left(\frac{F_s}{F_i}\right)^{1/3} Ai'\left(-\left[\frac{2m}{\hbar^2}\right]^{1/3}\frac{1}{F_s^{2/3}}\varepsilon\right)$$

$$\left. \cdot Bi\left(\left[\frac{2m}{\hbar^2}\right]^{1/3}\frac{1}{F_i^{2/3}}(\Phi_B - \varepsilon)\right)\right] \tag{3.39}$$

$$\beta_i = \alpha_s \pi \left[\left(\frac{F_s}{F_i}\right)^{1/3} Ai'\left(-\left[\frac{2m}{\hbar^2}\right]^{1/3}\frac{1}{F_s^{2/3}}\varepsilon\right)\right.$$

$$\cdot Ai\left(\left[\frac{2m}{\hbar^2}\right]^{1/3}\frac{1}{F_i^{2/3}}(\Phi_B - \varepsilon)\right) - Ai\left(-\left[\frac{2m}{\hbar^2}\right]^{1/3}\frac{1}{F_s^{2/3}}\varepsilon\right)$$

$$\left. \cdot Ai'\left(\left[\frac{2m}{\hbar^2}\right]^{1/3}\frac{1}{F_i^{2/3}}(\Phi_B - \varepsilon)\right)\right] \tag{3.40}$$

The barrier height (≈ 3.2 eV for electrons) is large compared with some characteristic carrier energy, for instance the first few energy levels in the potential well (≈ 250 meV). It is therefore reasonable to assume that near the Si/SiO$_2$ interface, only energies $\varepsilon < \Phi_B$ are significant. For those energies, we calculate the classical limit $\hbar \to 0$ in Eqs. (3.39) and (3.40). With the appropriate asymptotic expansions of the Airy functions Ai and Bi we obtain

$$\alpha_i = \frac{1}{2} \alpha_s \left(\frac{F_s}{F_i} \right)^{1/6} \left[\left(\frac{\Phi_B - \varepsilon}{\varepsilon} \right)^{1/4} \sin \left(\frac{2}{3} \left(\left[\frac{2m}{\hbar^2} \right]^{1/2} \frac{\varepsilon^{3/2}}{F_s} + \frac{1}{4} \pi \right) \right) \right.$$

$$\left. + \left(\frac{\varepsilon}{\Phi_B - \varepsilon} \right)^{1/4} \cdot \cos \left(\frac{2}{3} \left(\left[\frac{2m}{\hbar^2} \right]^{1/2} \frac{\varepsilon^{3/2}}{F_s} + \frac{1}{4} \pi \right) \right) \right]$$

$$\cdot \exp \left(\frac{2}{3} \left(\left[\frac{2m}{\hbar^2} \right]^{1/2} \frac{(\Phi_B - \varepsilon)^{3/2}}{F_s} \right) \right) \pi \qquad (3.41)$$

$$\beta_i = \frac{1}{2} \alpha_s \left(\frac{F_s}{F_i} \right)^{1/6} \left[\left(\frac{\Phi_B - \varepsilon}{\varepsilon} \right)^{1/4} \sin \left(\frac{2}{3} \left(\left[\frac{2m}{\hbar^2} \right]^{1/2} \frac{\varepsilon^{3/2}}{F_s} + \frac{1}{4} \pi \right) \right) \right.$$

$$\left. - \left(\frac{\varepsilon}{\Phi_B - \varepsilon} \right)^{1/4} \cdot \cos \left(\frac{2}{3} \left(\left[\frac{2m}{\hbar^2} \right]^{1/2} \frac{\varepsilon^{3/2}}{F_s} + \frac{1}{4} \pi \right) \right) \right]$$

$$\cdot \exp \left(-\frac{2}{3} \left(\left[\frac{2m}{\hbar^2} \right]^{1/2} \frac{(\Phi_B - \varepsilon)^{3/2}}{F_s} \right) \right) \pi \qquad (3.42)$$

In the classical limit, particles with energy $\varepsilon < \Phi_B$ cannot penetrate the barrier, so α_i and β_i have to approach zero. Because α_s is a finite number, we must therefore have

$$\left(\frac{\Phi_B - \varepsilon}{\varepsilon} \right)^{1/4} \sin \left(\frac{2}{3} \left(\left[\frac{2m}{\hbar^2} \right]^{1/2} \frac{\varepsilon^{3/2}}{F_s} + \frac{1}{4} \pi \right) \right) + \left(\frac{\varepsilon}{\Phi_B - \varepsilon} \right)^{1/4}$$

$$\cdot \cos \left(\frac{2}{3} \left(\left[\frac{2m}{\hbar^2} \right]^{1/2} \frac{\varepsilon^{3/2}}{F_s} + \frac{1}{4} \pi \right) \right) = 0 \qquad (3.43)$$

This equation determines the eigenvalues for the possible energy levels in the potential well. This is, of course, not the exact quantum-mechanical solution because it was obtained in the classical limit. For large energies, however, the classical approximation holds true and for the first few bound states of the potential well we are very close to the exact eigenvalues. If the barrier is large enough, which is the case for the MOSFET, the lowest eigenvalues will correspond to those of an infinite barrier. These are the zeros of the Airy function Ai. To solve Eq. (3.43), we try to find a function $\delta(\varepsilon, \Phi_B)$

$$\sqrt{(\varepsilon 2m/\hbar^2)} \frac{2}{3} \frac{\varepsilon}{F_s} + \frac{1}{4} \pi = n\pi - \delta(\varepsilon, \Phi_B) \qquad (3.44)$$

Inserting Eq. (3.44) into Eq. (3.43), we find

$$\delta(\varepsilon, \Phi_B) = a \tan(\sqrt{(\varepsilon/(\Phi_B - \varepsilon))}) \to \sqrt{(\varepsilon/\Phi_B)} \qquad (3.45)$$

Here, the last approximation holds in the limit $\varepsilon \ll \Phi_B$, which is the case for the first few energy levels, for instance. From Eqs. (3.44) and (3.45) we obtain the well-known result that the lowest energies in an inversion channel behave like

$$\varepsilon_n = Ry \left[\left(\frac{F_s a_0}{Ry} \right) \right]^{2/3} \left[\frac{3\pi}{8} (4n - 1) \right]^{2/3} \sim (F_s)^{2/3} \left[\frac{3\pi}{8} (4n - 1) \right]^{2/3}$$

(3.46)

and for the wave function φ_s in the inversion channel we obtain the following in the classical limit

$$\varphi_n(x) \sim \alpha_s \frac{1}{\sqrt{\pi}} \left[\left(\frac{2m}{\hbar^2} \right)^{1/3} \frac{(\varepsilon_n - F_s x)}{F_s^{2/3}} \right]^{-1/4} \sin \left(\frac{2}{3} \left(\frac{2m}{\hbar^2} \right)^{1/2} \right.$$
$$\left. \cdot \frac{(\varepsilon_n - F_s x)^{2/3} - \varepsilon_n^{2/3}}{F_s x} x - a \tan(\sqrt{(\varepsilon_n/(\Phi_B - \varepsilon_n)))} \right) \quad (3.47)$$

This equation simplifies considerably in the direct vicinity of the interface $x \le d_n \sim 2a_0 (Ry/F_s a_0)^{1/3} n^{2/3}$

$$\varphi_n(x) \sim \sin(q_n x + q_n x_0), \quad q_n = \sqrt{(\varepsilon_n 2m/\hbar^2)}, \quad x_0 = \sqrt{(\hbar^2/\Phi_B 2m)}$$

(3.48)

By replacing the wave functions in Eqs. (3.19) and (3.20) by Eq. (3.48), we finally obtain the following for the carrier and current densities in the classical limit.

$$n(R, T) = \sum_k \sin^2(k_t(R_t + x_0)) f(R, T, k) \tag{3.49}$$

$$j_1(R, T) = q \sum_k \frac{k_1}{m} \sin^2(k_t(R_t + x_0)) f(R, T, k) \tag{3.50}$$

$$j_t(R, T) = 0 \tag{3.51}$$

Here, the distribution function $f(R, T, k)$ is the solution of the Boltzmann equation on a semi-space. The current density component perpendicular to the interface is zero, which, in the conventional case, is the boundary condition for the transverse current at the interface j_t $(R_t = 0, T) = 0$. We will show below that, as in the bulk case, the distribution function is proportional to $\nabla_R \varepsilon_F$ in the linear response. Equation (3.51) implies $\nabla_{R_t} \varepsilon_F = 0$, which is satisfied near the surface in the conventional case. Because the lateral current flows in a thin layer of width $d \sim 2a_0 (Ry/F_s a_0)^{1/3}$ at the surface, we can also calculate the current density from

$$j(R, T) = q \sum_k \frac{k}{m} \sin^2(k_t(R_t + x_0)) f(R, T, k) \tag{3.52}$$

without any loss of generality, because the transverse component will vanish, owing to the boundary condition. We will come back to this problem in the next section, where we explicitly calculate the carrier and current densities.

To derive Eqs. (3.50) and (3.51), we considered a situation appropriate for the inversion channel of a MOSFET. The very large barrier between Si and SiO_2 allows us to consider the wave function in the barrier for the lowest energy levels in the limit $\hbar \to 0$. If the circumstances are such that the barrier height is comparable with the lowest bound states in the potential well, we have to consider the wave function in the barrier more precisely. This will be especially important in the event of a significant through-tunneling between adjoining potential wells. This can easily be incorporated if we consider both boundaries of the potential barrier for obtaining the tunneling transfer characteristic. In the classical limit, the tunnel contribution cannot be obtained as an additional current flow because the wave functions and the distribution function are real and the second term in the current equation Eq. (3.20) is therefore zero. The effect of tunneling has to be incorporated into the conventional formulation via a non-local particle generation term which accounts for the fact that particles vanish on one side of the classically forbidden region and reappear on the other side. This term must be added to the continuity equation and is related to the overlap of the wave functions from adjacent classically separated regions. For the potential from Eq. (3.32), we may obtain the following for the wave functions $\varphi(x)$ for $\varepsilon < \Phi_B$ at $x = -t_{\text{ins}}$

$$\varphi(x = -t_{\text{ins}}) \sim \exp \left(\frac{2}{3} \left[\frac{2m}{\hbar^2} \right]^{1/2} t_{\text{ins}} \right.$$
$$\left. \cdot \frac{(\Phi_B - \varepsilon - F_i t_{\text{ins}})^{3/2} - (\Phi_B - \varepsilon)^{3/2}}{F_i t_{\text{ins}}} \right) \qquad (3.53)$$

if $0 < \varepsilon < \Phi_B - F_i t_{\text{ins}}$, and

$$\varphi(x = -t_{\text{ins}}) \sim \exp \left(-\frac{2}{3} \left[\frac{2m}{\hbar^2} \right]^{1/2} \frac{(\Phi_B - \varepsilon)^{3/2}}{F_i} \right) \qquad (3.54)$$

for $\Phi_B - F_i t_{\text{ins}} < \varepsilon < \Phi_B$. In the limit of large oxide fields $F_i > \Phi_B/t_{\text{ins}}$, the first contribution vanishes and Eq. (3.54) describes the Fowler–Northeim tunneling with its characteristic field dependence. In the limit $F_i \to 0$, the second term has a negligible contribution and Eq. (3.53) describes the field-independent direct tunneling. Both contribute in the intermediate regime and both are field-dependent. However, the first term will be significant only for very thin oxides.

3.3 Surface Mobility

To calculate the channel mobility of a MOSFET, we must determine the distribution function $f(R, T, k)$ to evaluate Eqs. (3.49) and (3.50) or (3.52). We will only consider the case where the deviation from equilibrium is small, so that the analysis of the previous section holds. It is therefore permissible to solve the Boltzmann equation in the relaxation-time approximation. In the stationary case, we disregard the time dependence of the distribution function $f(R, k)$ and obtain the following from Eq. (2.31)

$$v_k \nabla_R f(R, k) - qE(R)\nabla_k f(R, k) = -\frac{f(R, k) - f_{equ}(R, k)}{\tau(\varepsilon_k)} \tag{3.55}$$

This equation must be solved for a semi-infinite space (Schrieffer 1955, Greene et al. 1960). For convenience, the coordinate perpendicular to the interface R_t will be called x. In the linear response regime, we will find a solution of a form similar to Eq. (2.32)

$$f(R, k) = f_{equ}(R, k) - q\frac{\partial f}{\partial \varepsilon} \Lambda(k_x, x)\tau(\varepsilon_k)v_k \cdot \nabla_R \varepsilon_F(R) \tag{3.56}$$

To account for the symmetry-breaking interface, we introduced the vertex function $\Lambda(k_x, x)$. In the bulk case that we studied in Chapter 2, it is identical to one. To find an equation for the vertex function Λ, we insert Eq. (3.56) into Eq. (3.55) and keep only linear terms in $\nabla_R \varepsilon_F$

$$\tau(\varepsilon_k)qE(R)\cdot\nabla_R\varepsilon_F(R) + v_k\cdot\nabla_R\varepsilon_F(R)\left\{ \tau(\varepsilon_k)qE_x(R)\frac{\partial}{\partial k_x}\Lambda(k_x, x)\right.$$

$$\left. - \tau(\varepsilon_k)v_{k_x}\frac{\partial}{\partial x}\Lambda(k_x, x) - \Lambda(k_x, x) + 1\right\} = 0 \tag{3.57}$$

We assume only that the current flow parallel to the surface $\nabla_R\varepsilon_F$ has no component in the x-direction. In the limit of a small applied bias, the lateral field will also be small and therefore the first term in Eq. (3.57) is of second order and should be neglected. Furthermore, Eq. (3.57) can be solved consistently only if we neglect a possible weak variation of E_x with respect to the lateral coordinates. This will be sufficiently satisfied in a device where the channel region is laterally homogeneous for a small drain bias.

$$\tau(\varepsilon_k)qE_x(x)\frac{\partial}{\partial k_x}\Lambda(k_x, x) - \tau(\varepsilon_k)v_{k_x}\frac{\partial}{\partial x}\Lambda(k_x, x) - \Lambda(k_x, x) + 1 = 0$$

$$\tag{3.58}$$

A change of variables

$$x \rightarrow \varepsilon_x = \frac{k_x^2}{2m} - qV(x) + qV_0 \tag{3.59}$$

$$\Lambda(k_x, x) \rightarrow \Lambda(k_x, \varepsilon_x) \tag{3.60}$$

gives an equation that can be integrated analytically.

$$\tau(\varepsilon_k)qE_x(x)\frac{\partial}{\partial k_x}\Lambda(k_x, \varepsilon_x) = \Lambda(k_x, x) - 1 \tag{3.61}$$

Here the variable x must be expressed as a function of ε_x and k_x by utilizing the inverse potential function.

$$x = (qV)^{-1}\left(\varepsilon_x - \frac{k_x^2}{2m} - qV_0\right) \tag{3.62}$$

The physical meaning of this new set of variables will be discussed below. Formal integration gives

$$\Lambda(k_x, \varepsilon_x) = 1 + c \cdot \exp\left(\int_0^{k_x} \frac{dk_x'}{\tau(\varepsilon_{k'})qE_x(x')}\right) \tag{3.63}$$

The prime indicates the variation with respect to the integration variable. The integration constant c has to be determined from the boundary conditions which have not yet been specified. Before we can discuss suitable boundary conditions, we have to discuss the physical consequences of the

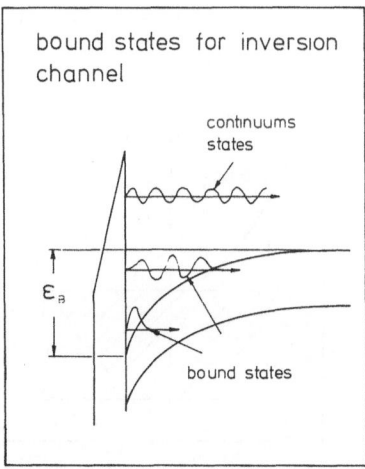

Fig. 22 The complete spectrum in an inversion channel has bound states in the potential well with depth ε_B and continuums states. The higher the bound state is located in the well the more it contributes to the carrier density far away from the interface

variable transformation Eq. (3.59). The variable ε_x corresponds to the classical energy of an electron in a one-dimensional potential field $-qV(x)$ as shown in Fig. 22.

The energy ε_x depends on both momentum k_x and space x. We have to distinguish different regions: the particle can be either confined in the potential well, which means $\varepsilon_x < \varepsilon_B = qV_0 - qV_\infty$ or has an energy larger than ε_B. In the (ε_x, k_x) plane, which we show in Fig. 23, we can therefore differentiate three regions:

$$\text{I. } 0 < \varepsilon_x < \varepsilon_B, \, k_x^2 < 2m\varepsilon_B$$

$$\text{II. } \quad \varepsilon_B < \varepsilon_x, \, k_x > 0$$

$$\text{III. } \quad \varepsilon_B < \varepsilon_x, \, k_x < 0$$

Similarly, we show the situation in the k_x, $V(x)$ plane in Fig. 24. The formal solution of Eq. (3.61) has to be restricted to each of the regions separately.

$$\Lambda^{\mathrm{I}}(k_x, \varepsilon_x) = 1 + c^{\mathrm{I}} \cdot \exp\left(\int_0^{k_x} \frac{dk'_x}{\tau(\varepsilon_{k'})qE_x(x')} \right) \tag{3.64.I}$$

$$\Lambda^{\mathrm{II}}(k_x, \varepsilon_x) = 1 + c^{\mathrm{II}} \cdot \exp\left(\int_0^{k_x} \frac{dk'_x}{\tau(\varepsilon_{k'})qE_x(x')} \right) \tag{3.64.II}$$

$$\Lambda^{\mathrm{III}}(k_x, \varepsilon_x) = 1 + c^{\mathrm{III}} \cdot \exp\left(\int_0^{k_x} \frac{dk'_x}{\tau(\varepsilon_{k'})qE_x(x')} \right) \tag{3.64.III}$$

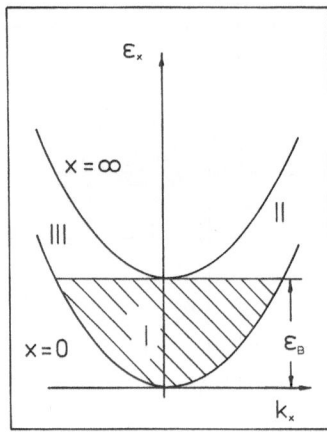

Fig. 23 ε_x, k_x phase space for a half space. Possible values of ε_x and k_x lie between the boundary parameters for $x = 0$ (surface) and $x = \infty$ (bulk). Region I is the phase space of bound states with $\varepsilon_x < \varepsilon_B$. Regions II and III contain the continuums states. They are connected by boundary conditions on the parabolas $x = 0$ and $x = \infty$ for fixed ε_x

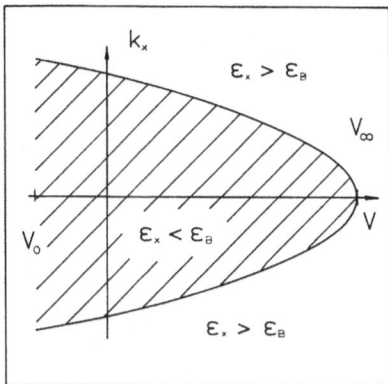

Fig. 24 k_x, V phase space for a half-plane

Equations (3.64) must be evaluated at the same point in the ε_x, k_x-plane. For constant energy surfaces, the k-integrals can be converted into an integration with respect to the potential V. From Eq. (3.59), we have

$$k_x = \text{sgn}(k_x) \cdot [2m(\varepsilon_x + qV - qV_0)]^{1/2} \qquad (3.65)$$

which gives the following, after differentiating with respect to V

$$dk_x = \text{sgn}(k_x)\frac{q}{2}(2m)^{1/2}[\varepsilon_x + qV - qV_0]^{-1/2}\, dV, \; \varepsilon_x = \text{const} \quad (3.66)$$

With the auxiliary function $G_{\varepsilon x}[V_1, V_2]$

$$G_{\varepsilon x}[V_1, V_2] = \tfrac{1}{2}(2m)^{1/2} \cdot \int_{V_2}^{V_1} \frac{dV'}{\tau(\varepsilon_{k'})E(V')}(\varepsilon_x + qV' - qV_0)^{-1/2}$$

$$(3.67)$$

where the k_x-dependence in the relaxation time should be replaced according to Eq. (3.65). Equations (3.64) will then read

$$\Lambda^{I}(k_x, \varepsilon_x) = 1 + c^{I} \cdot \exp\left(\text{sgn}(k_x) G_{\varepsilon x}\left[\frac{k_x^2}{2m} - \frac{1}{q}\varepsilon_x + V_0, \; -\frac{1}{q}\varepsilon_x + V_0 \right] \right)$$

$$(3.68.\text{I})$$

$$\Lambda^{II}(k_x, \varepsilon_x) = 1 + c^{II} \cdot \exp\left(G_{\varepsilon x}\left[\frac{k_x^2}{2m} - \frac{1}{q}\varepsilon_x + V_0, \; -\frac{1}{q}\varepsilon_x + V_0 \right] \right)$$

$$(3.68.\text{II})$$

$$\Lambda^{III}(k_x, \varepsilon_x) = 1 + c^{III} \cdot \exp\left(- G_{\varepsilon x}\left[\frac{k_x^2}{2m} - \frac{1}{q}\varepsilon_x + V_0, \; -\frac{1}{q}\varepsilon_x + V_0 \right] \right)$$

$$(3.68.\text{III})$$

We are now ready to formulate the boundary conditions. We do so on a constant energy surface ε_x which is represented as a straight line parallel to the k_x axis in Fig. 23. For all three regions, this surface intersects the surface parabola $x = 0$. It is clear that the boundary conditions have to contain information on the physical properties of the surface. This will be achieved by introducing the parameter p that accounts for the scattering properties of the surface. This parameter has the range $0 < p < 1$, which covers total absorptive to total reflective surfaces. The appropriate boundary condition at the surface $x = 0$, which corresponds to $k_x^0 = (2m\varepsilon_x)^{1/2}$, is therefore:

$$\text{region I:} \quad \Lambda^I(k_x^0, \varepsilon_x) = p\Lambda^I(-k_x^0, \varepsilon_x) \tag{3.69}$$

$$\text{region II/III:} \quad \Lambda^{II}(k_x^0, \varepsilon_x) = p\Lambda^{III}(-k_x^0, \varepsilon_x) \tag{3.70}$$

For the total absorptive surface $(p = 0)$, the vertex function in the first region will vanish at the surface. In the case of a specular reflective surface, it is an even function of k_x. For the regions II and III, the energy surface also cuts the parabola for $x \to \infty$, which corresponds to $k_x^\infty = (2m[\varepsilon_x - \varepsilon_B])^{1/2}$. In this case, we have the situation at a very remote distance from the interface. Surface properties should not be important here. We can therefore assume that the vertex function for incoming and outgoing particles is the same.

$$\text{region II/III:} \quad \Lambda^{II}(k_x^\infty, \varepsilon_x) = \Lambda^{III}(-k_x^\infty, \varepsilon_x) \tag{3.71}$$

The integration constants can now be determined from the boundary conditions Eqs. (3.69) to (3.71). After going back to the original variables k_x, x for the vertex function, we obtain

$$\Lambda^I(k_x, x) = 1 - (1 - p) \frac{\exp(G[V, V_0])}{1 - p \cdot \exp(-2G[V_0, V - k_x^2/q2m])},$$
$$k_x > 0 \tag{3.72}$$

$$\Lambda^I(k_x, x) = 1 - (1 - p)$$
$$\cdot \frac{\exp(-2G[V_0, V - k_x^2/q2m] - G[V, V_0])}{1 - p \cdot \exp(-2G[V_0, V - k_x^2/q2m])},$$
$$k_x < 0 \tag{3.73}$$

$$\Lambda^{II}(k_x, x) = 1 - (1 - p) \frac{\exp(G[V, V_0])}{1 - p \cdot \exp(2G[V_\infty, V_0])} \tag{3.74}$$

$$\Lambda^{III}(k_x, x) = 1 - (1 - p) \frac{\exp(2G[V_\infty, V_0] - G(V, V_0])}{1 - p \cdot \exp(2G[V_\infty, V_0])} \tag{3.75}$$

with $G[V_1, V_2]$

$$G[V_1, V_2] = \tfrac{1}{2}(2m)^{1/2} \cdot \int_{V_2}^{V_1} \frac{dV'}{\tau(\varepsilon_{k'})E(V')} (k_x^2/2m + qV' - qV)^{-1/2}$$

(3.76)

In regions II and III, we find $G[V_\infty, V_0] \sim -\varepsilon_B/E(V_\infty)$, which implies that it approaches $-\infty$ either if the field vanishes at infinity or if the binding energy is infinite. The latter is the case if we consider a constant field. With this, we finally obtain the following for the vertex function in regions II and III

$$\Lambda^{II}(k_x, x) = 1 - (1 - p)\exp(G[V, V_0])$$

(3.77)

$$\Lambda^{III}(k_x, x) = 1$$

(3.78)

To calculate density and current distribution, we have to use Eq. (3.56) with the vertex function Eqs. (3.72), (3.73), (3.77), and (3.78) in Eqs. (3.49) and (3.50). Only the first part of Eq. (3.56) contributes to the density distribution.

$$n(R) = \exp\left(\frac{qV(R) - q\varepsilon_F(R)}{k_B T}\right) \cdot \sum_k \sin^2[k_x(x + x_0)] \exp\left(-\frac{\varepsilon_k}{k_B T}\right)$$

$$= N \exp\left(\frac{qV(R) - q\varepsilon_F(R)}{k_B T}\right)\left(1 - \exp\left(-\left[\frac{x + x_0}{\lambda_{th}}\right]^2\right)\right)$$

(3.79)

Here, $\lambda_{th} = 1/\sqrt{2mk_B T}$ is the thermal wavelength and $N = (mk_B T/2\pi)^{3/2}$ the effective density of states. The effect of the wave function is contained in the third factor of Eq. (3.79). In the limit $\lambda_{th} \to 0$ ($\hbar \to 0$) we obtain the classical result. For the lateral current density we get

$$j_1(R) = -qn(R)\mu_{ch}(x)\nabla_{R_1}\varepsilon_F(R)$$

(3.80)

with the channel mobility μ_{ch}

$$\mu_{ch}(x) = \frac{q}{m}\frac{4}{\sqrt{\pi}}\int_0^\infty du\,u^3 \exp(-u^2)\int_{-\infty}^\infty dv\,\tau(u, v)\exp(-v^2)$$

$$\cdot \frac{\Lambda(v/\lambda_{th}, x)\sin^2\left(v\frac{x + x_0}{\lambda_{th}}\right)}{1 - \exp\left[-\left(\frac{x + x_0}{\lambda_{th}}\right)\right]}$$

(3.81)

Up to this point, we kept the energy dependence of the relaxation time τ. To proceed further analytically we have to replace $\tau(\varepsilon)$ by a suitable energy-independent average value $\langle \tau(\varepsilon) \rangle = \tau_{tr}$ (compare Eq. (2.38)).

$$\mu_{ch}(x) = \mu_{bulk}\alpha(x) \tag{3.82}$$

$$\alpha(x) = \frac{2}{\sqrt{\pi}} \int_{-\infty}^{\infty} dv \exp(-v^2) \Lambda(v/\lambda_{th}, x)$$

$$\cdot \frac{\sin^2\left(v\dfrac{x + x_0}{\lambda_{th}}\right)}{1 - \exp\left[-\left(\dfrac{x + x_0}{\lambda_{th}}\right)\right]} \tag{3.83}$$

All the information about how the interface influences the channel mobility is contained in $\alpha(x)$. The rest of this section will be devoted to an investigation of this function. Before we start out to do so, a comment on the current equation (3.80) is in order. Equation (3.80) describes only the lateral current density component. The transversal component is negligibly small in the channel region. As pointed out in the previous section, this is equivalent to the classical boundary condition $j_t(x = 0) = 0$. We will not go seriously wrong if we generalize Eq. (3.80) for the transversal current component too.

$$j(R) = -qn(R)\mu_{ch}(x)\nabla_R\varepsilon_F(R), \quad j_t(x = 0) = 0 \tag{3.84}$$

A different approach would have been to evaluate the current density from Eq. (3.52). This would give a matrix for the mobility that distinguishes the direction perpendicular and transversal to the surface. Because the transverse current is negligibly small, the kind of mobility used to calculate it is not significant. The situation is different in bulk. In the limit $x \to \infty$, we have $\mu_{ch} \to \mu_{bulk}$ so that both current density components are calculated with the same mobility. Equation (3.84) has the advantage that it can be used directly in the continuity equation to calculate the density distribution. The quasi-Fermi level is calculated from Eq. (3.79)

$$\varepsilon_F(R) = V(R) - u_0 \ln\left(1 - \exp\left(-\left[\frac{x + x_0}{\lambda_{th}}\right]^2\right)\right) + u_0 \ln\left(\frac{n(R)}{N}\right) \tag{3.85}$$

Replacing $\nabla_R\varepsilon_F$ in the current density Eq. (3.84) by the gradient of Eq. (3.85), we obtain the current density in the conventional drift–diffusion approximation

$$j(R) = -qn\mu_{ch}(x)\nabla_R V^*(R) + qD_{ch}(x)\nabla_R n(R) \tag{3.86}$$

However, the electrostatic potential V is replaced by the effective potential V^*

$$V^*(R) = V(R) - u_0 \ln\left(1 - \exp\left(-\left[\frac{x + x_0}{\lambda_{th}}\right]^2\right)\right) \tag{3.87}$$

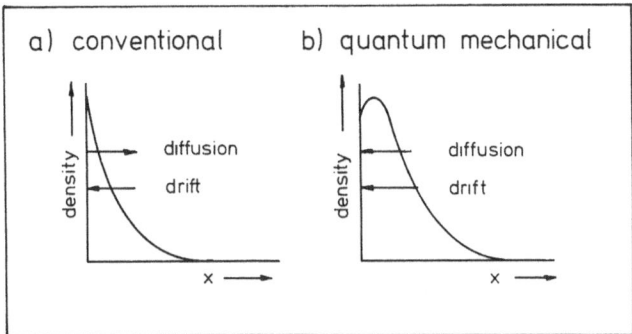

Fig. 25 Current density boundary condition at the semiconductor insulator interface and drift diffusion equation. (a) In the conventional case drift and diffusion component of the current density have opposite direction. (b) For a quantum mechanically modified density distribution drift and diffusion current density have the same direction at the surface

The physical interpretation is best done by inspecting Fig. 25, where we show the carrier density profile perpendicular to the interface.

The solid curve corresponds to the classical case. The boundary condition $j_t(x = 0) = 0$ forces a density distribution to adjust in such a way that field and diffusion currents cancel. The density maximum is therefore directly at the surface. The dashed line shows a density profile which would result if the correct boundary condition on the wave function had been taken into account. The maximum of the carrier density will then move away from the interface. With the classical drift–diffusion current equation, we never could satisfy the boundary condition $j_t(x = 0) = 0$ because field and diffusion currents have the same direction directly at the surface and would therefore be unable to cancel out. However, the correction term in the effective potential V^* provides a built-in field so that the total field current cancels the diffusion current at the interface. In the case of equilibrium (no source–drain bias), the Poisson–Schrödinger system is exactly solvable (Appendix II). In Fig. 26, we compare the results of such a calculation with the semi-classical and conventional methods.

Far away from the interface, all curves merge. For the quantum mechanical case, this is due only to the fact that all states (bound and continuous) of the corresponding Hamiltonian are accounted for and not only the first few bound states, as is usually done. The classical curve has a maximum at the interface. Both semi-classical and quantum-mechanical solutions have the same value at the interface and reach their maximum in close proximity to the interface. The maximum of the semi-classical approximation is located closer to the interface than that of the quantum-mechanical calculation. This is because the semi-classical approach is essentially an asymptotic approximation around the surface. Corrections appear to be more significant away from the surface. The value of the maximum density is approximately the same, but its position is shifted $x_{sc} = \frac{1}{2}x_{qm}$. As a first-order

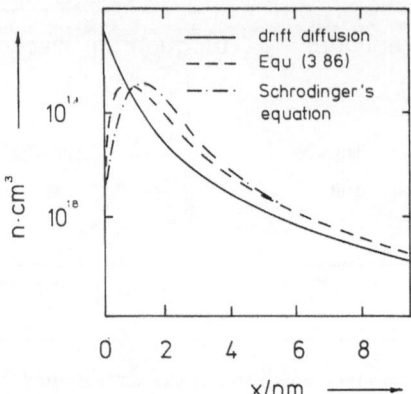

Fig. 26 Density distribution in an inversion channel of a MOSFET. Drift diffusion equation (solid line), modified drift diffusion equation Eq. (3.86) (dashed line), and exact solution of the quantum mechanical problem (dashed–dotted line)

approximation, the semi-classical approximation does the job. Its most significant advantage, however, is that it can be easily applied in a realistic device situation when a substantial current flows and there an inhomogeneous inversion channel exists. In this case, no exact quantum-mechanical treatment is currently possible. The modified current density equation Eq. (3.86) we incorporated together with the below specified channel mobility in the device simulator MINIMOS to obtain the terminal currents for a realistic MOSFET device under operating conditions (Hänsch et al. 1989). The numerical performance was comparable to that of the conventional approach and it can therefore serve as the starting point for a refined quantum mechanical treatment of carrier transport in MOSFET that meets the requirements of an efficient modeling tool.

We now will resume the discussion of the channel mobility $\mu_{ch}(x)$. All important features are contained in $\alpha(x)$, given by Eq. (3.83). It is useful to consider the contributions from regions I and II/III separately. If we normalize all energies in units of $k_B T$, all length in units of λ_{th}, and fields in units of u_0/λ_{th}, then we obtain the following for the contribution in region I

$$
\alpha^{\mathrm{I}}(x) = \frac{4}{\sqrt{\pi}} \int_0^{\sqrt{(V - V_\infty)}} dv \cdot \exp(-v^2) \left\{ 1 - (1 - p) \right.
$$
$$
\cdot \frac{\exp(-G[V_0, V - v^2])}{1 - p \exp(-2G[V_0, V - v^2])} \cosh(G[V, V - v^2])\left.\right\}
$$
$$
\cdot \frac{\sin^2(v(x + x_0))}{1 - \exp(-[x + x_0]^2)} \tag{3.88}
$$

and in region II/III we get

$$\alpha^{II/III}(x) = \frac{4}{\sqrt{\pi}} \int_{\sqrt{(V - V_\infty)}}^{\infty} dv \cdot \exp(-v^2)\{1 - \tfrac{1}{2}(1 - p)$$

$$\cdot \exp(G[V, V_0])\} \frac{\sin^2(v(x + x_0))}{1 - \exp(-[x + x_0]^2)} \qquad (3.89)$$

with

$$G[V_1, V_2] = \frac{\mu_0}{\mu_{bulk}} \int_{V_2}^{V_1} \frac{dV'}{E(V')} (v^2 - V + V')^{-1/2} \qquad (3.90)$$

Here, $\mu_0 = q\lambda_{th}^2/\hbar$ is the intrinsic mobility scale (we have shown $\hbar$ explicitly to have the usual dimension of the mobility). In the limit $x \to \infty$ ($V \to V_\infty$)α^I is zero. For $\alpha^{II/III}$ we get

$$\lim_{x \to \infty} \alpha^{II/III}(x) = \frac{4}{\sqrt{\pi}} \int_0^{\infty} dv \cdot \exp(-v^2) \frac{\sin^2(v(x + x_0))}{1 - \exp(-[x + x_0]^2)} = 1$$

$$(3.91)$$

This ensures that $\mu_{ch} \to \mu_{bulk}$ far away from the interface. Directly at the interface, $x = 0$, we find for $\alpha(0) = \alpha^I(0) + \alpha^{II/III}(0)$

$$\alpha(0) = \tfrac{1}{2}(1 + p) - \frac{2}{\sqrt{\pi}}(1 - p^2) \int_0^{\sqrt{(V_0 - V_\infty)}} dv \cdot \exp(-v^2)$$

$$\cdot \frac{\exp(-2G[V_0, V_0 - v^2])}{1 - p\exp(-2G[V_0, V_0 - v^2])} \frac{\sin^2(v(x + x_0))}{1 - \exp(-[x + x_0]^2)}$$

$$(3.92)$$

with

$$G[V_0, V_0 - v^2] = 2\frac{\mu_0}{\mu_{bulk}} \int_0^{v} \frac{dv'}{E(V_0 - v^2 + v'^2)} \qquad (3.93)$$

An interesting feature of Eqs. (3.92) and (3.93) is that the field contributes to the mobility reduction at the interface in a non-local way through Eq. (3.93). We can also calculate the limit for a zero field. In this limit, we find $G \to \infty$ and the integral in Eq. (3.92) therefore vanishes.

$$\lim_{E \to 0} \alpha(0) = \tfrac{1}{2}(1 + p) \qquad (3.94)$$

Equation (3.94) describes the intrinsic mobility reduction at the interface due to "surface roughness". This effect is much less pronounced than that of a finite transversal field. Without specifying a potential profile, further analysis is very cumbersome. We will therefore evaluate Eqs.. (3.88)–(3.90) and (3.92)–(3.93) for a special field profile. A linear and an exponentially decreasing potential are analytically tractable. The latter one is more realistic for a device situation. Given the surface field E_0, such a potential reads

$$V(x) = E_0 æ \exp\left(-\frac{x}{æ}\right) + V_\infty \tag{3.95}$$

We leave the evaluation of the integrals as an exercise to the reader and quote only the results. We obtain the following for $\alpha(z)$

$$\alpha(z) = 1 - \alpha'(z) - \alpha''(z) \tag{3.96}$$

with

$$\alpha'(z) = (1-p)\frac{4}{\sqrt{\pi}}\int_0^{\sqrt{\Psi}} dv \cdot \exp(-v^2)$$

$$\cdot \frac{\exp\left(-\dfrac{c}{(\Psi - v^2)^{1/2}} \operatorname{arctg}\left(\left[\dfrac{\Psi_0}{\Psi - v^2} - 1\right]^{1/2}\right)\right)}{1 - p\exp\left(-\dfrac{2c}{(\Psi - v^2)^{1/2}} \operatorname{arctg}\left(\left[\dfrac{\Psi_0}{\Psi - v^2} - 1\right]^{1/2}\right)\right)}$$

$$\cdot \cosh\left(\frac{c}{(\Psi - v^2)^{1/2}} \operatorname{arctg}\left(\left[\frac{\Psi}{\Psi - v^2} - 1\right]^{1/2}\right)\right)$$

$$\cdot \frac{\sin^2(v(x + x_0))}{1 - \exp(-[x + x_0]^2)} \tag{3.97}$$

$$\alpha''(x) = (1-p)\frac{2}{\sqrt{\pi}}\int_{\sqrt{\Psi}}^{\infty} dv \cdot \exp(-v^2)\exp\left(-\frac{c}{(v^2 - \Psi)^{1/2}}\right)$$

$$\cdot \left\{\operatorname{Arth}\left(\left[\frac{\Psi}{v^2 - \Psi} + 1\right]^{1/2}\right) - \operatorname{Arth}\left(\left[\frac{\Psi_0}{v^2 - \Psi} + 1\right]^{1/2}\right)\right\}$$

$$\cdot \frac{\sin^2(v(x + x_0))}{1 - \exp(-[x + x_0]^2)} \tag{3.98}$$

Here we have used the abbreviations $\Psi = V - V_\infty$ and $c = 2æ\mu_0/\mu_{\text{bulk}}$. We obtain the constant field case by taking the limit $æ \to \infty$, $V_\infty \to -\infty$ and $æE_0 = -V_\infty$

$$\alpha(x) = 1 - (1-p)\frac{4}{\sqrt{\pi}}\int_0^{\infty} dv \cdot \exp(-v^2)$$

$$\cdot \frac{\exp(-c'(v^2 + E_0 x)^{1/2})}{1 - p\exp(-2c'(v^2 + E_0 x)^{1/2})} \cosh(c'v)$$

$$\cdot \frac{\sin^2(v(x + x_0))}{1 - \exp(-[x + x_0]^2)} \tag{3.99}$$

$$c' = 2\mu_0/\mu E_0$$

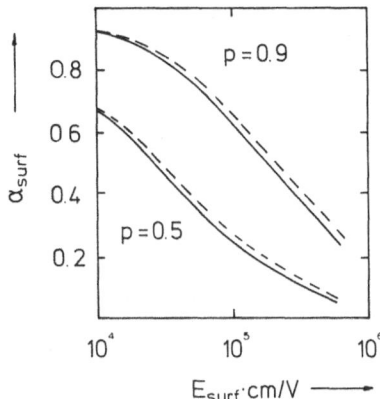

Fig. 27 Mobility reduction versus surface field for different surface scattering parameters p.
The solid line assumes a constant electric field in the vicinity of the interface and the dahsed
line an exponentially decreasing field with the same surface field strength

To match a realistic situation, we parameterized the exponential potential
so that it coincides with a potential distribution obtained from a device
simulator for the extension of the channel width $d \sim 2a_0(Ry/qE_0a_0)^{1/3} \sim$
5 nm. This corresponds to a potential drop $\Delta V = V_0 - V_\infty$ of a few hundred
mV over the characteristic length $\ae = \Delta V/E_0$. In Fig. 27, we show the
variation of $\alpha(0)$ with the surface field. The surface scattering parameter is
$p = 1$ for an ideal surface. The Si/SiO$_2$ interface is very close to an ideal
interface, so we chose $p = 0.9$ and $p = 0.5$.
Figure 27 shows the constant field case and that for the exponential
decreasing potential. Both are essentially the same. The influence of non-
local corrections is not significant. Figure 28 is more interesting. Here we
show $\alpha(x)$ for zero field and assuming a surface field $E_0 = 6 \cdot 10^5$ V/cm. The
scattering parameter was chosen to be $p = 0.8$ and $p = 0.5$.
In the close vicinity of the surface both field distributions yield approximately the same result. The constant field case approaches the bulk value
very slowly. The exponential field follows the behavior of the field qualitat-
ively, although it is not a local function of the field. If this were the case,
there would be no difference if the limit $E \to 0$ is taken at the surface
($\alpha = \frac{1}{2}(1 + p)$) or in bulk ($\alpha = 1$). To fit the exponential field case, we tried
an expression that depends only on the local field

$$\alpha_{\mathrm{fit}}(x) = \frac{\dfrac{1+p}{2}}{1 + \frac{1}{2}(1-p)\left(\dfrac{E(x)}{E_{\mathrm{ref}}}\right)^{1/2(1+p)}}, \quad E_{\mathrm{ref}} = 2\frac{\mu_0}{\mu_{\mathrm{bulk}}} \cdot \frac{u_0}{\lambda_{\mathrm{th}}} \quad (3.100)$$

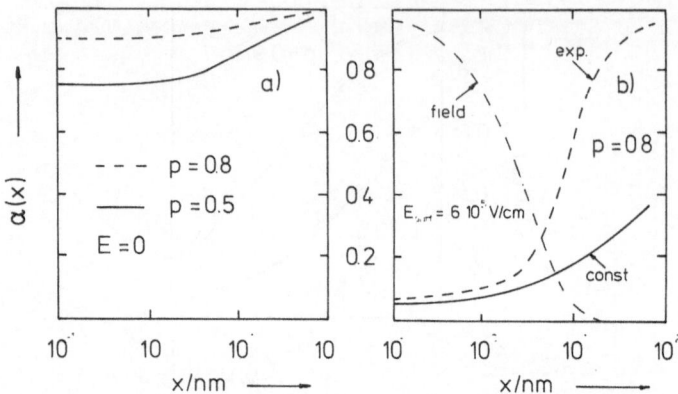

Fig. 28 Mobility reduction versus distance from the interface. (a) Zero field case for surface scattering parameter $p = 0.8$ (dashed line) and $p = 0.5$ (solid line). (b) Exponentially decreasing electric field (dashed line) with surface field $E_{surf} = 6 \cdot 10^5$ V/cm and constant field (solid line) with the same surface field strength. The dashed-dotted line shows the spatial variation of the exponentially decaying field

At the surface, we obtain a reasonable fit but the distance dependence is poor towards the bulk. It can, in fact, never approach the correct bulk value $\alpha = 1$ for non-zero field.

This can be corrected if we introduce an explicit x dependence that turns the effects of the interface off and switches the correct bulk value on.

$$\alpha_{fit}(x) = \frac{1 + \left(\dfrac{1 + p}{2} - 1\right)\exp\left(-\left[\dfrac{x}{\lambda_{th}}0.1\right]^2\right)}{1 + \frac{1}{2}(1 - p)\left(\dfrac{E(x)}{E_{ref}}\right)^{1/2(1+p)}\exp\left(-\left[\dfrac{x}{\lambda_{th}}0.1\right]^2\right)} \tag{3.101}$$

Equation (3.101) agrees well with the exact result, as can be seen in Fig. 29. A mobility model based on Eq. (3.101) would read

$$\mu_{ch}(E_x, x) = \frac{\mu_{bulk} + (\mu_{surf} - \mu_{bulk})F(x)}{1 + \frac{1}{2}(1 - p)\left(\dfrac{E(x)}{E_{ref}}\right)^{1/2(1-p)}F(x)},$$

$$F(x) = \exp\left(-\left[\dfrac{x}{\lambda_{th}}0.1\right]^2\right) \tag{3.102}$$

A channel mobility of this type was suggested by Selberherr and has repeatedly proved to be very accurate (Selberherr et al. 1990).

We find that the channel mobility can indeed be represented to a very good approximation by a local-field model for the channel region. Near the interface ($F(x) \sim 1$) we find a less pronounced influence of the bulk mobility, especially its doping dependence. To maintain global consistency,

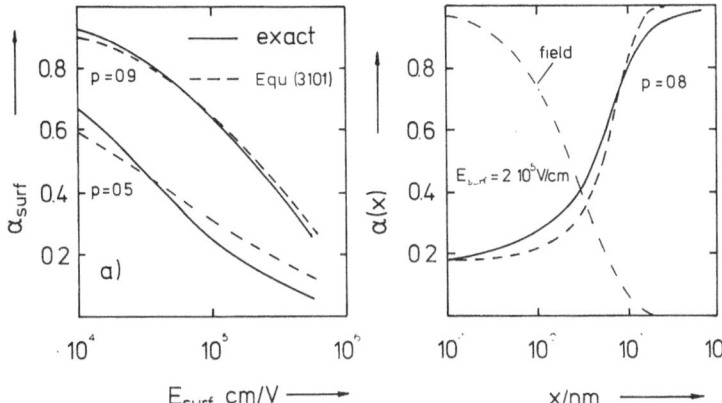

Fig. 29 Comparison of exact results (solid line) with the fit formula Eq. (3.101) (dashed line). (a) Mobility reduction versus surface field. (b) Mobility reduction versus distance from the interface for exponentially decaying electric field with surface field strength $E_{surf} = 2 \cdot 10^5$ V/cm. The dashed–dotted line shows the spatial variation of the electric field

however, an explicit spatial dependence must be introduced that ensures the correct transition from the channel into the bulk region.

References

Cooper J. A., Nelson D. F. (1981): IEEE Electr. Device Lett. *EDL2*, 171.

Greene R. F., Frankel D. R., Zemel J. (1960): Phys. Rev. *118*, 967.

Hänsch W., Mahan G. D. (1983): Phys. Rev. *B28*, 1902.

Hänsch W., Vogelsang Th., Kircher R., Orlowski M. (1989): Solid State Electr. *32*, 839.

Hiroki A., Odanaka S., Ohe K., Esaki H. (1987): IEEE Electr. Device Lett. *EDL8*, 231.

Hsing C. T., Kennedy D. P., Sutherland A. D., Van Vliet K. M. (1979): Physica Status Solidi *56a*, 129.

Kadanoff L. P., Baym G. (1962): Quantum Statistical Mechanics. Benjamin, London.

Laux S. E., Warren A. C. (1986): IEDM 86 Technical Digest 567.

Mahan G. D. (1981): Many-Particle Physics. Plenum, New York.

Schrieffer J. R. (1955): Phys. Rev. *97*, 641.

Schwarz S. A., Russek S. E. (1983): IEEE Trans. Electr. Devices. *ED30*, 1634.

Selberherr S., Hänsch W., Seavey M., Slotboom J. W. (1990): Solid State Electr., to be published.

Stern F. (1972): Phys. Rev. *35*, 4891.

Stern F., Howard W. E. (1967): Phys. Rev. *163*, 816.

Sze S. M. (1981): Physics of Semiconductor Devices (2nd Edition). John Wiley, New York.

Takagi S., Iwase M., Toriumi A. (1988): IEDM 88 Technical Digest 398.

Übensee H., Paasch G., Zöllner J. P., Fiedler Th., Grobsch G. (1985): Physica Status Solidi *130b*, 387.

Walker A. J., Woerlee P. H. (1988): J. Phys. Colloque *C4*, 265.

Yamaguchi K. (1979): IEEE Trans. Electr. Devices *ED26*, 1068.

Yamaguchi K. (1983): IEEE Trans. Electr. Devices *ED30*, 658.

High Energetic Carriers 4

4.1 Introduction

High energetic carriers are electrons (holes) that occupy energetic states far above (below) the conduction (valence) band edge. An example of the characteristic energy is the threshold energy for impact ionization ε_{imp} which is of the order of the bandgap E_G. A carrier possessing this energy can transfer its energy via collision to an electron in the valence band and move it into the conduction band. In this process, which is called impact ionization, an electron–hole pair is created and the original high energetic carrier is scattered into a state of considerably lower energy. Another energy range of interest is the conduction and valence-band barriers between bulk silicon ($E_G = 1.12\,\text{eV}$) and the large-gap oxide SiO_2 ($E_G \approx 9\,\text{eV}$). These range from 3.2 eV to about 5 eV for the conduction and valence bands, respectively. Carriers with comparable energies can enter the oxide and accumulate electrical charge in traps. This eventually leads to a failure of the device. To minimize this oxide charging is a major concern of device design. We will treat it in the next chapter. We use the substrate and gate currents of a MOSFET for the experimental verification of high energetic carriers. Both are very indirect measures of device degradation due to oxide damage because the damage is caused by filling up traps in the oxide, which requires an understanding of the trap mechanisms. Furthermore, the substrate and gate currents are very sensitive to uncertainties in device processing. This is particularly so in the accuracy of the doping profiles which determine the electric field in the device.

Our previous analysis in Chapters 2 and 3 was concerned with taking averages over the distribution function. These averages are only slightly sensitive to the exact energy band structure of the material. It was therefore at least reasonable to use the effective mass approximation for that analysis. The situation is different for high energetic particles. A simple look at the band structure of Si shows that for an electron energy of approximately 1 eV, the band structure looks like anything but the effective mass approximation. Matters do not improve if we approach even higher energies.

Furthermore, it is well known that the coupling between the different equivalent regions in the Brillouin zone becomes increasingly important for such high energies. Therefore a simple analysis of the distribution of the high energetic particles seems vain. On the other hand, modeling requires a simple formulation for utilizing in a numerical code. This leaves us with a real dilemma! As greater computer power becomes available, Monte Carlo methods are gaining increasing popularity to solve this problem (Jacoboni and Lugli 1989). Here, we also face problems with the high energy particles which are not easy to overcome. In principle, Monte Carlo approaches to the high energetic particles involve two different groups of problems.

The first one concerns the physics of high energy particles. To find a reasonably correct description of the impact ionization process it must without question be self-consistently included in the calculation to obtain a physically relevant result. As mentioned above, impact ionization is an effect of the electron–electron interaction. Monte Carlo approaches are essentially independent particle methods. Trajectories of separate particles are monitored in a self-consistently determined field. As we learned in Chapter 1, the concept of a particle has a limited range of validity. Let us recall that a particle exists if the imaginary part of the self-energy is sufficiently small compared with other characteristic energies in the system (if the real part of the self-energy is also of slow variation). Self-energies for the high energies considered here are currently not well known in Si. A rough estimate, however, shows that the collision broadening could be as high as a few hundred millivolts, which certainly restricts the particle concepts in which energies transferred through optical phonon scattering are of the order of a hundred millivolts or less. Granted that the particle concept holds, there remains the problem of calculating the scattering cross-section of the impact ionization process. Simplified versions can be found in the literature, such as the Keldysh formula (Keldysh 1959). The key point is, however, that the interaction strength for this process depends on the overlap of the conduction and valence band wave functions and on the special form of the interaction between the carriers. For the former, we have to account for the conduction-band wave functions for the high energy states and the complex band structure of the holes. For the interaction, a screened Coulomb potential can often be used, which is itself a functional of the carrier density. All these points must be included in a Monte Carlo study of high energetic carriers. There are certainly some features which are less important. However, to take full advantage of the Monte Carlo approach as a unique tool to discriminate the different influences in a very transparent way, they all have to be included on a sound physical foundation in the first place.

The second class of problems are numerical in nature. Owing to limited memory and CPU resources, Monte Carlo methods are restricted to be performed with a limited number of particles. This limitation drastically reduces the number of particles found in the higher energies. To find statistically relevant results, the number of particles in the higher energetic states has to be increased artificially. Problems involving this statistic enhancement are discussed in another monograph in this series (Jacoboni and Lugli 1989).

All in all, we encounter the same problem in Monte Carlo methods: as long as we are dealing with averages of the distribution function, the Monte Carlo approach can serve as a powerful tool for analyzing the physics of the device. However, as soon as we direct our attention to high energy particles, many uncertainties remain.

Following the spirit of the previous chapters, we would once again like to present some precise results which are of certain relevance for modeling high energetic carriers. We will not even attempt to find a correct description but try to give a guideline for physical reasoning. At best, we can ask the questions: in what mode do the high energy particles move: is ballistic motion dominant or scattering limited ensemble motion? What influence can we expect from such concepts as a soft or abrupt threshold? How do we get particles into the oxide? Many answers can be found in the literature. In the following sections, we will provide a consistent view on these based on the study of the Boltzmann equation. The results we present are not correct in a puristic sense, but will reveal some physical implications which are important for effective modeling.

4.2 Impact Ionization Scattering Strength

In the impact ionization process, a particle–hole pair is created. It must therefore be included in the generation/recombination term of the continuity equation of the electrons and holes, respectively. However, it must be noted that this process does not really create particles in the physical sense, but only transfers them from the valence to the conduction band, or vice versa. This is a new feature which we have not yet considered. It introduces a mutual interaction between electrons in the conduction band and holes in the valence band and renders the one-band approach invalid. Other such coupling mechanisms are Shockley–Read–Hall and Auger recombination (Selberherr 1984). Shockley–Read–Hall recombination can be interpreted as an impurity-assisted recombination–generation process (Shockley and Read 1952, Hall 1952) in which carriers are transferred from the conduction to the valence band, or vice versa, via impurities located energetically somewhere in the middle of the bandgap. Auger recombination (Beattie and Landsberg 1959, Haug 1970) is a process based on the Coulomb interaction among the carriers. In a collision between two carriers, one of them picks up sufficient energy to overcome the energy gap. Many possible variations

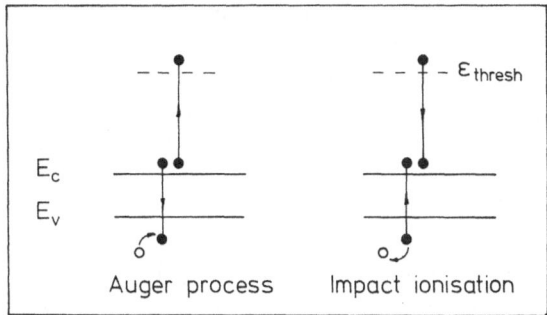

Fig. 30 Auger process and impact ionization. E_c and E_v are the conduction and valence
band edges and $\varepsilon_{\text{thresh}}$ is the threshold energy characteristic for the process

of this process contribute to the total generation/recombination rate. Auger
recombination becomes important in high non-equilibrium carrier concentrations. We will not consider Shockley–Read–Hall and Auger recombination in detail in this text. However, the impact ionization process is very
closely related to Auger recombination. It could be considered as an inverse
Auger process. In an Auger process, the initially colliding particles generally
have a low energy. In the collision process, one of the partners is transferred
across the bandgap, where it immediately recombines, and the other to a
high energetic state in the original band. We illustrate this in the band
diagram of Fig. 30a for the collision of two electrons in the conduction
band. There is a minimum energy in which we have to find the carrier
remaining in the conduction band. Impact ionization considers the process
in which a high energetic carrier in the conduction band, for example, kicks
an electron from the valence into the conduction band. The final states of
this process, illustrated in Fig. 30b, are then two generally low energetic
electrons in the conduction band and an additional hole in the valence
band. This process is possible only if the initial high energetic carrier has an
energy greater than the threshold ε_{imp}. This threshold energy corresponds
to the minimum energy in which we can find the post-collision high
energetic particle in the Auger process.

In the rigorous formulation of carrier transport, a distinction between
Auger recombination and impact ionization is therefore not possible. Both
should be contained in a proper scattering term of the Boltzmann equation.
From Eq. (2.1), we have for the one-band model

$$\frac{\partial}{\partial T} f(R, T, k) + v(k)\nabla_R f(R, T, k) - qE(R, T)\nabla_k f(R, T, k) =$$

$$\sum_{k'} [w(k, k')(1 - f(R, T, k))f(R, T, k')$$

$$- w(k', k)f(R, T, k)(1 - f(R, T, k'))] \qquad (4.1)$$

where $w(k, k')$ are the scattering terms given by Eq. (2.2). As a remainder, the first and second contributions of the right-hand side of Eq. (4.1) were interpreted as scattering events that scatter particles into and out of the phase–space volume around (R, k), respectively. The detailed balance relationship, Eq. (2.6), ensures that the scattering term vanishes in equilibrium. In a modification of Eq. (4.1) to include impact ionization as well as Auger recombination, the scattering term has to be altered. To this end, we have to adopt the two-band picture. We will distinguish electrons (e) in the conduction (c) band and holes (h) in the valence (v) band by their distribution functions $f_{ce}(R, T, k)$ and $f_{vh}(R, T, k) = 1 - f_{ve}(R, T, k)$. For the sake of simplicity, we will consider only the process shown in Fig. 30. Many other processes are possible due to the complex band structure of the material. We find for the Boltzmann equation

$$\frac{\partial}{\partial T} f_{ce}(R, T, k) + v_c(k)\nabla_R f_{ce}(R, T, k) - qE(R, T)\nabla_k f_{ce}(R, T, k) =$$

$$= \sum_{k'} [w_{cc}(k, k')(1 - f_{ce}(R, T, k))f_{ce}(R, T, k')$$

$$- w_{cc}(k', k)f_{ce}(R, T, k)(1 - f_{ce}(R, T, k'))]$$

$$+ 2\pi \sum_{k', k_1, k_1'} |v_{ccvc}(k, k'; k_1, k_1')|^2 \delta(\varepsilon_{ck'} + \varepsilon_{vk_1} - \varepsilon_{ck} - \varepsilon_{ck_1'})$$

$$\cdot [(1 - f_{ce}(R, T, k))f_{ce}(R, T, k')(1 - f_{vh}(R, T, k_1))(1 - f_{ce}(R, T, k_1'))$$

$$- f_{ce}(R, T, k)(1 - f_{ce}(R, T, k'))f_{vh}(R, T, k_1)f_{ce}(R, T, k_1')] \qquad (4.2)$$

With the energy-conserving δ-function, it is easily shown that the last contribution vanishes in equilibrium. The inter-band Coulomb interaction matrix element v_{ccvc} will be discussed below. A similar equation would hold for the holes in the valence band. To make any progress, we have to simplify Eq. (4.2) further. Although both impact ionization and Auger recombination are contained simultaneously, their influence dominates in different situations. For relatively low carrier concentrations, we can neglect the Auger recombination if the electric field is sufficiently high. Pair generation by impact ionization will dominate. For a MOSFET, this approximation usually holds to a good degree because impact ionization is important in the pinch-off region where the field is high and the carrier concentrations are low. Whenever we meet a situation with both high field and large carrier concentration, complex interference will occur between the Auger effect and impact ionization. In such cases, it might not be appropriate to model them separately. For low carrier concentrations, it is also justified to assume a non-degenerate carrier gas. This implies $f \ll 1$ and we have approximately $1 - f \sim 1$. This gives the following for the inter-band scattering term of Eq. (4.2)

$$2\pi \sum_{k',k_1,k'_1} |v_{ccvc}(k,k';k_1,k'_1)|^2 \delta(\varepsilon_{ck'} + \varepsilon_{vk_1} - \varepsilon_{ck} - \varepsilon_{ck'_1})$$

$$\cdot [f_{ce}(R,T,k') - f_{ce}(R,T,k)f_{vh}(R,T,k_1)f_{ce}(R,T,k'_1))] \qquad (4.3)$$

In equilibrium, Eq. (4.3) still vanishes in accordance with the theorem of detailed balance. The first term is responsible for pair generation by impact ionization and the second for Auger recombination or pair recombination by collision. In non-equilibrium, the contribution of the first term will dominate in high fields and low carrier concentrations. We can therefore neglect the second term and get the following for the Boltzmann equation (4.1)

$$\frac{\partial}{\partial T} f_{ce}(R,T,k) + v_c(k)\nabla_R f_{ce}(R,T,k) - qE(R,T)\nabla_k f_{ce}(R,T,k) =$$

$$= \sum_{k'} [w_{cc}(k,k')f_{ce}(R,T,k') - w_{cc}(k',k)f_{ce}(R,T,k)]$$

$$+ 2\pi \sum_{k',k_1,k'_1} |v_{ccvc}(k,k';k_1,k'_1)|^2$$

$$\cdot \delta(\varepsilon_{ck'} + \varepsilon_{vk_1} - \varepsilon_{ck} - \varepsilon_{ck'_1})f_{ce}(R,T,k') \qquad (4.4)$$

Integrating over the moments of Eq. (4.4), we can again derive an appropriate set of transport equations suitable for modeling. The new feature in these equations compared to that derived in Chapter 2, comes from the impact ionization term. It is clear that this term will have the most pronounced influence in the continuity equation because it describes the generation of electrons in the conduction band, which was neglected so far. Therefore the right-hand side of Eq. (2.104) will no longer be zero. However, an interesting question is: what will happen to the other equations? There will be additional contributions to the momentum and energy relaxation times τ_v and τ_ε as well as a new channel of energy dissipation which will enter the energy conservation (2.106). For the continuity equation we obtain

$$\frac{\partial}{\partial T} n(R,T) + \nabla_R v_n(R,T) = G_n(R,T) \qquad (4.5)$$

with the generation term

$$G_n(R,T) = 2\pi \sum_{\substack{k_1,k'_1 \\ k_2,k'_2}} |v_{ccvc}(k'_1,k_1;k_2,k'_2)|^2$$

$$\cdot \delta(\varepsilon_{ck_1} + \varepsilon_{vk_2} - \varepsilon_{ck'_1} - \varepsilon_{ck'_2})f_{ce}(R,T,k_1) \qquad (4.6)$$

where we have added the index n to indicate electrons in the conduction band. We will discuss the influence on the other equations later. For the

Coulomb inter-band matrix element, we have

$$v_{ccvc}(k'_1, k_1; k_2, k'_2) = \int dx' \int dx\, \varphi^*_{ck'_1}(x)\varphi_{ck_1}(x)$$
$$\cdot v(x - x')\varphi_{vk_2}(x')\varphi^*_{ck'_2}(x') \tag{4.7}$$

where $\varphi_{ck}(x)$ and $\varphi_{vk}(x')$ are the Bloch wave functions of electrons and holes, respectively.

$$\varphi_{ck}(x) = \frac{1}{\sqrt{\Omega}} \cdot \exp(ikx)u_{ck}(x)$$
$$\tag{4.8}$$
$$\varphi_{vk}(x) = \frac{1}{\sqrt{\Omega}} \cdot \exp(-ikx)u_{vk}(x)$$

Here Ω is the volume of the Brilliouin zone. As we shall soon see, it is essential at this point to keep the features of the periodic lattice. In Eq. (4.7), we use the Fourier transform of the matrix element $v(x - x')$

$$v_{ccvc}(k'_1, k_1; k_2, k'_2) = \sum_q v_q \int dx\, \exp(-iqx)\varphi^*_{ck'_1}(x)\varphi_{ck_1}(x)$$

$$\int dx'\, \exp(iqx')\varphi_{vk_2}(x')\varphi^*_{ck'_2}(x')$$

$$= \sum_q v_q \int dx\, \exp(-i[q + k'_1 - k_1]x)$$

$$\cdot u^*_{ck'_1}(x)u_{ck_1}(x)\int dx'\, \exp(i[q - k_2 - k'_2]x')$$

$$\cdot u_{vk_2}(x')u^*_{ck'_2}(x')$$

The dominant part of the integrals comes from k-values for which the argument of the oscillatory exponential is small or zero. We will keep only these terms and therefore the inter-band matrix element will be reduced to

$$v_{ccvc}(k'_1, k_1; k_2, k'_2) = v_{|k_1 - k'_1|}\, \delta_{k_1 - k'_1, k_2 + k'_2}$$
$$F_{cc}(k'_1, k_1)F_{vc}(k'_2, k_2) \tag{4.9}$$

with the overlap matrix elements

$$F_{cc}(k'_1, k_1) = \frac{1}{\Omega} \int dx\, u^*_{ck'_1}(x)u_{ck_1}(x) \tag{4.10}$$

$$F_{vc}(k'_2, k_2) = \frac{1}{\Omega} \int dx\, u_{vk_2}(x)u^*_{ck'_2}(x)$$

In the effective mass approximation, these overlap matrix elements are identical to one and Eq. (4.9) is essentially exact. In real material, however, they can differ considerably from one. These overlap matrix elements will therefore renormalize the interaction strength. This has to be considered in a more exact treatment, which may be a Monte Carlo approach, for instance. It is beyond the scope of this text to evaluate the properties of F_{cc} and F_{vc}. Further on, we will therefore approximate them by the constant numbers $F_{cc} \approx 1$ and $F_{vc} \approx 0.1$ which are used in a different context (Antoncik and Landsberg 1963, Schorch and Abram 1988) and represent the fact that conduction wave functions overlap perfectly while the overlap between the conduction and valence-band wave functions is small. If we use a screened Coulomb potential for the interaction between the carriers, similar to Eq. (2.43), we finally obtain the following for the generation rate $G_n(R, T)$

$$G_n(R, T) = 2\pi \left(4\pi \frac{q^2}{\kappa} \right)^2 \sum_{\substack{k_1, k_1' \\ k_2, k_2'}} \left(\frac{|F_{cc} F_{vc}|}{|k_1 - k_1'|^2 + k_c^2} \right)^2 \delta_{k_1 - k_1', k_2 + k_2'}$$
$$\cdot \delta(\varepsilon_{ck_1} + \varepsilon_{vk_2} - \varepsilon_{ck_1'} - \varepsilon_{ck_2'}) \cdot f_{ce}(R, T, k_1) \qquad (4.11)$$

This is brought into a more convenient form by using the impact ionization scattering strength $1/\tau_{imp}(k_1)$

$$G_n(R, T) = \sum_{k1} \frac{1}{\tau_{imp}(k_1)} f_{ce}(R, T, k_1) \qquad (4.12)$$

$$\frac{1}{\tau_{imp}(k_1)} = 2\pi \left(4\pi \frac{q^2}{\kappa} \right)^2 \sum_{k, q'} \left(\frac{|F_{cc} F_{vc}|}{k^2 + k_c^2} \right)^2$$
$$\cdot \delta(\varepsilon_{ck_1} + \varepsilon_{vk + q'} - \varepsilon_{ck + k_1} - \varepsilon_{c - q'}) \qquad (4.13)$$

To obtain Eq. (4.13) from Eq. (4.11), we performed the summation over k_2 using the momentum conservation, and introduced the new integration variables $k = k_1 - k_1'$ and $q = -k_2'$. If we re-install the k-dependence of the overlap integrals, Eq. (4.13) would give an accurate description of the impact ionization scattering strength for a general material. No restriction was yet made for the band structure. To proceed, it is necessary to introduce a specific band structure. In the rest of this section, we will investigate different means of evaluating the scattering rate τ_{imp}. Our concern will not be to provide an evaluation which is exact for a specific material. As outlined above, this is impossible without a detailed knowledge of band structure and wave functions. We will rather concentrate in comparing two different evaluation schemes to evaluate Eq. (4.13), which will lead us to the concept of an abrupt or soft threshold behavior of the impact ionization process. In the next section, we will find an expression for the distribution function f_{ce} which is needed to calculate the generation rate G_n.
An exact analytical evaluation of Eq. (4.13) can be performed in the two-band effective mass approximation where the conduction and valence

bands are given by an energy sphere with effective masses m_c and m_v, and separated by an energy gap E_G. For the energy dispersion we then have

$$\varepsilon_{ck} = E_c + \frac{k^2}{2m_c} \tag{4.14}$$

$$\varepsilon_{vk} = E_v - \mu \frac{k^2}{2m_c}; \quad \mu = \frac{m_c}{m_v} \tag{4.15}$$

$$E_G = E_c - E_v = \frac{k_G^2}{2m_c} \tag{4.16}$$

An energy paraboloid with rotational symmetry is also manageable. Because this will neither yield anything new for our purposes nor be of any relevance for a realistic situation as far as impact ionization is concerned, we will consider only the simplest case. We perform the angular q-integration first.

$$\sum_q \delta(\varepsilon_{ck_1} + \varepsilon_{vk+q} - \varepsilon_{ck+k_1} - \varepsilon_{c-q}) = 2m_c \sum_q \delta(k_1^2 - \mu(k+q)^2$$

$$- (k+k_1)^2 - q^2 - k_G^2) = \frac{m_c}{(2\pi)^2} \frac{1}{\mu k} \int_0^\infty q dq \int_{-2\mu kq}^{2\mu kq} du$$

$$\cdot \delta(k_G^2 + (1+\mu)k^2 + (1+\mu)q^2 + 2k \cdot k_1 - u)$$

$$= \frac{m_c}{(2\pi)^2} \frac{1}{\mu k} \int_0^\infty q dq [\theta(k_G^2 + (1+\mu)k^2$$

$$+ (1+\mu)q^2 + 2\mu kq + 2k \cdot k_1) - \theta(k_G^2 + (1+\mu)k^2$$

$$+ (1+\mu)q^2 - 2\mu kq + 2k \cdot k_1)] \tag{4.17}$$

Next, we perform the angular k-integration

$$\sum_k \frac{1}{(k^2 + k_c^2)^2} \sum_q \delta(\varepsilon_{ck1} + \varepsilon_{vk+q} - \varepsilon_{ck+k_1} - \varepsilon_{c-q})$$

$$= \frac{m_c}{(2\pi)^4} \frac{1}{2\mu k_1} \int_0^\infty dk \frac{1}{(k^2 + k_c^2)^2} \int_0^\infty q dq \int_{-2kk_1}^{2kk_1} du$$

$$\cdot [\theta(k_G^2 + k^2 + q^2 + \mu(k+q)^2 + u)$$

$$- \theta(k_G^2 + k^2 + q^2 + \mu(k-q)^2 + u)] \tag{4.18}$$

$$= \frac{m_c}{(2\pi)^4} \frac{1}{2\mu k_1} \int_0^\infty dk \frac{1}{(k^2 + k_c^2)^2} \int_0^\infty q dq [4\mu kq - g^+(q)\theta(g^+(q))$$

$$+ g^-(q)\theta(g^-(q))] \tag{4.19}$$

$$g^+(q) = k_G^2 + k^2 + q^2 + \mu(k+q)^2 - 2kk_1$$

$$g^-(q) = k_G^2 + k^2 + q^2 + \mu(k-q)^2 - 2kk_1 \tag{4.20}$$

For the remaining integrations we have to consider the bounds of the integrals very carefully. In the region where both auxiliary functions g^+ and g^- are positive the integrand vanishes by virtue of the relationship

$$g^+(q) - g^-(q) = 4\mu kq \tag{4.21}$$

This also proves $g^+(q) > g^-(q)$ for $q > 0$. Keeping all variables fixed except q, $g^-(q)$ will first change sign if we come from sufficiently large q values. In the region $g^+(q) > 0$ and $g^-(q) < 0$ the integrand will be

$$4\mu kq - g^+(q) = -g^- > 0, \qquad q > 0 \tag{4.22}$$

Moving further to lower q-values the function $g^+(q)$ will change sign. In the region $g^+(q) < 0$ and $g^-(q) < 0$ the integrand is

$$4\mu kq > 0, \qquad q > 0 \tag{4.23}$$

Both g^+ and g^- have one zero for positive q and one zero for negative q values. Because of their symmetry we find $q_{1,2}^- = -q_{1,2}^+$. Because only positive q are relevant for the integration of Eq. (4.20) we have to consider only the positive zeros. We illustrate the functions $g^+(q)$ and $g^-(q)$ in Fig. 31. For fixed k and k_1 we have two parallels with minimum at negative and positive q for $g^+(q)$ and $g^-(q)$, respectively.
For the zeros of $g^-(q)$ we find

$$Q_{1,2} = \frac{\mu}{1+\mu}k \pm \left| \frac{2kk_1}{1+\mu} - \frac{1+2\mu}{(1+\mu)^2}k^2 - \frac{1}{1+\mu}k_G^2 \right|^{1/2} \tag{4.24}$$

The zeros are real only if the argument of the square root is assured to be positive. This is the case if the following relationship holds

$$0 < \left| k - \frac{1+\mu}{1+2\mu}k_1 \right|^2 < \frac{(1+\mu)^2}{(1+2\mu)^2}k_1^2 - \frac{1+\mu}{1+2\mu}k_G^2 \tag{4.25}$$

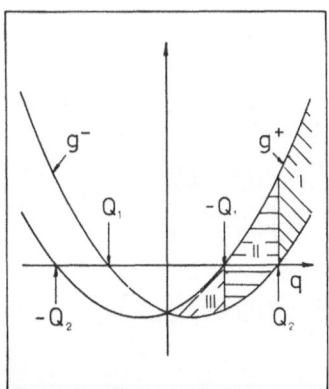

Fig. 31 The limits for the q-integration of Eq. (4.19)

Reading this from the very left to the right we find a lower bound for k_1

$$k_1^2 > \frac{1 + 2\mu}{1 + \mu} k_G^2 = k_{imp}^2 \tag{4.26}$$

which immediately gives the threshold energy ε_{imp}

$$\varepsilon_{imp} = \frac{1 + 2\mu}{1 + \mu} E_G \tag{4.27}$$

We find that the threshold for the impact ionization process is larger than the band gap. This comes from the fact that in the collision process not only the total energy is conserved but also the momentum. The bounds for the k-integration we obtain by the zeros of the argument of the square root in Eq. (4.24)

$$K_{1,2} = \frac{1 + \mu}{1 + 2\mu} k_1 \pm \frac{1 + \mu}{1 + 2\mu} (k_1^2 - k_{imp}^2)^{1/2} \tag{4.28}$$

Equation (4.25) is fulfilled if k is restricted to the values

$$K_1 < k < K_2 \tag{4.29}$$

For the q-integration we have to split up the integrand into three domains. In the first domain, $q > Q_2$, the integrand vanishes identically. In the second region, $Q_2 > q > -Q_1$, it is given by Eq. (4.22) and in the third, $-Q_1 > q > 0$, by Eq. (4.23). We finally obtain for the remaining integrals

$$\sum_k \frac{1}{(k^2 + k_c^2)^2} \sum_q \delta(\varepsilon_{ck_1} + \varepsilon_{vk+q} - \varepsilon_{ck+k_1} - \varepsilon_{c-q})$$

$$= \frac{m_c}{(2\pi)^4} \frac{\theta(k_1 - k_{imp})}{2\mu k_1} \int_{K_1}^{K_2} dk \frac{1}{(k^2 + k_c^2)^2}$$

$$\cdot \left[\int_{-Q_1}^{Q_2} q dq \{-g^-(q)\} + 4\mu k \int_0^{-Q_1} dq q^2 \right] \tag{4.30}$$

$$= \frac{1}{(2\pi)^4} \frac{2m_c}{3k_1} (1 + 2\mu)^{-3/2} \left(\frac{1 + 2\mu}{1 + \mu} \right)^2$$

$$\cdot \theta(k_1 - k_{imp}) \int_{-(k_1^2 - k_{imp}^2)^{1/2}}^{(k_1^2 - k_{imp}^2)^{1/2}} dx$$

$$\cdot \frac{(x + k_1)}{((x + k_1)^2 + (k_c(1 + 2\mu)/(1 + \mu))^2)^2} ((k_1^2 - k_{imp}^2) - x^2)^{3/2} \tag{4.31}$$

In the near threshold region $(k_1^2 - k_{imp}^2) \ll k_1^2$ we have $x \ll k_1$ in the integrand. We can therefore neglect the x dependence compared with the k_1

dependence. This leaves a simple analytical integration,

$$\sum_k \frac{1}{(k^2 + k_c^2)^2} \sum_q \delta(\varepsilon_{ck_1} + \varepsilon_{vk+q} - \varepsilon_{ck+k_1} - \varepsilon_{c-q})$$

$$= \frac{m_c}{8(2\pi)^3} (1 + 2\mu)^{-3/2} \left(\frac{1 + 2\mu}{1 + \mu}\right)^2$$

$$\cdot \frac{\theta(k_1 - k_{imp})}{(1 + (k_c/k_1 \cdot (1 + 2\mu)/(1 + \mu))^2)^2} \left(\frac{k_1^2 - k_{imp}^2}{k_1^2}\right)^2 \qquad (4.32)$$

To obtain the impact-ionization scattering rate, we have to insert Eq. (4.32) into Eq. (4.13)

$$\frac{1}{\tau_{imp}(k_1)} = \frac{Ry}{\kappa^2 \hbar} \frac{m_c}{m_0} |F_{cc} F_{vc}|^2 (1 + 2\mu)^{-3/2} \left(\frac{1 + 2\mu}{1 + \mu}\right)^2 \theta(k_1 - k_{imp})$$

$$\cdot \frac{1}{(1 + (k_c/k_1 \cdot (1 + 2\mu)/(1 + \mu))^2)^2} \left(\frac{k_1^2 - k_{imp}^2}{k_1^2}\right)^2 \qquad (4.33)$$

In Eq. (4.33), we re-installed the explicit appearance of Planck's number and expressed the electron charge by the atomic Rydberg. Assuming that the μ-dependent contributions will result in a factor of order unity, the order of magnitude for the impact ionization scattering rate is given by

$$\frac{1}{\tau_{imp}(k_1)} \approx \frac{Ry}{\kappa^2 \hbar} \frac{m_c}{m_0} |F_{cc} F_{vc}|^2 \theta(k_1 - k_{imp}) \left(\frac{k_1^2 - k_{imp}^2}{k_1^2}\right)^2 \qquad (4.34)$$

If we put some characteristic numbers for Si into Eq. (4.34), we obtain the following, converting the k-dependence into an energy.

$$\frac{1}{\tau_{imp}(\varepsilon)} \approx 10^{12} \left(\frac{\varepsilon - \varepsilon_{imp}}{\varepsilon}\right)^2 \theta(\varepsilon - \varepsilon_{imp}) s^{-1} \qquad (4.35)$$

In the limit of large energies $\varepsilon \gg \varepsilon_{imp}$, the scattering rate is comparable to that of phonons and impurities. This result was also obtained by Kane (1967), who performed a very elaborate calculation, including the exact band structure in Si. However, the net scattering rate that enters the semiconductor equations is given by integrals like Eq. (4.12). If the average energy of the carrier system is much smaller than the threshold energy, the distribution function decays exponentially at such high energies. Therefore only a small energy range near the threshold will contribute to the integral. Assuming a threshold energy of approximately 1.8 eV, the scattering rate for impact ionization will amount to $\tau_{imp} > 10$ ps in a range of 200 meV away from the threshold. This is about two orders of magnitude larger than the other scattering rates in the system. It thus seems justified to neglect the influence of impact ionization as compared to other scattering processes in the system as long as the average carrier energy is small compared to the impact ionization threshold energy. To be more explicit, we will calculate an effective scattering rate. To do so, we have to use a result from the next section. We will show there that in the limit of high energies, the distribution function can be represented by an exponential of the form

$$f(\varepsilon) \approx (\varepsilon/\beta)^\nu \exp(-\varepsilon/\beta); \quad \varepsilon \gg \omega_{opt} \tag{4.36}$$

where β can be interpreted as a high energy temperature and ω_{opt} is the characteristic energy of the optical phonons. In the high-field limit, we have $\nu \to 1$. In general we have $\beta \neq k_B T_c$, where T_c is the average carrier temperature. The effective scattering rate is defined by

$$\frac{1}{\tau_{imp}} = \frac{10^{12}}{\beta^2} \int_{\varepsilon_{imp}}^{\infty} d\varepsilon (\varepsilon - \varepsilon_{imp})^2 \varepsilon^{-1} \exp(-\varepsilon/\beta) \, s^{-1} \tag{4.37}$$

$$= 2 \cdot 10^{12} (\varepsilon_{imp}/\beta)^{1/2} \exp(-\varepsilon_{imp}/2\beta) W_{-3/2, -1}(\varepsilon_{imp}/\beta) \, s^{-1} \tag{4.38}$$

$$= 2 \cdot 10^{12} \, \beta/\varepsilon_{imp} \exp(-\varepsilon_{imp}/\beta) \left[1 + O(\beta/\varepsilon_{imp}) \right] s^{-1} \tag{4.39}$$

To go from Eq. (4.38) to (4.39) we used the asymptotic expansion of the Whitaker function $W_{n,m}$, which is justified as long as $\beta \ll \varepsilon_{imp}$. For $\beta = 300$ meV, we have $\tau_{imp} = 1$ ns, which makes the above estimate look conservative. Therefore, it seems justified to neglect contributions from impact ionization whenever they have to be compared with other scattering rates in the system. We can neglect its contributions in all but the continuity equation.

Four independent particle states are involved in the impact ionization process. Two for the initial states of the colliding particles and two for their final states. Owing to energy and momentum conservation, there are certain restrictions on the permitted states. The explicit calculation presented above took full account of these restrictions. As a result, we obtained a threshold energy that is, in general, not equal to the bandgap but larger. For a realistic band structure, the exact calculation could be very tedious. It is therefore worth looking for a simplified procedure. A clue is given by the fact that the order of magnitude is given by the k-independent term of the interaction matrix element and the energy dependence emerges from the integration in k-space. Disregarding the momentum conservation, it is nothing other than the convolution of the densities of states of the collision partners (Kane 1967). This would result in an integral of the form

$$\frac{1}{\tau_{imp}(\varepsilon_1)} = V^2 \int_{E_c}^{\infty} d\varepsilon_1' \, \Omega_c(\varepsilon_1') \int_{-\infty}^{E_v} d\varepsilon_2 \Omega_v(\varepsilon_2)$$

$$\cdot \int_{E_c}^{\infty} d\varepsilon_2' \Omega_c(\varepsilon_2') \delta(\varepsilon_{c1} + \varepsilon_2 - \varepsilon_1' - \varepsilon_2') \tag{4.40}$$

with the densities of states

$$\Omega_i(\varepsilon) = \sum_k \delta(\varepsilon - \varepsilon_{ik}) \tag{4.41}$$

for the conduction ($i = c$) and valence ($i = v$) band. The effective interaction is given by

$$V^2 = (4\pi)^3 \frac{Ry}{\kappa^2 m_0 k_0^4} |F_{cc} F_{vc}|^2 \Omega_0 \tag{4.42}$$

where Ω_0 is the volume of the unit cell and k_0 a characteristic momentum transfer. Performing the ε_2'-integration, we obtain

$$\frac{1}{\tau_{imp}(\varepsilon_1)} = V^2 \int_{E_c}^{\infty} d\varepsilon_1' \Omega_c(\varepsilon_1') \int_{-\infty}^{E_v} d\varepsilon_2 \Omega_v(\varepsilon_2)$$

$$\cdot \Omega_c(\varepsilon_{c1} + \varepsilon_2 - \varepsilon_1') \theta(\varepsilon_1 + \varepsilon_2 - \varepsilon_1' - E_c) \tag{4.43}$$

From this we get the restrictions for the other integration variables. The integral gives a non-zero contribution only if the argument of the θ-function is positive.

$$\varepsilon_1 + \varepsilon_2 - \varepsilon_1' - E_c > 0 \tag{4.44}$$

From this we find

$$E_v > \varepsilon_2 > \varepsilon_1' - \varepsilon_1 + E_c \tag{4.45}$$

as the range for ε_2 and also

$$E_v - E_c + \varepsilon_1 > \varepsilon_1' > E_c \tag{4.46}$$

which gives a restriction for ε_1

$$\varepsilon_1 > 2E_c - E_v \tag{4.47}$$

We can now write down Eq. (4.43) with the appropriate boundaries for the integrals

$$\frac{1}{\tau_{imp}(\varepsilon_1)} = V^2 \int_{E_c}^{\infty} d\varepsilon_1' \Omega_c(\varepsilon_1') \int_{E_c + \varepsilon_1' - \varepsilon_1}^{E_v} d\varepsilon_2 \Omega_v(\varepsilon_2) \Omega_c(\varepsilon_{c1}$$

$$+ \varepsilon_2 - \varepsilon_1') \theta(\varepsilon_1 - 2E_c + E_v) \tag{4.48}$$

We evaluate this integral as we did above in the effective mass approximation. We get the following for the densities of state if we take the energy origin as the valence band edge

$$\Omega_c(\varepsilon) = \frac{(2m_c)^{3/2}}{4\pi^2} (\varepsilon - E_G)^{1/2} \tag{4.49}$$

$$\Omega_v(\varepsilon) = \frac{(2m_v)^{3/2}}{4\pi^2} (-\varepsilon)^{1/2} \tag{4.50}$$

and Eq. (4.48) turns into

$$\frac{1}{\tau_{\text{imp}}(\varepsilon)} = V^2 \frac{(2m_c)^3 (2m_v)^{3/2}}{(4\pi^2)^3} \int_0^{\varepsilon - E_G} d\varepsilon' \sqrt{\varepsilon'} \int_0^{\varepsilon - E_G - \varepsilon'}$$

$$\cdot d\varepsilon'' \sqrt{\varepsilon''} \sqrt{(\varepsilon - E_G - \varepsilon' - \varepsilon'')} \theta(\varepsilon - E_G) \qquad (4.51)$$

$$= V^2 \frac{(2m_c)^3 (2m_v)^{3/2}}{(4\pi^2)^3} \frac{\pi}{8} \int_0^{\varepsilon - E_G} d\varepsilon' \sqrt{\varepsilon'} (\varepsilon - E_G - \varepsilon')^2 \theta(\varepsilon - E_G)$$

$$(4.52)$$

$$= V^2 \frac{(2m_c)^3 (2m_v)^{3/2}}{(4\pi^2)^3} \frac{2\pi}{105} \theta(\varepsilon - E_G)(\varepsilon - E_G)^{7/2} \qquad (4.52)$$

With the effective interaction Eq. (4.42) we finally get the following for the impact ionization scattering strength

$$\frac{1}{\tau_{\text{imp}}(\varepsilon)} = \sigma \theta(\varepsilon - E_G) \left(\frac{\varepsilon - E_G}{E_G} \right)^{7/2} \qquad (4.53)$$

with

$$\sigma = \frac{4}{\pi} \frac{Ry}{\hbar} \frac{|F_{cc} F_{vc}|^2}{105\kappa^2} \left(\frac{m_c}{m_0} \right) \mu^{-3/2} \left(\frac{k_G}{k_0} \right)^4 \Omega_0 k_G^3 \qquad (4.54)$$

Because we no longer have the phase space restriction due to momentum conservation, the threshold energy turns out to be the energy gap. Furthermore, we also obtain a different threshold behavior for the energy dependence. It is interesting to compare the threshold behaviors of equations (4.35) and (4.54). If we assume $k_0 \approx k_G$, typical parameters for Si yield $\sigma \approx 10^{11} s^{-1}$. In Fig. 32a we show a linear plot of the threshold behavior and in Fig. 32b a logarithmic one. Owing to the higher exponent, we observe a smoother onset of the scattering rate in Eq. (4.54) than in (4.35). The more abrupt nature of Eq. (4.35) is clearly seen in the logarithmic plot which is more appropriate in connection with the exponential variation of the distribution function.

The smoother threshold behavior was due to the phase space available for the colliding carriers being less restricted. In our approach, this was achieved artificially by disregarding the momentum conservation. It might, however, happen that collision processes appear where phonons are involved. Especially if these are acoustical phonons, they can carry momentum away without contributing significantly to the energy balance. For such a situation, the second way of calculating the impact ionization scattering rate would be a suitable approximation. The effective interaction will then, of course, include contributions from the phonons. In the indirect band gap semiconductor Si, it is very likely that phonon-assisted processes will contribute to the impact-ionization scattering rate (Lochmann 1977, Takashima 1981).

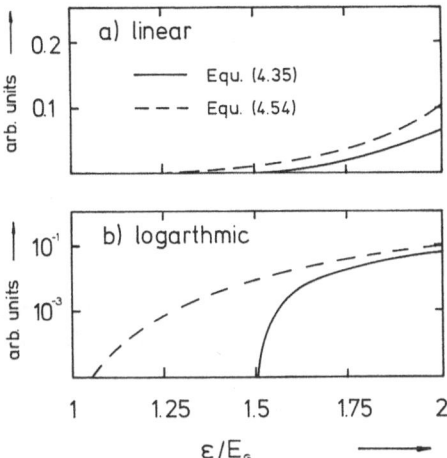

Fig. 32 Threshold behavior for the impact ionization cross section on linear (a) and logarithmic scale (b). Restricted phase space due to momentum and energy conservation (solid line). Only energy conservation (dashed line)

4.3 Distribution Function

After having discussed the impact-ionization scattering strength in the previous section, we will now turn our attention to determining the distribution function. According to Eqs. (4.12) and (4.33) the distribution function is needed for energies above the threshold energy ε_{imp}. A straight-forward procedure would be to solve the Boltzmann equation, for instance by Monte Carlo methods. This is, however, very cumbersome and time-consuming as already outlined in the first section of this chapter. For modeling in the realm of the drift–diffusion approximation, we need a compact analytical expression for the generation rate that permits easy implementation in a numerical code. On the other hand, it must correctly describe the physics. Various approaches found in the literature cover pure ballistic motion (Chynoweth 1959, Thurgate and Chan 1985, Kunert *et al.* 1985, Meinerzhagen and Engl 1988) which implies weak scattering, to ensemble motion (Henning *et al.* 1987, Fukuma and Lui 1987, Hänsch and Schwerin 1989) where scattering provides a stationary state in the electric field. The former goes back to Shockley's lucky electron (Schockley 1958) and would give a very anisotropic distribution of high energy carriers along the electric field lines. The latter is described by Wolff's diffusion mode (Wolff 1954) of high energy carriers and was already considered by the Russian scientists Davydov (1937), Druyvesteyn and Penning (1940) and Conwell (1967). It results in an isotropic distribution function. Both

Shockley's lucky electron, Eq. (4.55), and Wolff's diffusion mode, Eq. (4.56), show a characteristic field dependence

$$f(\varepsilon) \sim \exp\left(-\frac{\varepsilon}{F\lambda}\right) \tag{4.55}$$

$$f(\varepsilon) \sim \exp\left(-\frac{\varepsilon\delta}{F\lambda F\lambda}\right) \tag{4.56}$$

Here we have assumed a constant field E and $F = qE$. The energy δ is of the order of the optical phonon energy ω_{opt} and λ is the mean free path between collisions. Baraff (1962) was the first to solve the Boltzmann equation exactly for the high energy tail. His numerical result shows both lucky electron and diffusion mode behavior in limiting cases. Keldysh (1965) was able to find an analytical solution of the Boltzmann equation in the limit $\varepsilon \gg \omega_{opt}$, which is the relevant limit for impact ionization. Unlike Baraff, Keldysh was able to find a solution for finite temperatures. The next substantial contribution was the lucky-drift mode motion by Ridley (1983). In this mode, the electrons experience rapid momentum-changing collisions, whereas energy-changing collisions are less frequent. The lucky-drift electron therefore drifts along in the field with less frequent energy-changing collisions. The lucky-drift concept is somehow a link between pure ballistic motion, as in Shockley's lucky electron, and Wolff's diffusion mode. Below, we will compare the lucky-drift concept with an exact solution of the Boltzmann equation for the case of a constant field. To obtain this solution, we will be guided by Keldysh's work but will avoid its sophisticated algebra. We start with the stationary Boltzmann equation, assuming acoustical phonons as elastic scatterers and optical phonons with a constant energy ω_{opt} as inelastic scatterers

$$v(k)\nabla_R f(R, k) - qE(R)\nabla_k f(R, k) =$$

$$\sum_{k'} [w(k, k')f(R, k') - w(k', k)f(R, k)] \tag{4.55}$$

with the elastic scattering rate $w_{ac}(k, k')$

$$w_{ac}(k, k') = 2\pi|m_{k-k'}|^2\delta(\varepsilon_k - \varepsilon_{k'}) \tag{4.56}$$

and the inelastic scattering rate $w_{opt}(k, k')$

$$w_{opt}(k, k') = 2\pi|m_{k-k'}|^2[n_B(\omega_{opt})\delta(\varepsilon_k - \varepsilon_{k'} - \omega_{opt})$$

$$+ (n_B(\omega_{opt}) + 1)\cdot\delta(\varepsilon_k - \varepsilon_{k'} + \omega_{opt})] \tag{4.57}$$

The interaction matrix element $|m_{k-k'}|^2$ for elastic scattering due to acoustic phonons is, according to Eqs. (2.50) to (2.53):

$$|m_{k-k'}|^2 = C^2\frac{k_B T}{\rho s^2} = M_{el}^2 \tag{4.58}$$

We also assume a constant matrix element for optical phonon scattering. This approximation is justified if coupling to polar optical phonons is negligible.

$$|m_{k-k'}|^2 = m_{\text{opt}}^2 \tag{4.59}$$

Using Eqs. (4.56) and (4.57) with Eqs. (4.58) and (4.59) in the scattering term of the Boltzmann equation we get

$$v(k)\nabla_R f(R, k) - qE(R)\nabla_k f(R, k) = \Sigma(\varepsilon_k, f_0) - \frac{1}{\tau(\varepsilon_k)} f(R, k) \tag{4.60}$$

Here, $\tau(\varepsilon)$ is the total collision time and corresponds to the time between events that scatter the particle out of the phase-space volume around k, R

$$\frac{1}{\tau(\varepsilon)} = \frac{1}{\tau_{\text{ac}}(\varepsilon)} + \frac{1}{\tau_{\text{opt}}(\varepsilon)} \tag{4.61}$$

with

$$\frac{1}{\tau_{\text{ac}}(\varepsilon)} = 2\pi M_{\text{ac}}^2 \sum_k \delta(\varepsilon - \varepsilon_k) = 2\pi M_{\text{ac}}^2 \Omega(\varepsilon) \tag{4.62}$$

and

$$\frac{1}{\tau_{\text{opt}}(\varepsilon)} = 2\pi m_{\text{opt}}^2 [n_B(\omega_{\text{opt}}) \sum_k \delta(\varepsilon - \varepsilon_k - \omega_{\text{opt}}) + (n_B(\omega_{\text{opt}}) + 1)$$

$$\cdot \sum_k \delta(\varepsilon - \varepsilon_k + \omega_{\text{opt}})]$$

$$= 2\pi m_{\text{opt}}^2 [n_B(\omega_{\text{opt}}) \Omega(\varepsilon - \omega_{\text{opt}}) + (n_B(\omega_{\text{opt}}) + 1)\Omega(\varepsilon + \omega_{\text{opt}})]$$

$$= 2\pi m_{\text{opt}}^2 [2n_B(\omega_{\text{opt}}) + 1]\Omega(\varepsilon) \tag{4.63}$$

To get the last line in Eq. (4.63), we have assumed that the density of states $\Omega(\varepsilon)$ varies only little over an energy range ω_{opt}. The functional $\Sigma(\varepsilon, f_0)$ depends only on the energy ε and an average of the distribution function f over a surface with constant energy ε. Its physical interpretation is a scattering time for bringing particles into the phase-space volume around k, R. This contribution is absent in ballistically oriented approaches like the lucky electron. With Eqs. (4.62) and (4.63) it can be expressed as

$$\Sigma(\varepsilon, f_0) = \frac{1}{\tau_{\text{ac}}(\varepsilon)} \frac{f_0(R, \varepsilon)}{\Omega(\varepsilon)} + \frac{1}{\tau_{\text{opt}}(\varepsilon)}$$

$$\cdot \frac{\exp\left(-\frac{\omega_{\text{opt}}}{2k_B T}\right) f_0(R, \varepsilon - \omega_{\text{opt}}) + \exp\left(+\frac{\omega_{\text{opt}}}{2k_B T}\right) f_0(R, \varepsilon + \omega_{\text{opt}})}{\Omega(\varepsilon) \, 2\cosh\left(\frac{\omega_{\text{opt}}}{2k_B T}\right)}$$

$$\tag{4.64}$$

where f_0 is given by

$$f_0(R, \varepsilon) = \sum_k f(R, k)\delta(\varepsilon - \varepsilon_k) \qquad (4.65)$$

It will also be useful to consider the velocity average

$$f_1(R, \varepsilon) = \sum_k v(k)f(R, k)\delta(\varepsilon - \varepsilon_k) \qquad (4.66)$$

If the impact ionization scattering rate $1/\tau_{imp}(k)$ is a function of energy alone, then only $f_0(R, \varepsilon)$ is needed to calculate the generation rate. We therefore aim to derive an equation for $f_0(R, \varepsilon)$. To this end, we first multiply the Boltzmann equation Eq. (4.60) by $\delta(\varepsilon - \varepsilon_k)$ and perform a sum over k. The result of this manipulation is

$$\left(-qE_i(R)\frac{\partial}{\partial\varepsilon} + \frac{\partial}{\partial R_i}\right)f_{1,i}(R, \varepsilon) = \Omega(\varepsilon)\Sigma(\varepsilon, f_0) - \frac{1}{\tau(\varepsilon)}f_0(R, \varepsilon)$$

$$(4.67)$$

The second step is to multiply Eq. (4.60) by $v(k)\delta(\varepsilon - \varepsilon_k)$ and to again perform the k summation. We will then obtain the equation

$$qE_i(R)e_j\sum_k f(R, k)\frac{\partial}{\partial k_i}\left[v_j(k)\delta(\varepsilon - \varepsilon_k)\right]$$

$$+ e_j\frac{\partial}{\partial R_i}\sum_k f(R, k)v_i(k)v_j(k)\delta(\varepsilon - \varepsilon_k) = -\frac{1}{\tau(\varepsilon)}f_1(R, \varepsilon) \qquad (4.68)$$

which turns in the effective mass approximation into

$$q\frac{E(R)}{m}\left(1 - \frac{2}{3}\varepsilon\frac{\partial}{\partial\varepsilon}\right)f_0(R, \varepsilon) - \frac{2}{3m}\varepsilon\nabla_R f_0(R, \varepsilon) = -\frac{1}{\tau(\varepsilon)}f_1(R, \varepsilon)$$

$$(4.69)$$

If we use the identity $\varepsilon = \frac{1}{2}mv^2(\varepsilon)$ and define the mean free path $\lambda = \tau(\varepsilon)v(\varepsilon)$, we can express $f_1(R, \varepsilon)$ as a functional depending on $f_0(R, \varepsilon)$

$$f_1(R, \varepsilon) = -\frac{1}{3}v(\varepsilon)\left\{-qE(R)\lambda\frac{\partial}{\partial\varepsilon} + \lambda\nabla_R\right\}f_0(R, \varepsilon)$$

$$- q\frac{\tau(\varepsilon)}{m}E(R)f_0(R, \varepsilon) \qquad (4.70)$$

With the generalized differential operator

$$D = -qE(R)\lambda\frac{\partial}{\partial\varepsilon} + \lambda\nabla_R \qquad (4.71)$$

equations (4.67) and (4.70) read

$$\frac{1}{\lambda} D_i f_{1,i}(R, \varepsilon) = \Omega(\varepsilon)\Sigma(\varepsilon, f_0) - \frac{1}{\tau(\varepsilon)} f_0(R, \varepsilon) \tag{4.72}$$

$$f_1(R, \varepsilon) = -\frac{1}{3} v(\varepsilon) D f_0(R, \varepsilon) - \frac{qE(R)\lambda}{mv(\varepsilon)} f_0(R, \varepsilon) \tag{4.73}$$

We now insert Eq. (4.73) into Eq. (4.72) and multiply the resulting equation by $\tau(\varepsilon)$

$$-\frac{1}{3v(\varepsilon)} Dv(\varepsilon)\, Df_0(R, \varepsilon) - \frac{1}{v(\varepsilon)} D\frac{1}{v(\varepsilon)} \frac{qE(R)\lambda}{m} f_0(R, \varepsilon)$$

$$= \tau(\varepsilon)\Omega(\varepsilon)\Sigma(\varepsilon, f_0) - f_0(R, \varepsilon) \tag{4.74}$$

By evaluating the hierarchy of differentiations, we finally obtain the following, after some algebra which is left to the reader:

$$-\frac{1}{3} D^2 f_0(R, \varepsilon) - \frac{1}{3}\left[\frac{qE(R)\lambda}{\varepsilon} D + \frac{3}{2}\lambda \cdot \operatorname{div}\left(\frac{qE(R)\lambda}{\varepsilon} \right) \right] f_0(R, \varepsilon)$$

$$+\frac{1}{4}\left(\frac{qE(R)\lambda}{\varepsilon} \right)^2 f_0(R, \varepsilon) = \tau(\varepsilon)\Omega(\varepsilon)\Sigma(\varepsilon, f_0) - f_0(R, \varepsilon)$$

$$= -\frac{\tau(\varepsilon)}{\tau_{\mathrm{opt}}(\varepsilon)} f_0(R, \varepsilon) + \frac{\tau(\varepsilon)}{\tau_{\mathrm{opt}}(\varepsilon)}$$

$$\cdot \frac{\exp\left(-\dfrac{\omega_{\mathrm{opt}}}{2k_B T} \right) f_0(R, \varepsilon - \omega_{\mathrm{opt}}) + \exp\left(+\dfrac{\omega_{\mathrm{opt}}}{2k_B T} \right) f_0(R, \varepsilon + \omega_{\mathrm{opt}})}{2\cosh\left(\dfrac{\omega_{\mathrm{opt}}}{2k_B T} \right)} \tag{4.75}$$

The last expression for the right hand side of Eq. (4.75) was obtained by using Eq. (4.64). For a field strength of 4.10^5 V/cm and a total mean free path $\lambda \approx 90$ Å, we have a characteristic energy picked up in the field between collisions of $qE\lambda \approx 360$ meV. This must be compared with a typical threshold energy $\varepsilon_{\mathrm{imp}} \approx \frac{3}{2} E_G = 1.8$ eV. We can therefore use $qE\lambda/\varepsilon_{\mathrm{imp}} \ll 1$ to obtain the correct asymptotic limit. More difficulties are caused by the third term on the left hand side of Eq. (4.75). Utilizing the Poisson equation, we get the following for this contribution

$$\lambda \cdot \operatorname{div} \frac{qE(R)\lambda}{\varepsilon} \approx \frac{\lambda^2 q^2}{\kappa\varepsilon} N_A = 2\frac{a_0}{\lambda} \frac{\lambda^3 N_A}{\kappa} \frac{Ry}{qE\lambda} \frac{qE\lambda}{\varepsilon_{\mathrm{imp}}} \approx 3\cdot 10^{-3} \frac{qE\lambda}{\varepsilon_{\mathrm{imp}}} \tag{4.76}$$

For this estimation, we assumed a channel doping of approximately 10^{17} cm^{-3}. We also neglected the contributions of the carrier densities in the space charge. This is justified in the pinch-off region where carrier densities are small compared to the doping. Under the usual circumstances in a MOSFET, we can neglect the third term on the left-hand side of Eq. (4.75). This is certainly also true for the fourth term in the asymptotic limit for large energies. Equation (4.75) therefore reduces to

$$D^2 f_0(R, \varepsilon) + \frac{qE(R)\lambda}{\varepsilon} Df_0(R, \varepsilon) = 3\frac{\lambda}{\lambda_{opt}} f_0(R, \varepsilon) - 3\frac{\lambda}{\lambda_{opt}}$$

$$\cdot \frac{\exp\left(-\dfrac{\omega_{opt}}{2k_B T}\right) f_0(R, \varepsilon - \omega_{opt}) + \exp\left(+\dfrac{\omega_{opt}}{2k_B T}\right) f_0(R, \varepsilon + \omega_{opt})}{2\cosh\left(\dfrac{\omega_{opt}}{2k_B T}\right)}$$

$$(4.77)$$

For the general case of a space-dependent electric field, we will study Eq. (4.77) later in this section. For now we will consider only the constant-field case in which the distribution function has no explicit spatial dependence, which gives

$$D = -qE\lambda \frac{\partial}{\partial \varepsilon} \tag{4.78}$$

With the new variable $y = \varepsilon/qE\lambda$ Eq. (4.77) reads

$$f_0''(y) - \frac{1}{y} f_0'(y) = 3\frac{\lambda}{\lambda_{opt}} f_0(y) - 3\frac{\lambda}{\lambda_{opt}}$$

$$\cdot \frac{\exp\left(-\dfrac{\omega_{opt}}{2k_B T}\right) f_0\left(y - \dfrac{\omega_{opt}}{qE\lambda}\right) + \exp\left(+\dfrac{\omega_{opt}}{2k_B T}\right) f_0\left(y + \dfrac{\omega_{opt}}{qE\lambda}\right)}{2\cosh\left(\dfrac{\omega_{opt}}{2k_B T}\right)}$$

$$(4.79)$$

where the prime indicates a derivative with respect to y. To solve Eq. (4.79), we try the Ansatz

$$f(y) = y^v \exp(-ys) \tag{4.80}$$

where v and s are unknown numbers. Inserting Eq. (4.80) into Eq. (4.79), we obtain the following if we again neglect contributions $O(y^{-2})$

$$s^2 = 3\frac{\lambda}{\lambda_{opt}} \left[1 - \frac{\cosh\left(\dfrac{\omega_{opt}}{2k_B T} - \dfrac{\omega_{opt}}{qE\lambda} s\right)}{\cosh\left(\dfrac{\omega_{opt}}{2k_B T}\right)}\right] \tag{4.81}$$

and

$$v = \cfrac{s}{2s - 3\cfrac{\lambda}{\lambda_{opt}}\cfrac{\omega_{opt}}{qE\lambda}\cfrac{\sinh\left(\cfrac{\omega_{opt}}{2k_B T} - \cfrac{\omega_{opt}}{qE\lambda} s\right)}{\cosh\left(\cfrac{\omega_{opt}}{2k_B T}\right)}} \tag{4.82}$$

We observe that the electric field in Eqs. (4.81) and (4.82) is scaled by the energy of the optical phonon. The physics behind this is that the energy gain in the electric field between collisions is compared with the typical energy loss during a collision event. Because the cosh function is always greater than one, we can find the following from Eq. (4.81) as the upper bound for s

$$s < s_{max} = \left(3\frac{\lambda}{\lambda_{opt}}\left[1 - \frac{1}{\cosh\left(\dfrac{\omega_{opt}}{2k_B T}\right)}\right]\right)^{1/2} \tag{4.83}$$

which makes it a number of order one in the low temperature limit $2k_B T < \omega_{opt}$. In the high temperature limit, $2k_B T \gg \omega_{opt}$ s_{max} approaches zero like $\omega_{opt}/k_B T$. In Fig. 33, we show a graphical representation of the left and right hand side of Eq. (4.81). The intersection for non-zero s of the two

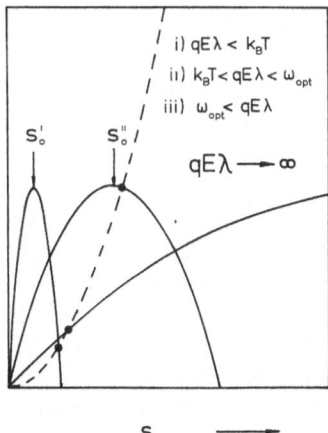

Fig. 33 The solution of Eq. (4.81) is the intersection ($\bullet$) of the solid line with the dashed line ($= s^2$)

curves gives the solution. The maximum of the right hand side is reached for $s_0 = qE\lambda/2k_B T$ and has the two zeros $s_1 = 0$ and $s_2 = 2s_0$.
The behavior of s is of interest for $qE\lambda$ in comparison to $k_B T$ and ω_{opt}. We can distinguish three important cases.

(i) $k_B T \gg qE\lambda$; small field
We find $s_0 < s < s_2$. s is therefore of the order $qE\lambda/k_B T$. To find the leading contribution for s, the right-hand side of Eq. (4.81) must vanish because the square on the left hand side is of second order and can be neglected. This gives the following for s and v

$$s = \frac{qE\lambda}{k_B T}; \quad v = 0 \qquad\qquad (4.84)$$

(ii) $k_B T < qE\lambda \ll \omega_{opt}$; moderate field
The location of the intersection of the two curves in Fig. 33 is close to s_{max} as the field, and so s_0, moves towards higher values. We get an approximately field-independent solution for s because we are in the region where the parabola has its steepest slope.

$$s = s_{max}; \quad 0 < v < 1 \qquad\qquad (4.85)$$

(iii) $qE\lambda \gg \omega_{opt}$; strong field
As the field continues to increase, s_0 moves further to greater s values, but the intersection of two curves will move towards zero again. Because the parabola has a decreasing slope, we will have a more pronounced field dependence of s. Because s is bounded, we can perform an expansion of the cosh-function in Eq. (4.81) and obtain the following term in leading order for s

$$s = 3\frac{\omega_{opt}}{qE\lambda_{opt}} \tanh\left(\frac{\omega_{opt}}{2k_B T}\right); \quad v = 1 \qquad\qquad (4.86)$$

In Figs. 34 and 35, we show an evaluation of Eqs. (4.81) and (4.82) for the characteristic cases discussed above. We assume $\omega_{opt} = 60$ meV and plot s and v versus the energy $\varepsilon = qE\lambda$. In Fig. 34 we show s and v for room temperature for an energy range 5 meV $< \varepsilon <$ 500 meV.
In Fig. 35 we show the interesting case of a moderate to strong perturbation of the carrier system in the low-temperature regime $k_B T = 1.0$ meV in the energy range 2.5 meV $< \varepsilon <$ 500 meV.
Comparing Figs. 34 and 35, we find for the low-temperature case an extended plateau for which the function $s(E, T)$ is a constant. For the higher temperatures, this plateau degenerates into a pronounced maximum of the bell-shaped curve for s. The reason for this behavior is that only for sufficiently low temperatures can case (ii) be realized. The energy gain of the carriers in the field is smaller than the energy of the optical phonon and the coupling to the optical phonons is therefore weak and we have almost

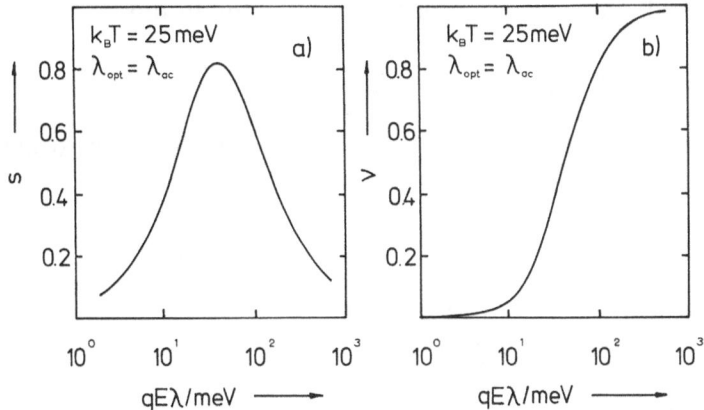

Fig. 34 The field dependence of the functions s (a) and v (b) at room temperature assuming equal scattering rates for optical and acoustical phonons. For the energy of the optical phonon 60 meV is used. The electric field is scaled by the total mean free path

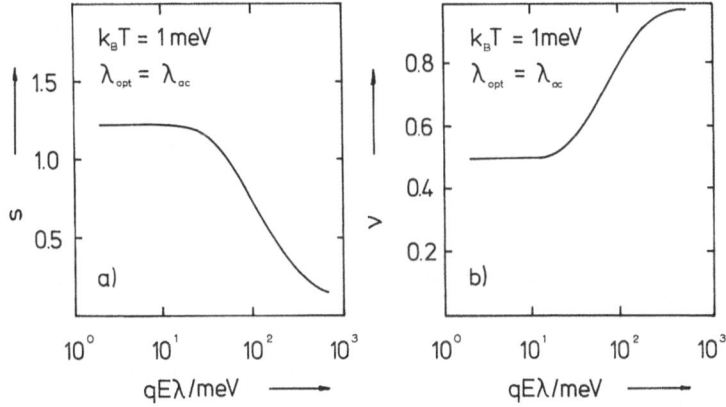

Fig. 35 The field dependence of the functions s (a) and v (b) at low temperature assuming equal scattering rates for optical and acoustical phonons. The electric field is scaled by the total mean free path

ballistic behavior. As soon as the energy gain in the field approaches ω_{opt}, optical phonon scattering becomes stronger and the carrier will no longer move in purely ballistic mode. The onset of inelastic scattering will bring the carrier into the lucky-drift mode, which we will discuss later. Let us summarize the findings we get for the distribution function Eq. (4.80) in the limits discussed in Eqs. (4.84) to (4.86)

(i) Maxwellian distribution near thermal equilibrium

$$f(\varepsilon) \approx \exp\left(-\frac{\varepsilon}{k_B T}\right); \quad k_B T \gg qE\lambda \tag{4.87}$$

(ii) Shockley's lucky electron is accelerated ballistically to achieve an energy ε in the electric field without collisions

$$f(\varepsilon) \approx \exp\left(-\frac{\varepsilon}{qE\lambda^*}\right); \quad k_B T < qE\lambda \ll \omega_{\mathrm{opt}} \tag{4.88}$$

with the effective mean free path $\lambda^* = \lambda/s_{\max}$
(iii) Wolff's diffusion model

$$f(\varepsilon) \approx \exp\left(-\frac{\varepsilon}{qE\lambda} \frac{\delta}{qE\lambda}\right); \quad \omega_{\mathrm{opt}} \ll qE\lambda \tag{4.89}$$

with a characteristic energy δ

$$\delta = 3\omega_{\mathrm{opt}} \frac{\lambda}{\lambda_{\mathrm{opt}}} \tanh\left(\frac{\omega_{\mathrm{opt}}}{2k_B T}\right) \tag{4.90}$$

Thus our approach to solving the Boltzmann equation covers different modes of carrier motion in a very natural way and provides the correct limit for $E \to 0$.

The lucky-drift model (Ridley 1983) is a suitable alternative for calculating the population of high energetic states. This model does not suffer from the weak scattering limit of the purely ballistic lucky-electron model because it accounts for both momentum and energy-relaxing collisions: a carrier can achieve its energy ballistically without a momentum-relaxating collision at all or after a single scattering event it will drift in the field without experiencing an energy relaxation to obtain the energy ε. The two characteristic times in this process are the momentum scattering rate τ_m and the energy scattering rate τ_ε. For the momentum scattering rate, we have the following in line with Eq. (2.50)

$$\frac{1}{\tau_m(\varepsilon_k)} = 2\pi \sum_{k'} |m_{k-k'}|^2 \left[n_B(\omega_{k-k'})\delta(\varepsilon_k - \varepsilon_{k'} - \omega_{k-k'}) \right.$$

$$\left. + (n_B(\omega_{k-k'}) + 1) \cdot \delta(\varepsilon_k - \varepsilon_{k'} + \omega_{k-k'}) \right] (1 - \cos\theta_{kk'}) \tag{4.91}$$

and the energy scattering rate is the difference between the energy transfer from phonon emission and absorption divided by the carrier energy

$$\frac{1}{\tau_\varepsilon(\varepsilon_k)} = 2\pi \sum_{k'} |m_{k-k'}|^2 \frac{\omega_{k-k'}}{\varepsilon_k} \left[(n_B(\omega_{k-k'}) + 1)\delta(\varepsilon_k - \varepsilon_{k'} + \omega_{k-k'}) \right.$$

$$\left. - n_B(\omega_{k-k'}) \cdot \delta(\varepsilon_k - \varepsilon_{k'} - \omega_{k-k'}) \right] \tag{4.92}$$

We evaluate Eqs. (4.91) and (4.92) for optical phonons, use the same assumption as above and keep the phonon energy ω_{opt} constant. In the limit $\varepsilon \gg \omega_{opt}$, we then obtain

$$\frac{1}{\tau_m(\varepsilon_k)} = 2\pi(2n_B(\omega_{opt}) + 1)\sum_{k'}|m_{k-k'}|^2\,\delta(\varepsilon_k - \varepsilon_{k'})\,(1 - \cos\theta_{kk'})$$

(4.93)

$$\frac{1}{\tau_\varepsilon(\varepsilon_k)} = 2\pi\frac{\omega_{opt}}{\varepsilon_k}\sum_{k'}|m_{k-k'}|^2\,\delta(\varepsilon_k - \varepsilon_{k'})$$

(4.94)

After performing the momentum summation, this yields the following relationship between τ_m and τ_ε

$$\frac{1}{\tau_\varepsilon(\varepsilon)} = \frac{3}{2}\frac{\omega_{opt}}{\varepsilon}\tanh\left(\frac{\omega_{opt}}{2k_BT}\right)\frac{1}{\tau_m(\varepsilon)}$$

(4.95)

Equation (4.95) implies that for sufficiently high energies $\varepsilon \gg \omega_{opt}$ we have $\tau_m \ll \tau_\varepsilon$. The carrier dynamics is now described by the following equations of motions for the momentum and energy balance

$$\Delta p = F\Delta t - p\frac{\Delta t}{\tau_m(t)}$$

(4.96)

$$\Delta\varepsilon = Fv_d(t)\Delta t - \varepsilon\frac{\Delta t}{\tau_\varepsilon(t)}$$

(4.97)

For $\Delta t < \tau_m$, momentum transfer to the phonons can be neglected and we have pure ballistic motion obeying Newton's law

$$\dot{p} = F$$

(4.98)

The lucky-drift case is dominant for $\tau_m < \Delta t < \tau_\varepsilon$. We can neglect the energy transfer and get

$$\dot{\varepsilon} = Fv_d(t)$$

(4.99)

In the limit $\Delta t > \tau_m$, the stationary state of Eq. (4.96) is reached and we have

$$v_d(t) = \frac{F\tau_m(t)}{m(t)}$$

(4.100)

The probability for a lucky electron to proceed from t_0 to t is

$$P_{LE}(t) = P_{LE}(t_0)\exp\left(-\int_{t_0}^{t}\frac{d\tau}{\tau_m(\tau)}\right)$$

(4.101)

Converting Eq. (4.101) into energy space, we get the following by utilizing Eq. (4.98)

$$P_{LE}(\varepsilon) = P_{LE}(\varepsilon_0)\exp\left(-\int_{\varepsilon_0}^{\varepsilon}\frac{d\varepsilon'}{F(\partial\varepsilon/\partial p)\tau_m(\varepsilon')}\right) \tag{4.102}$$

Here $\partial\varepsilon/\partial p$ is the group velocity of the electrons $v_p(\varepsilon')$. Similarly, we have the following for the probability of a lucky-drift electron

$$P_{LD}(t) = P_{LD}(t_0)\exp\left(-\int_{t_0}^{t}\frac{d\tau}{\tau_\varepsilon(\tau)}\right) \tag{4.103}$$

which can be converted into energy space by using Eq. (4.99) where we express the drift velocity by its stationary value $v_d(\varepsilon)$ from Eq. (4.100)

$$P_{LD}(\varepsilon) = P_{LD}(\varepsilon_0)\exp\left(-\int_{\varepsilon_0}^{\varepsilon}d\varepsilon'\frac{m(\varepsilon')}{F^2\tau_\varepsilon(\varepsilon')\tau_m(\varepsilon')}\right) \tag{4.104}$$

with Eq. (4.95) we can simplify the integrand in the exponential of Eq. (4.104).

$$\frac{m(\varepsilon)}{\tau_\varepsilon(\varepsilon)\tau_m(\varepsilon)} = 3\frac{\omega_{opt}}{\lambda_m(\varepsilon)^2}\tanh\left(\frac{\omega_{opt}}{2k_BT}\right) \tag{4.105}$$

To derive Eq. (4.105), we used $\varepsilon = \frac{m(\varepsilon)}{2}v^2(\varepsilon)$ and $\lambda_m(\varepsilon) = \tau_m(\varepsilon)v(\varepsilon)$. With Eqs. (4.102) and (4.104), we found the probability to find a lucky electron or a lucky-drift electron at that energy. For a lucky-drift electron it is possible that it initially started as a lucky electron and will then, after a collision, continue as a lucky-drift electron. If we assume that one collision is sufficient to convert a lucky electron into a lucky-drift electron, this process will have the probability

$$P_{LD}(\varepsilon) = P_{LE}(\varepsilon_0)\int_{\varepsilon_0}^{\varepsilon}\frac{d\varepsilon'}{F\lambda_m(\varepsilon')}\exp\left(-\int_{\varepsilon_0}^{\varepsilon'}\frac{d\varepsilon'}{Fv(\varepsilon')\tau_m(\varepsilon')}\right)$$

$$\cdot\exp\left(-\int_{\varepsilon'}^{\varepsilon}d\varepsilon'\frac{m(\varepsilon'')}{F^2\tau_\varepsilon(\varepsilon'')\tau_m(\varepsilon'')}\right) \tag{4.106}$$

The total probability is then the sum of the lucky electron contribution Eq. (4.102) and the lucky-drift electron from Eq. (4.106)

$$P_{tot}(\varepsilon) = P_{LE}(\varepsilon) + P_{LD}(\varepsilon) \tag{4.107}$$

In the effective mass approximation, the mean free path λ_m is constant and the integrals can be evaluated easily.

$$P_{tot}(\varepsilon) = P_{LE}(\varepsilon_0) \left[\exp\left(-\frac{\varepsilon - \varepsilon_0}{F\lambda_m}\right) \right.$$

$$\left. + \frac{\exp\left(-\dfrac{\varepsilon - \varepsilon_0}{F\lambda_m}\dfrac{3\omega_{opt}}{F\lambda_m}\tanh\left(\dfrac{\omega_{opt}}{2k_B T}\right)\right) - \exp\left(-\dfrac{\varepsilon - \varepsilon_0}{F\lambda_m}\right)}{1 - \dfrac{3\omega_{opt}}{F\lambda_m}\tanh\left(\dfrac{\omega_{opt}}{2k_B T}\right)} \right]$$

$$(4.108)$$

It is reasonable to assume that the particles will start from a distribution of initial values ε_0. A proper average over this distribution should therefore be performed. The largest weight of such a distribution will be centered around the average carrier energy $\varepsilon_{av} = 3/2k_B T_c$ and we could therefore replace ε_0 by ε_{av}. As long as this average energy is small compared to the impact ionization threshold, we can neglect it relative to ε.

$$P_{tot}(\varepsilon) \approx \exp\left(-\frac{\varepsilon}{F\lambda_m}\right) + \frac{\exp\left(-\dfrac{\varepsilon}{F\lambda_m}\dfrac{\delta'}{F\lambda_m}\right) - \exp\left(-\dfrac{\varepsilon}{F\lambda_m}\right)}{1 - \dfrac{\delta'}{F\lambda_m}}$$

$$(4.109)$$

$$\delta' = 3\omega_{opt}\tanh\left(\frac{\omega_{opt}}{2k_B T}\right) \qquad (4.110)$$

In the low-temperature limit $k_B T \ll \omega_{opt}$, the tanh-function approaches one. For moderate field strength $k_B T < F\lambda_m \ll \omega_{opt}$, we obtain the limit of the lucky electron

$$P_{tot}(\varepsilon) \approx \exp\left(-\frac{\varepsilon}{F\lambda_m}\right) \qquad (4.111)$$

If we consider the strong field case, we have Wolff's diffusion mode

$$P_{tot}(\varepsilon) \approx \exp\left(-\frac{\varepsilon}{F\lambda_m}\frac{3\omega_{opt}}{F\lambda_m}\tanh\left(\frac{\omega_{opt}}{2k_B T}\right)\right) \qquad (4.112)$$

which is exactly the same as Eq. (4.89) if we only consider scattering by optical phonons, which means $\delta = \delta'$ in Eq. (4.90). In Figs. 36 to 38 we compare the solution of the Boltzmann equation with the lucky-drift

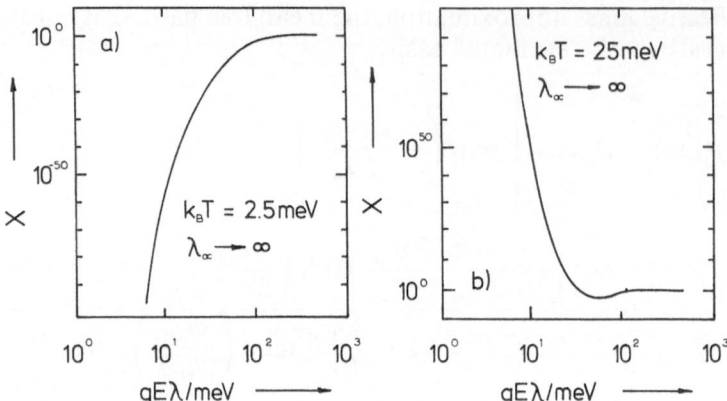

Fig. 36 The field dependence of the ratio of the solution of Boltzmann's equation and the lucky drift model at low (a) and room (b) temperature neglecting the influence of acoustical phonons

model. We plot the quantity $\chi(\varepsilon)$ versus the field energy $\varepsilon = qE\lambda$ at an impact ionization threshold $\varepsilon_{\mathrm{imp}} = 1.8$ eV

$$\chi(\varepsilon) = \frac{\exp\left(-\dfrac{\varepsilon_{\mathrm{imp}}}{\varepsilon} s(\varepsilon, T)\right)}{P_{\mathrm{tot}}(\varepsilon, T)} \tag{4.113}$$

In Fig. 36 only the influence of optical phonon scattering is considered. In the high-field regime, both expressions coincide. In the region of moderate field strength where the lucky electron mode dominates, the lucky-drift model has a considerably larger particle number than the solution of Boltzmann's equation. The reason is that the effective mean free path for the considered temperatures $k_B T < \omega_{\mathrm{opt}}$ is greater in the lucky drift model

$$\lambda_{BE}^{\max} = \lambda^* = \lambda_{LE} \left/ \left(\left[3\left[1 - \frac{1}{\cosh\left(\dfrac{\omega_{\mathrm{opt}}}{2k_B T}\right)}\right]\right]\right)^{1/2}\right. \tag{4.114}$$

In Fig. 37 we show the situation if the mean free path in the lucky drift model is adjusted to be the same as the maximum mean free path obtained from the solution of the Boltzmann equation $\lambda_{LE} \to \lambda^*$. We find that for moderate and large fields the lucky drift model coincides with the solution of the Boltzmann equation.

For small fields the particle number estimated from the solution of the Boltzmann equation is much larger than that from the lucky drift model. This is an artifact of the lucky drift model which is not applicable for very

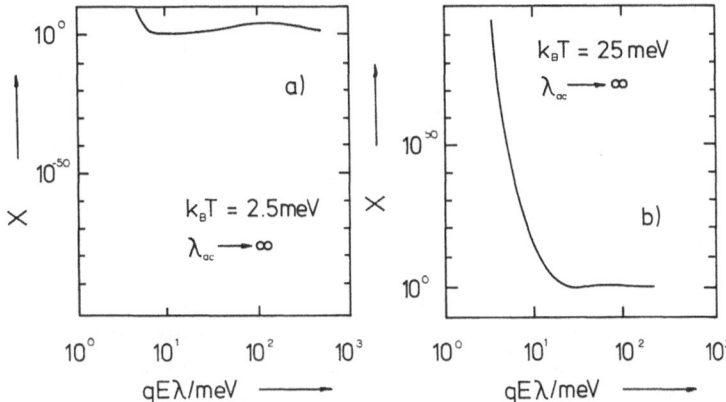

Fig. 37 The field dependence of the ratio of the solution of Boltzmann's equation and the lucky drift model at low (a) and room (b) temperature neglecting the influence of acoustical phonons. The ballistic mean free path in the lucky drift model is adjusted to the maximal mean free path of the solution of the Boltzmann equation, compare Eq. (4.114)

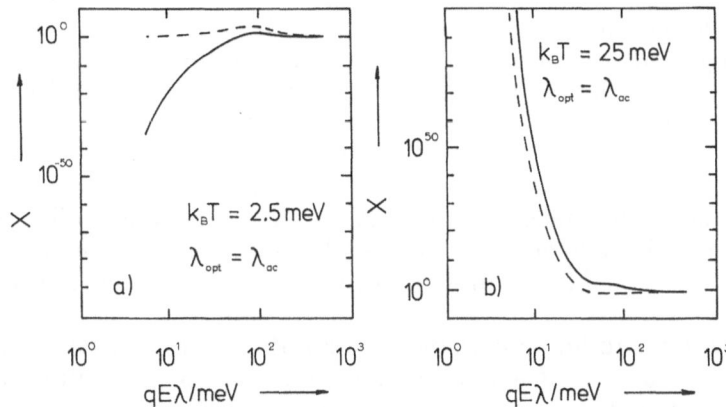

Fig. 38 The field dependence of the ratio of the solution of Boltzmann's equation and the lucky drift model at low (a) and room (b) temperature assuming equal scattering rates for optical and acoustical phonons. Solid line—no adjustment of the ballistic mean free path in the lucky drift model. Dashed line—the ballistic mean free path in the lucky drift model is adjusted to the maximal mean free path of the solution of the Boltzmann equation, cf. Eq. (4.114)

small fields because it does not produce the correct behavior in equilibrium. In Fig. 38 we show the influence of acoustical phonon scattering. The interference of optical and acoustical phonons is taken into account in the lucky drift model by $\delta' \to \delta$. This is justified if one assumes that acoustical phonons have the dominant influence in the momentum relaxation scattering processes. On the other hand we can include in the elastical channel all other contributions that scatter only elastically, such as impurities for instance, and therefore contribute to the momentum relaxation. For moderate fields we now have $\lambda_{BE}^{max} = \lambda^* = \lambda_{LE}/s_{max}$ with s_{max} from Eq. (4.83) with $\lambda/\lambda_{opt} = 0.5$. The discrepancy observed at room temperature has now completely disappeared. It remains, however, at lower temperatures which is again attributed to the different mean free path in the lucky drift model compared to the maximum mean free path of the solution of Boltzmann's equation. If we again adjust the maximum mean free path to coincide in lucky drift and the solution of the Boltzmann equation $\lambda_{LE} \to \lambda^*$ both give similar results. H remains however the discrepancy approaching thermal equilibrium.

In the previous paragraphs we studied the stationary Boltzmann equation to obtain a solution for the distribution of high energy particles in a constant electric field. Although this case is of fundamental interest, the reality does not usually provide such a simple situation, especially in a MOSFET. The electric field can change rapidly over the distance of several mean free paths λ. To apply the constant field solution, Eq. (4.80), the carriers have to relax into a stationary state with the local electric field. This is possible only if the spatial variation of this field is moderate over the mean free path. If this is not the case, the solution of the Boltzmann equation will no longer respond to the local electric field. An ensemble of carriers subjected to a field will relax to a stationary state after several collisions with phonons. For the constant field case, we have implicitly assumed that the spatial region which we considered is large enough for such a relaxation to take place. In a field with considerable spatial variation therefore, we have to account for this relaxation. Because this relaxation is very closely related to a transient phenomenon, we will start our investigation from the time-dependent Boltzmann equation. Our goal for this investigation is to find a possible generalization of the analysis performed above for the case of a non-homogeneous electric field. For the time-dependent Boltzmann equation, we obtain the following, similarly to Eqs. (4.72) and (4.73)

$$\left[1 + \tau \frac{\partial}{\partial T}\right] f_0 + \frac{1}{v} D_i f_{1,i} = \tau \Omega \Sigma [f_0] \tag{4.115}$$

$$\left[1 - \tau \frac{\partial}{\partial T}\right] f_1 = -\frac{1}{3} v D f_0 - f_0 \frac{qE\lambda}{mv} \tag{4.116}$$

Here we omitted the variables for simplicity. Formally we can express the brackets on the left-hand side of Eqs. (4.115) and (4.116) as the operators

T^+ and T^-, respectively.

$$T^\pm = 1 \pm \tau \frac{\partial}{\partial T} \qquad (4.117)$$

acting on the components f_0 and f_1 of the distribution function. We now multiply both equations with the corresponding inverse operator from the left and get

$$f_0 + (T^+)^{-1} \left[\frac{1}{v} D_i f_{1,i} \right] = (T^+)^{-1} \tau \Omega \Sigma [f_0]$$
$$f_1 = -(T^-)^{-1} \frac{1}{3} v D f_0 - (T^-)^{-1} \frac{qE\lambda}{mv} f_0 \qquad (4.118)$$

These equations are completely equivalent to Eq. (4.72) and (4.73). We can follow exactly the same arguments that lead to Eq. (4.77) and obtain in leading order

$$f_0 - \frac{1}{3} D^2 (T^+)^{-1} (T^-)^{-1} f_0 = (T^+)^{-1} \tau \Omega \Sigma [f_0] \qquad (4.119)$$

In this order, we omit the pre-factor of the exponential. In the ideal case of an infinite scattering rate $\tau = 0$, relaxation would be instantaneous to the local field and we could apply the results for a constant field derived above. To obtain a more realistic description, we will consider the limit of very strong coupling, which means $\tau \to 0$. In this limit, we will linearize Eq. (4.119) in τ. In this linearized theory we have, for T^+ and T^-

$$T^+ T^- = 1 + O\left(\tau^2 \frac{\partial^2}{\partial T^2} \right) \approx 1; \quad T^+ \approx (T^-)^{-1}; \quad T^- \approx (T^+)^{-1} \qquad (4.120)$$

and therefore Eq. (4.119) turns into

$$f_0 - \frac{1}{3} D^2 f_0 = \tau \Omega \Sigma [T^- f_0] \qquad (4.121)$$

Equation (4.121) will serve as the basis for incorporating the relaxation phenomena discussed above. Of course, for a stationary solution there will be no explicit time dependence of f. However, adjusting to the field situation will involve a motion in phase space according to the collision events which transfer energy and momentum to the environment. We will therefore replace the time derivative by

$$T^- = 1 - \tau \frac{\partial}{\partial T} \to 1 - \tau \dot{\varepsilon} \frac{\partial}{\partial \varepsilon} - \tau \dot{R} \frac{\partial}{\partial R} \qquad (4.122)$$

Applying T^- to f_0 yields the following, if the collision rate is sufficiently high

$$T^- f_0 (R, \varepsilon) \approx f_0(R - \tau v, \varepsilon - \tau \dot{\varepsilon}) \tag{4.123}$$

The order of magnitude of $\dot{\varepsilon}$ will be $\varepsilon/\tau_\varepsilon$, with the energy scattering rate from Eq. (4.92). If we use its estimate given in Eq. (4.95), we obtain

$$\dot{\varepsilon} \approx \frac{3}{2} \frac{\omega_{\text{opt}}}{\tau_m(\varepsilon)} \tanh\left(\frac{\omega_{\text{opt}}}{2k_B T}\right) \approx \frac{\omega_{\text{opt}}}{\tau_m(\varepsilon)} \tag{4.124}$$

Because $\tau \ll \tau_m$ for a short-range interaction (cf. discussion of Eq. (2.49)), which we assumed by keeping the matrix element constant, the correction term to the energy is much less than the optical phonon energy. Within the accuracy we are interested in, we can therefore neglect this contribution. We will discuss the contribution to the spatial variable later. The explicit form of Eq. (4.121) is now

$$\left(qE(R)\lambda\frac{\partial}{\partial\varepsilon} - \lambda\nabla R\right)^2 f_0(R, \varepsilon) - 3\frac{\lambda}{\lambda_{\text{opt}}} f_0(R, \varepsilon) =$$

$$-3\frac{\lambda}{\lambda_{\text{opt}}} \frac{\exp\left(-\dfrac{\omega_{\text{opt}}}{2k_B T}\right) f_0(R - \tau v, \varepsilon - \omega_{\text{opt}}) + \exp\left(+\dfrac{\omega_{\text{opt}}}{2k_B T}\right) f_0(R - \tau v, \varepsilon + \omega_{\text{opt}})}{2\cosh\left(\dfrac{\omega_{\text{opt}}}{2k_B T}\right)} \tag{4.125}$$

A change of variables

$$\varepsilon \to \varepsilon - qV(R) = \Phi \tag{4.126}$$

$$\nabla_R \to \nabla_R + qE(R)\frac{\partial}{\partial\varepsilon}$$

brings Eq. (4.125) into the form

$$\lambda^2 \nabla_R^2 f_0(R, \Phi) - 3\frac{\lambda}{\lambda_{\text{opt}}} f_0(R, \Phi) =$$

$$-3\frac{\lambda}{\lambda_{\text{opt}}} \frac{\exp\left(-\dfrac{\omega_{\text{opt}}}{2k_B T}\right) f_0(R - \tau v, \Phi - \omega_{\text{opt}}) + \exp\left(+\dfrac{\omega_{\text{opt}}}{2k_B T}\right) f_0(R - \tau v, \Phi + \omega_{\text{opt}})}{2\cosh\left(\dfrac{\omega_{\text{opt}}}{2k_B T}\right)} \tag{4.127}$$

This equation is very closely related to the inhomogeneous Helmholtz equation

$$\nabla_R^2 f + k^2 f = \rho \tag{4.128}$$

which in infinite space is solved by the following integral (Morse and Feshbach 1953)

$$f(R) = -\int \frac{dR'^3}{4\pi} \frac{\exp(ik|R - R'|)}{|R - R'|} \rho(R') \qquad (4.129)$$

or if the problem is described sufficiently in two dimensions

$$f(R) = -\int \frac{dR'^2}{2\pi} K_0(|k||R - R'|)\rho(R') \qquad (4.130)$$

Here $K_0(z)$ is the modified Bessel function of order zero. We find a formal solution of Eq. (4.127), for example for the two-dimensional case, to be

$$f_0(R, \Phi) = \frac{3}{4\pi} \frac{\lambda_{opt}}{\lambda} \int \frac{dR'^2}{\lambda_{opt}^2} K_0(|k||R - R'|)$$

$$\cdot \frac{\exp\left(-\frac{\omega_{opt}}{2k_B T}\right) f_0(R' - \tau v, \Phi' - \omega_{opt}) + \exp\left(+\frac{\omega_{opt}}{2k_B T}\right) f_0(R' - \tau v, \Phi' + \omega_{opt})}{\cosh\left(\frac{\omega_{opt}}{2k_B T}\right)}$$

$$(4.131a)$$

with $k = (3/\lambda_{opt}\lambda)^{1/2}$. The prime in the integrand indicates a variation with the integration variable. Going back to the original variables, we get

$$f_0(R, \varepsilon) = \frac{3}{4\pi} \frac{\lambda_{opt}}{\lambda} \int \frac{dR'^2}{\lambda_{opt}^2} K_0(|k||R - R'|) \frac{1}{\cosh\left(\frac{\omega_{opt}}{2k_B T}\right)}$$

$$\cdot \left[\exp\left(-\frac{\omega_{opt}}{2k_B T}\right) f_0(R' - \tau v, \varepsilon - qV' + qV - \omega_{opt}) \right.$$

$$\left. + \exp\left(+\frac{\omega_{opt}}{2k_B T}\right) f_0(R' - \tau v, \varepsilon - qV' + qV + \omega_{opt}) \right]$$

$$(4.131b)$$

We solve this integral equation approximately by first-order iteration. However, the problem is to find an appropriate initial solution. In the first place, it is interesting to study the solution for a constant field. In this case, the distribution function will have no explicit spatial dependence and the potential V is a linear function along the x-axis, for example. We perform the redundant space integration analytically and obtain the following from

Eq. (4.131)

$$f_0(\varepsilon) = \frac{3}{4}\frac{1}{\lambda}\int_{-\infty}^{\infty}\frac{dx}{\lambda_{opt}|k|}\exp(-|k||x|)$$

$$\cdot \frac{\exp\left(-\dfrac{\omega_{opt}}{2k_BT}\right)f_0(\varepsilon - qEx - \omega_{opt}) + \exp\left(+\dfrac{\omega_{opt}}{2k_BT}\right)f_0(\varepsilon - qEx + \omega_{opt})}{\cosh\left(\dfrac{\omega_{opt}}{2k_BT}\right)}$$

(4.132)

with the formulation

$$f_0 = \exp\left(-\frac{\varepsilon}{qE\lambda}s\right) \qquad (4.133)$$

the previous equation turns into

$$1 = \frac{3}{2}\frac{1}{\lambda}\int_{-\infty}^{\infty}\frac{dx}{\lambda_{opt}|k|}\exp\left(-|k||x|+\frac{x}{\lambda}s\right)\frac{\cosh\left(\dfrac{\omega_{opt}}{2k_BT}-\dfrac{\omega_{opt}}{qE\lambda}s\right)}{\cosh\left(\dfrac{\omega_{opt}}{2k_BT}\right)}$$

(4.134)

After the remaining integration is performed, this yields

$$s^2 = 3\frac{\lambda}{\lambda_{opt}}\left[1 - \frac{\cosh\left(\dfrac{\omega_{opt}}{2k_BT}-\dfrac{\omega_{opt}}{qE\lambda}s\right)}{\cosh\left(\dfrac{\omega_{opt}}{2k_BT}\right)}\right] \qquad (4.135)$$

This is precisely Eq. (4.81), which we derived previously. We now come back to the integral equation (4.131). As a first guess, we could take the homogeneous case distribution function as the initial solution for the iteration. The resulting expression is still simple enough to be handled in a modeling environment. But it has a built-in physical inconsistency: because it depends only on the local electric field, it would not reproduce the correct limit in thermal equilibrium. A device, for instance a simple pn-junction, still contains significant electric fields even in thermal equilibrium. In order to find a solution to this problem, we will consider the generalized differential operator D defined in Eq. (4.71).

$$D = -qE(R)\lambda\frac{\partial}{\partial\varepsilon} + \lambda\nabla_R \qquad (4.136)$$

This operator is related to the generalized driving force F for the particle current, which is as follows according to Eq. (2.180)

$$F(R) = E(R) + \frac{1}{n(R)} \nabla_R [u_T(R) n(R)] \tag{4.137}$$

the connection between D and F is obtained by performing the following average over the distribution function

$$q F(R) \lambda = \frac{2}{3} \frac{\int d\varepsilon \, \Omega(\varepsilon) \varepsilon D f(R, \varepsilon)}{\int d\varepsilon \, \Omega(\varepsilon) f(R, \varepsilon)} \tag{4.138}$$

Equation (4.138) is an implicit equation for D in terms of F. A possible solution is found by inspection

$$D = -q F \lambda \frac{\partial}{\partial \varepsilon} \tag{4.139}$$

In thermal equilibrium, we have no current flow and therefore $F = 0$. If in the solution for the homogeneous case we replace the electric field E by F, this distribution function will assume correct behavior in thermal equilibrium in the general case of a non-homogeneous field distribution. Over the distance τv, the distribution function relaxes to the electric field distribution. In the same spirit as we replaced the electric field by its average equivalent F, we now replace τv by $\lambda_\varepsilon = \tau_\varepsilon v_{\text{sat}}$, which is the average distance between energy-relaxing collisions.

$$
\begin{aligned}
f_0^{(1)}(R, \varepsilon) \sim & \int \frac{dR'^2}{\lambda^2} K_0(|k| \, |R - R'|) \left(1 - \frac{\lambda_{\text{opt}}}{3\lambda} s(R' - \lambda_\varepsilon)^2 \right) \\
& \cdot \exp\left(-\frac{q V(R) - q V(R')}{q F(R' - \lambda_\varepsilon) \lambda} s(R' - \lambda_\varepsilon) \right) \\
& \cdot \exp\left(-\frac{\varepsilon}{q F(R' - \lambda_\varepsilon) \lambda} s(R' - \lambda_\varepsilon) \right)
\end{aligned} \tag{4.140}
$$

If we further assume that the electric field does not change significantly over the mean free path λ_ε, we can expand the potential difference to obtain

$$
\begin{aligned}
V(R) - V(R') = \approx & - E(R' - \lambda_\varepsilon)(R - R') \\
& - 2(R - R') \cdot \nabla_{R'} \lambda_\varepsilon E(R' - \lambda_\varepsilon) + O(\lambda_\varepsilon^2)
\end{aligned} \tag{4.141}
$$

In the leading order, we neglect the second term of Eq. (4.141) and replace the potential difference in Eq. (4.140).

$$
f_0^{(1)}(R, \varepsilon) \sim \int \frac{dR'^2}{\lambda^2}\, K_0(|k|\,|R - R'|)\left(1 - \frac{\lambda_{\mathrm{opt}}}{3\lambda}\, s(R' - \lambda_\varepsilon)^2\right)
$$

$$
\cdot \exp\left(\frac{qE(R' - \lambda_\varepsilon)\,(R - R')}{qF(R' - \lambda_\varepsilon)\lambda}\, s(R' - \lambda_\varepsilon)\right)
$$

$$
\cdot \exp\left(-\frac{\varepsilon}{qF(R' - \lambda_\varepsilon)\lambda}\, s(R' - \lambda_\varepsilon)\right) \tag{4.142}
$$

With Eq. (4.142) we have the desired result. It is correct in the strong coupling limit $\lambda_\varepsilon \to 0$. Although in reality we have a finite mean free path, the asymptotic limit, which we consider here, is justified in the realm of the drift–diffusion approximation. The component f_0 of the distribution function as defined in Eq. (4.65) has the dimension of the density of states, while according to Eq. (4.142) f_0 is a dimensionless number. This contradiction is easily resolved by the observation that the distribution function is determined only up to a multiplicative factor. This can be chosen appropriately. It is related to the normalization of the distribution function. We will discuss this point in the next section where we consider the implications of Eq. (4.142) on modeling the phenomena of high energetic carriers.

With Eq. (4.142) we have derived an expression for the population of the high energy states in the carrier distribution function. We compared this solution for the case of a constant electric field with the lucky-drift model. We found reasonable agreement for the strong fields at elevated temperatures. A possible discrepancy was observed for the low temperature limit $k_B T \ll \omega_{\mathrm{opt}}$ and was due to the different mean free path in the ballistic regime. It is impossible to say *a priori* which of these models is correct. The solution of the Boltzmann equation assumes that scattering is sufficiently strong to maintain a stationary state in the field. Lucky electron and lucky-drift models assume ballistic behavior of the carriers, at least in part, which implies a weak scattering limit. An independent means of comparison can be achieved by calculating the shape of the distribution function by Monte Carlo methods. We performed such a comparison for the same model specifications as used above. All parameters were calculated consistently from first-principle scattering processes used in the Monte Carlo calculation. This leaves no fitting parameter. However, we have not yet addressed the problem of normalization of the distribution function. This must be done to obtain the absolute scale on which f varies. Because this normalization factor varies very slowly compared to the exponential, we consider it to be a constant number. This number is chosen such that the Monte Carlo and analytically obtained results coincide for a specified energy. In Figs. 39 and 40, we show the energy dependence of the distribution function for the energy interval $0\,\mathrm{eV} < \varepsilon < 2.5\,\mathrm{eV}$ in a constant field.

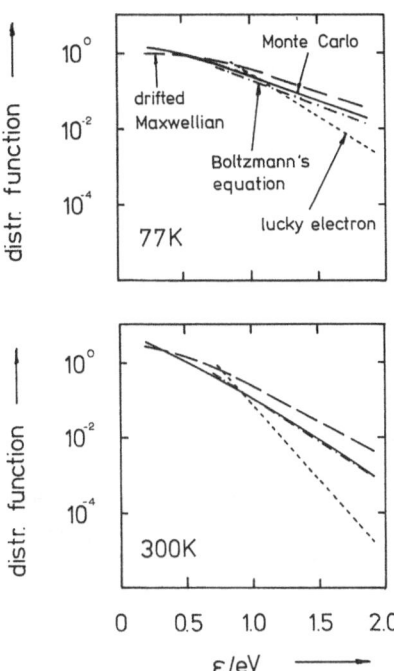

Fig. 39 Comparison of Monte Carlo calculation (solid line), asymptotic solution of Boltzmann's equation (dashed–dotted line), lucky electron (short-dashed line) and drifted Maxwellian (long-dashed line) at 77 K and 300 K. The drifted Maxwellian was adjusted to give the same average energy as obtained from the Monte Carlo calculation. Normalization of the lucky electron model and the asymptotic solution of Boltzmann's equation was achieved by assuming equal values at $\varepsilon = 0.9\,\text{eV}$ for 300 K. All mean free paths (λ, λ_{op}, λ_{ac}) were provided from the Monte Carlo calculation. No additional fitting was done

The shape of the distribution function was obtained without additional fitting. A comparison for the temperatures $T = 300$ K and $T = 77$ K is shown in Fig. 39.

Figure 40 shows a comparison between Monte Carlo calculation, our analytical approach and the lucky-electron mode for $T = 300$ K.

In Fig. 41 we show the field dependence of the distribution function for a constant energy $\varepsilon = 1.8$, 1.4 and 1.0 eV. The values are normalized to the distribution function for the highest field considered.

From Figs. 39 to 40 we can learn that the analytical solution of the Boltzmann equation is in very good agreement with the Monte Carlo calculations for the shape of the distribution functon. In Fig. 43, we compare Eq. (4.142) for a one-dimensional non-homogeneous field profile with the corresponding Monte Carlo results. The field profile was taken from MINIMOS for a typical MOSFET under strong avalanche conditions and is shown in Fig. 42. For the comparison, we normalized the result

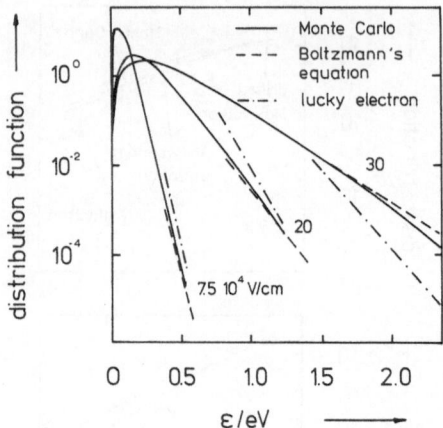

Fig. 40 Comparison of Monte Carlo calculation (solid line), asymptotic solution of Boltzmann's equation (dashed line), and lucky electron (dashed–dotted) line for different electric fields at room temperature. Normalization was achieved by adjusting values at $\varepsilon = 1.3\,\mathrm{eV}$ for $E = 3\cdot10^{5}\,\mathrm{V/cm}$

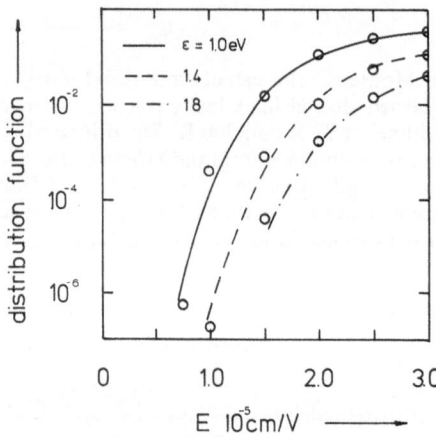

Fig. 41 Comparison of the field dependence of Monte Carlo calculation (open circles) and asymptotic solution of Boltzmann's equation for different energies $\varepsilon = 1\,\mathrm{eV}$ (solid line), $\varepsilon = 1.4\,\mathrm{eV}$ (dashed line), and $\varepsilon = 1.8\,\mathrm{eV}$ (dashed–dotted line). Normalization was achieved by adjusting values at $\varepsilon = 1.0\,\mathrm{eV}$ and $E = 3\cdot10^{5}\,\mathrm{V/cm}$

from Eq. (4.142) to the Monte Carlo result so that they coincide for an energy $\varepsilon = 1.0\,\mathrm{eV}$ at a position $y = 1.1\,\mu\mathrm{m}$.
We find a striking agreement of the Monte Carlo result with the analytical solution found in Eq. (4.142). In particular, we observe the retarded relaxation of the distribution function to the maximum of the electric field. All in

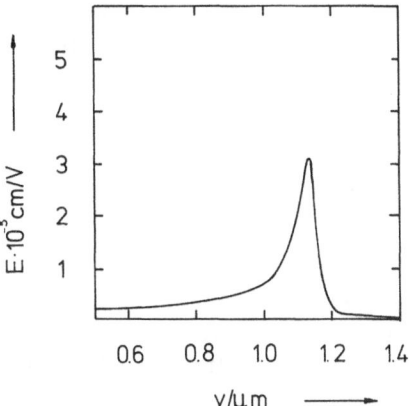

Fig. 42 Lateral electric field of a MOSFET. The effective channel length is $L_{eff} = 0.8 \; \mu m$ and the oxide thickness $t_{ins} = 20$ nm. The profile was taken through the point of maximal lateral electric field

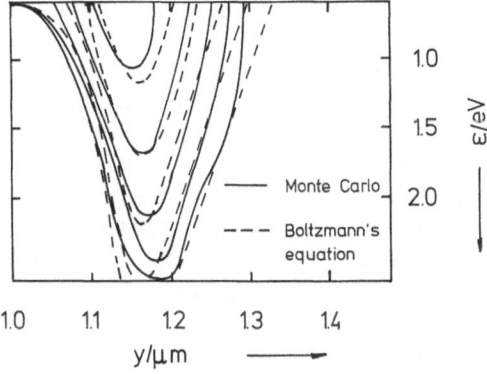

Fig. 43 Comparison of Monte Carlo calculation (solid line) and asymptotic solution of Boltzmann's equation for the non-homogeneous electric field of Fig. 42. Normalization was achieved by equal values at $y = 1.1 \; \mu m$ and $\varepsilon = 1$ eV

all, the Monte Carlo results show that our analytical approach gives reasonable results, at least for the model specifications. To apply these results to a realistic situation, a full account of the band structure must be included. Any analytical approach is doomed to fail in this case. However, we can speculate that the band structure will in the first place affect the mean free paths related to the scattering process that enter the formulation and not the essential field dependence. If we replace the mean free paths by appropriately fitted values, the analytical approach will also hold under realistic conditions. In the next section, we will discuss the application of Eq. (4.142) for calculating the impact ionization rate and gate oxide injection.

4.4 Impact Ionization Coefficient and Gate Oxide Injection

In the previous two sections of this chapter, we discussed features relating to the calculation of the scattering rate for impact ionization and the high-energy tail of the carrier distribution function. We will now investigate how these results can be utilized for modeling the impact ionization rate and oxide injection in a realistic device setting. The very first point of concern relates to the model specifications. Our analysis relied heavily on the effective mass approximation, which is a poor approximation for the high energetic carriers, as already mentioned in the introduction. We can, however, put forward some reasoning how a more realistic band structure can be included in the expression for the distribution function such as Eq. (4.142). A major advantage of the effective mass approximation is that the mean free path due to phonon scattering is approximately energy independent. This is no longer the case if the correct band structure is considered. The basic physics described in the formulation based on the effective mass approximation will most probably not change. We therefore argue that with properly chosen mean free paths we have a reasonable approximation of the real physics. For different energy ranges, these mean free paths would be different. The mean free path will most likely decrease if the energy increases. With this approach, we retain the simplicity of Eq. (4.142). We found a solution of the Boltzmann equation in Section 3 for infinitely extended bulk material. In a MOSFET, everything happens in close vicinity to the interface. To account for the effect of this interface, we must look for a solution restricted to a semi-infinite domain. This problem is similar to that addressed in Chapter 3 for the channel mobility. There, a complex algebra was required to find a suitable solution. A key factor was the boundary condition for the momentum transfer at the interface. There is experimental evidence that the impact ionization coefficient near the interface is reduced compared with the bulk (Thurgate and Chan 1985, Slotboom *et al.* 1987). This points in the direction of a modified distribution function very close to the surface. There are arguments in favor of this. In the first place, the surface provides an additional scattering channel due to coupling between surface excitations and bulk carriers in the close vicinity of the interface. This will eventually lead to a reduction of the population of high energetic carriers. Secondly, the density, and henceforth the distribution function will be reduced due to the boundary condition of the wave function. To account for these effects, we propose the boundary condition

$f_0(y, x = 0, \varepsilon) \approx 0$ for the high energetic states. Of course, we do not suggest that the distribution function vanishes at the surface. This rather implies that the population of high energetic states is drastically reduced at the surface as compared with the bulk. A closed solution is still possible for this boundary condition (Morse and Feshbach 1953). On the basis of the theory of the Green's functions for the Helmholtz equation (4.128), we have to replace the kernel of the integral in Eq. (4.142) by

$$K_0(|k||R - R'|) \rightarrow K_0(|k| |R - R'|_-) - K_0(|k| |R - R'|_+) \quad (4.143)$$

Here we have

$$|R - R'|_\pm = [(y - y')^2 + (x \pm x')^2]^{1/2} \quad (4.144)$$

Up to this point, we proposed arguments as to how we might generalize our findings to match the situation in a real device. One important point was left out in this discussion. To find the population of the high energetic states in the previous section, we concentrated primarily on finding the correct field dependence or shape of the distribution function. Its absolute value is related to its normalization. To find the correct normalization, we have to know the complete distribution function. Because we have investigated only energies that were large compared with the average carrier energy, they contribute insignificantly to the normalization integral. The correct normalization is therefore beyond the scope of our work. The correctly normalized distribution function could be represented, for example, by

$$f_0(R, \varepsilon) \approx n(R)\gamma(R, \varepsilon) f_0^{(1)}(R, \varepsilon); \quad \varepsilon \gg \varepsilon_{av} = \frac{3}{2} k_B T_c(R) \quad (4.145)$$

Here $\gamma(R, \varepsilon)$ is an unknown function of dimension ε^{-1}. In the case of a Maxwellian-shaped distribution function, $\gamma(R, \varepsilon)$ would be

$$\gamma(R, \varepsilon) = \frac{2}{\sqrt{\pi}} \left(\frac{\varepsilon}{k_B T_c(R)} \right)^{1/2} \frac{1}{k_B T_c(R)} \quad (4.146)$$

A direct calculation of the distribution function in the way done in Monte Carlo calculations, for instance, shows that the distribution function in high electric fields is not Maxwellian. However, the high energy tail is always Maxwellian for the effective mass approximation, as demonstrated in the previous section. In Fig. 44 we show the shape of the distribution function as obtained from a Monte Carlo calculation for two different electric fields and compare it with a Maxwellian which gives the same average carrier energy.

The auxiliary function $\gamma(R, \varepsilon)$ will also for the exact distribution function vary on a slower scale than the approximative exponentially decreasing tail. Therefore it is justified to replace $\gamma(R, \varepsilon)$ by an appropriately chosen constant γ_0 as we have demonstrated already above.

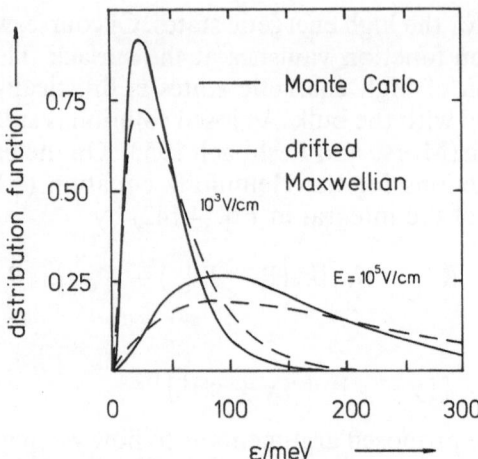

Fig. 44 Comparison of Monte Carlo calculation (solid line) and drifted Maxwellian (dashed line) for different field strength

4.4.1 Impact Ionization Generation Rate

To obtain the generation rate due to impact ionization, we now combine the results from the previous two sections. If we neglect the anisotropy corrections in the scattering rate τ_{imp}, it becomes only a function of energy, as we showed in the previous section. In this case, the generation rate is directly related to f_0. We find the following from Eq. (4.12) for the stationary state

$$G_n(R) = \int d\varepsilon \, \frac{1}{\tau_{imp}(\varepsilon)} f_0(R, \varepsilon) \tag{4.151}$$

As long as the high-energy temperature $qE\lambda/s$ is small compared to the threshold energy ε_{imp}, the integral is, to a good approximation, represented in the threshold approximation as already shown in Section 2 of this chapter (see discussion after Eq. (4.37)).

$$G_n(R) = n(R) \frac{1}{\tau_{imp}^0(R)} f_0^{(1)}(R, \varepsilon_{imp}) \tag{4.152}$$

The effective scattering rate $1/\tau_{imp}^0$ is related to the following integral

$$\frac{1}{\tau_{imp}^0(R)} = \int_{\varepsilon_{imp}}^{\infty} d\varepsilon \gamma(R, \varepsilon) \frac{1}{\tau_{imp}(\varepsilon)} \frac{f_0^{(1)}(R, \varepsilon)}{f_0^{(1)}(R, \varepsilon_{imp})} \tag{4.153}$$

and will vary slowly compared to the exponential variation of f_0. Impact ionization is important if the electric field is sufficiently high $qE\lambda \gg k_B T$, so that the carriers will move with saturation velocity. With the effective mean free path for impact ionization $\lambda_{\mathrm{imp}}^0 = \tau_{\mathrm{imp}}^0 v_{\mathrm{sat}}$ we can reduce Eq. (4.152) to the well-established expression for the generation rate

$$G_n(R) = |v_n(R)|\alpha_n(R) \tag{4.154}$$

with the impact ionization coefficient

$$\alpha_n(R) = \frac{1}{\lambda_{\mathrm{imp}}^0} f_0^{(1)}(R, \varepsilon_{\mathrm{imp}}) \tag{4.155}$$

Here we have replaced the slowly varying function $\lambda_{\mathrm{imp}}^0(R)$ by a constant that is adjusted to the experimental situation. For an order of magnitude estimate, we find $\lambda_{\mathrm{imp}}^0 = 10^3 \ldots 10^2$ nm, which corresponds to our earlier estimate that the effective scattering strength for impact ionization is one to two orders of magnitude smaller than the typical phonon scattering rate. Following the same arguments as above, we can repeat the same analysis for the generation rate in the valence band. Here, two holes collide and generate an electron–hole pair. By adding both conduction and valence band contributions of the generation rate, we obtain the continuity equation for the stationary case

$$\nabla_R v_n(R) = |v_n(R)|\alpha_n(R) - |v_p(R)|\alpha_p(R) \tag{4.156}$$

Because the particles that appear in the conduction band must vanish in the valence band, the hole continuity equation reads

$$\nabla_R v_p(R) = -|v_n(R)|\alpha_n(R) + |v_p(R)|\alpha_p(R) \tag{4.157}$$

This completes the discussion of the generation rate due to impact ionization.

4.4.2 Gate Oxide Injection

To calculate the number of carriers that can enter the gate oxide, we have to sample all carriers with an energy greater than the barrier height which hit the interface with a velocity component perpendicular to it. The interface is located at $x = 0$ and the barrier height is $\varepsilon_{\mathrm{inj}}$. The main quantity we are interested in is the number of particles that cross the interface per unit area and time. For this current density, we can write

$$v_{\mathrm{inj},x}(R) = \sum_k v_x(k) f(R, k); \quad \varepsilon_k > \varepsilon_{\mathrm{inj}}|_{x=0} \tag{4.158a}$$

Converting the k-summation into an energy integration, we can relate the injected current density to the distribution function $f_{1x}(R, \varepsilon)$ that was defined in Eq. (4.66)

$$v_{\mathrm{inj},x}(R) = \int_{\varepsilon_{\mathrm{inj}}}^{\infty} d\varepsilon\, f_{1x}(R, \varepsilon)|_{x=0} \tag{4.158b}$$

With Eq. (4.70), f_1 is expressed in terms of f_0. Utilizing the boundary condition on $f_0|_{x=0} = 0$ we then get

$$v_{inj,\,x}(R) = -\frac{1}{3}\lambda_{inj} \cdot \int_{\varepsilon_{inj}}^{\infty} d\varepsilon \, v(\varepsilon)\frac{\partial}{\partial x} f_0(R, \varepsilon)|_{x=0} \tag{4.159}$$

which finally gives us, with Eq. (4.145)

$$v_{inj,\,x}(R) = -\frac{1}{3}\lambda_{inj} n(R) \cdot \int_{\varepsilon_{inj}}^{\infty} d\varepsilon \gamma(R, \varepsilon)v(\varepsilon)\frac{\partial}{\partial x} f_0^{(1)}(R, \varepsilon)|_{x=0} \tag{4.160}$$

The subscript inj indicates that the mean free path has to be taken at $\varepsilon = \varepsilon_{inj}$. Equation (4.160) is conveniently evaluated in a similar way to the threshold approximation employed earlier.

$$v_{inj,x}(R) = -\frac{1}{3} v_{inj}^0 n(R) \lambda_{inj} \frac{\partial}{\partial x} f_0^{(1)}(R, \varepsilon_{inj})|_{x=0} \tag{4.161}$$

Here we have introduced an effective velocity v_{inj}^0, which is a proper average of $v(\varepsilon)$. From Eq. (4.161) we can learn that the injection process is related to the diffusion of high energy carriers towards the surface. For an order of magnitude estimate for v_{inj}, we can use $v_{inj} \approx v_{th}\sqrt{(\varepsilon_{inj}/k_BT)}$; here $v_{th} \approx 10^7$ cm/s is the thermal velocity of electrons in silicon at room temperature. With a barrier height of approximately 3 eV we get $v_{inj} \approx 10^8$ cm/s. A somewhat lower value is more likely because of the larger effective mass in the higher-energy states.

A shortcoming of the effective mass approximation is that we must adjust the mean free paths to the corresponding energy range. To find the solution of Eq. (4.55), we neglected the contribution of the impact ionization process. This is justified as long as energies near the threshold ε_{imp} are considered, as was discussed in the first section of this chapter (see discussion after Eq. (4.35)). For the injection current, we consider an energy range far above the threshold energy and the impact ionization process contributes significantly to the scattering in the system. It must therefore be included in the solution of Eq. (4.55). This is accomplished by adding the last term of the right-hand side of Eq. (4.4) to the right-hand side of Eq. (4.55). The calculation is straightforward and gives a slightly modified equation for s.

$$s^2 = 3\frac{\lambda}{\lambda_{opt}}\left[1 - \frac{\cosh\left(\dfrac{\omega_{opt}}{2k_BT} - \dfrac{\omega_{opt}}{qE\lambda}s\right)}{\cosh\left(\dfrac{\omega_{opt}}{2k_BT}\right)}\right] + \frac{\lambda_{imp}}{\lambda_{opt}} \tag{4.162}$$

Here λ_{imp} is the mean free path for the impact ionization process at an energy ε_{inj}. It is of the order of the phonon mean free path and should not be confused with the effective scattering rate λ_{imp}^0 in which only carriers close to the threshold energy contribute. With Eqs. (4.155) and (4.161), we have related both the impact ionization coefficient and the current density injected into the gate oxide to the same distribution function. In the next chapter we will use these models to study the degradation phenomena of a MOSFET.

References

Antoncik E., Landsberg P. T. (1963): Proc. Phys. Soc. *82*, 337.
Baraff G. (1962): Phys. Rev *128*, 2507.
Beattie A. R., Landsberg P. T. (1959): Proc. Roy. Soc. *A249*, 16.
Chynoweth A. G. (1959): Phys. Rev. *109*, 1537.
Conwell E. M. (1967): High Field Transport in Semiconductors. In: Solid State Physics Suppl. 9, ed. F. Seitz, D. Thornbull, H. Ehrenreich. Academic Press, New York.
Davydov B. I. (1937). JETP 7, 1069.
Druyvesteyn M. J., Penning F. M. (1940): Rev. Modern Phys. *12*, 87.
Fukuma M., Lui W. W. (1987): IEEE Electr. Device Lett. *EDL8*, 214.
Hall R. N. (1952): Phys. Rev. *87*, 387.
Haug A. (1972): Theoretical Solid State Physics Vol. II. International Series of Monographs in Natural Philosophy Vol. 36. Pergamon Press, Oxford.
Hänsch W., Schwerin A. (1989): J. Appl. Phys. *66*, 1435.
Henning A. K., Chan N. C., Watt J. T., Plummer J. (1987): IEEE Trans. Electr. Devices *ED34*, 64.
Jacoboni C., Lugli P. (1989): The Monte Carlo Method for Semiconductor Device Simulation. In Computational Microelectronics (ed. S. Selberherr). Springer, Wien.
Kane E. O. (1967): Phys. Rev. *159*, 624.
Keldysh L. V. (1959): JETP *37*, 713 [Sov. Phys. JETP *10*, 509 (1960)].
Keldysh L. V. (1965): JETP *48*, 1692 [Sov. Phys. JETP *21*, 1135].
Kunert R., Werner C., Schütz A. (1985): IEEE Trans. Electr. Devices *ED32*, 1057.
Lochmann W. (1977): Physica Status Solidi *40*, 285.
Meinerzhagen B., Engl W. L. (1988): IEEE Trans. Electr. Devices *ED35*, 689.
Morse P. M., Feshbach H. (1953): Methods of Theoretical Physics. McGraw-Hill, New York.
Ridley B. K. (1983): J. Phys. *C16*, 3373.
Scharoch P., Abram R. A. (1988): Semicond. SCI. Technol. *3*, 973.
Selberherr S. (1984): Analysis and Simulation of Semiconductor Devices. Springer, Wien.
Shockley W. (1958): Solid State Electr. *2*, 1537.
Shockley W., Read W. T. (1952): Phys. Rev. *87*, 835.
Slotboom J. W., Streuker G., Davis G., Hartog P. (1987): IEDM 87, Technical Digest.
Takashima M. (1981): Phys. Rev. *B23*, 6625.
Thurgate T., Chan N. (1985): IEEE Trans. Electr. Devices *ED32*, 400.
Wolff P. A. (1954): Phys. Rev. *95*, 1415.

Degradation 5

5.1 Introduction

Modeling aims at two goals: first it provides a tool for an engineer to optimize a device for an existing technology; secondly it should help to show the electrical behavior of a device which is not yet available. Both goals make different demands on the tool. While accuracy and efficiency are needed in the former, the latter focuses on qualitative features and should be very flexible. In the previous chapters, we discussed the physics of carrier transport in a device, using the MOSFET as a guiding example. The work we have done so far provides the physics necessary to determine the internal field and carrier distributions of the device and henceforth also its terminal currents. Casting this physics into numerical code form will enable the engineer to do his job. Provided the correct doping profiles are known, the engineer can tailor a device to suit his needs. In the first place, this would involve the question: how should I implement the process flow so that the device has the specifications required for its functioning in an electric circuit? Usually the threshold voltage, saturation currents at operating bias, subthreshold swing, transconductance, etc. are specified by the circuit environment. Many realizations satisfy the given specifications. We can single out variations that are more compatible with the process flow than others because the transistor is just one building block in the large system that constitutes the chip. To find the best choice among the remaining variations, the long-term stability of the device is considered: once the ideal device satisfies the circuit specifications, how likely is it that these might change during operation? This drift of device-specific parameters is called degradation. In a device optimization cycle the engineer tries to find the device that shows minimum degradation.

There are many reasons why a device may degrade, but the most important is hot-carrier degradation (Hänsch *et al.* 1988). Hot-carrier degradation includes all degradation effects related to the very high electric field in the device that enables carriers to overcome the potential barrier between Si and the gate insulator SiO_2. Once the carriers have entered the gate oxide, they may get trapped and build up charge that interferes with the controlling gate field. Because the maximum electric field in the device is the primary cause of hot carrier degradation, efforts have been directed to find device structures that reduce the electric field. The most successful example to reduce the lateral electric field in a MOSFET device is to replace the abrupt *pn*-junction on source and drain by a graded one. This structure is called lightly doped drain or LDD and is now standard for all submicron NMOS devices. We show the drain side of such a device in Fig. 45. The clue is to use two implants for the formation of the source and drain region instead of a single one.

The first implant labeled n^- is a comparatively low dose and low energy phosphorus implant which provides a properly graded *pn*-junction. The second high dose and high energy arsenic implant n^+ serves as the proper

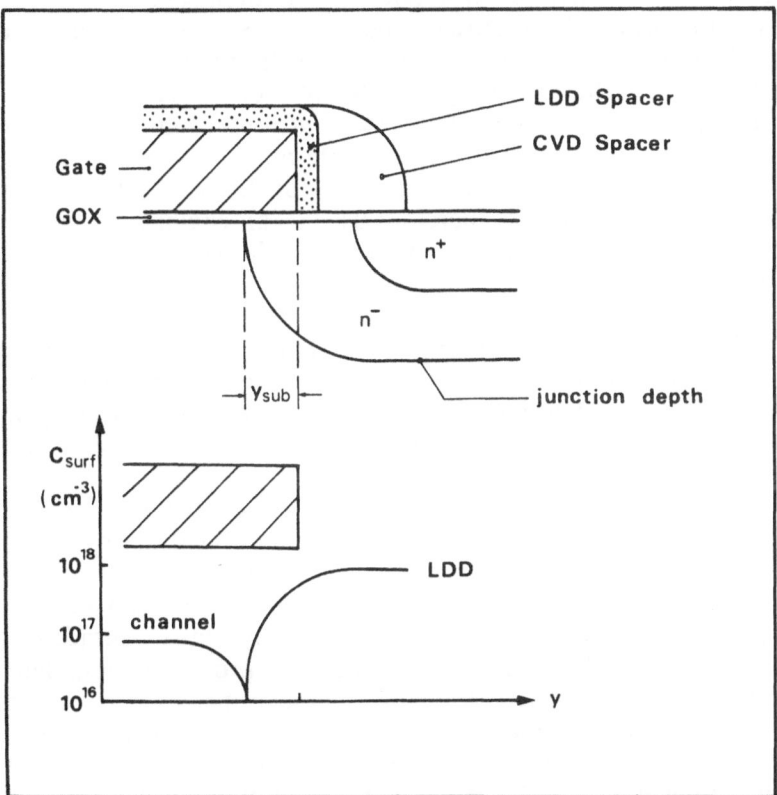

Fig. 45 Construction of the LDD source/drain profile. The reliability and device performance depend crucially on the choice of parameters, as there are: LDD dose and energy (n^-), HDD dose and energy (n^+), LDD and CVD (HDD) spacer to mask implantation, junction depth and sub-diffusion (y_{sub})

contact region as in the conventional device. It is separated from the first implant by a spacer large enough to prevent the formation of the channel *pn*-junction after diffusion. To find the optimal device structure requires a fine tuning of all controllable parameters in this structure (Orlowski *et al.* 1989a, b).

The basic dilemma we face is that under operating conditions the specified device lifetime is approximately ten years, owing to hot carrier drift. For a technology under development, we cannot wait this long to decide whether the device structure is reliable or not. So devices are electrically stressed to accelerate the degradation process. Under the assumption that only the total number of oxide charges is important, these results then allow an extrapolation to operating conditions where degradation happens at a much lower speed.

A closer investigation shows that not only the maximum electric field but also its position relative to the carrier distribution, especially the current density, is important. A measure for the overlap of current density and field is the substrate current which is related directly to the impact ionization generation term in the continuity equation (4.156). Because the electric field enters exponentially, the generation rate contributes significantly only at a very local level. In contrast to the other terminal currents, therefore, the substrate current can serve as an indicator for a local quantity: the maximum electric field. It is therefore used as an engineering guideline to characterize the hot carrier degradation of a device (Takeda and Suzuki 1983, Hu *et al.* 1985). There is, however, experimental evidence that devices with a larger maximum field or substrate current show less degradation than those with lower substrate current. This indicates that the field or substrate current criterion is not sufficient. After all, the change of the transistor characteristics due to hot carrier degradation is caused primarily by the amount and position of the damage in the oxide. The electric field is only the cause and not the actual effect.

In this chapter, we will discuss how modeling can contribute to the problem of hot-carrier degradation beyond the maximum field approach. To this end, we can separate two different approaches which are distinguished in their level of sophistication. In the next section, we will study the first approach which is best characterized by analyzing a degraded sample (Schwerin *et al.* 1987). After the degradation process is completed, the experiment shows characteristic changes in the properties of the device such as a shift in threshold voltage or a decrease in transconductance. Modeling can now be used to determine the kind of damage responsible for these changes. We will discuss the prominent degradation mechanisms and their influence on the device characteristics. We must consider different kinds of traps that have an influence on both the potential distribution in the device and the scattering rate of the mobility due to Coulomb centers near the Si/SiO_2 interface. It is important in this approach to include as many device

features as possible and it is almost certainly fatal to look at only one
feature and draw conclusions from it. In this regard, it can be considered as
a jigsaw puzzle where different pieces are joined to find the total picture. We
will discuss different characterization schemes for degraded devices which
include both conventional electrical and charge-pumping measurements
and their modeling. The second and more ambitious approach is then
presented in section three: modeling the degradation process itself. The
second step is impossible without the first because the complexity of the
problem requires a detailed analysis to single out its prominent features. To
calculate the degradation process, we have to solve two problems. (1) How
do the carriers enter the gate oxide region and how do we determine their
distribution there? and (2) How do we describe the oxide damage? Both
constitute difficult problems which cannot be solved rigorously. More than
in the previous chapters, the lack of precisely obtainable results will literally
force us to model the essential physics. Therefore the contents of Section 3
of this chapter in particular have to be understood more as an outline of
an approach than a finished treatment. There is certainly considerable
room for improvement here, but the material gathered so far is strongly
encouraging.

5.2 Analyzing a Degraded MOSFET

Various methods are available to determine the damage caused by carriers
injected into the gate oxide. In principle, we can distinguish between
investigations performed on MOS capacitors and those on MOSFETs. The
former usually inject carriers created by driving the device into deep
depletion. Carriers from the substrate are then accelerated towards the
Si/SiO_2 interface and eventually gain enough energy in the transverse field
to overcome the barrier at the interface. The injection process is homogeneously distributed over the SiO_2/Si interface. These experiments are therefore known as homogeneous injection experiments. Another possibility for
a homogeneous injection experiment is to inject minority carriers into the
substrate of a MOSFET under strong inversion. Here the existence of the
inversion layer pins the surface potential so that, the fields in the gate oxide
and Si-substrate are easily controlled, therefore the situation in the MOS
structure is well-defined. A second class of experiments is performed on
MOSFETs under operating conditions. Here, of course, the device bias is
chosen so that there is a significant probability of injection. The carriers
injected into the gate oxide are now provided by the channel carriers that
are heated up in the lateral source-drain field of the device. Typical bias
conditions are maximum substrate (n-MOSFET) or maximum gate current
(p-MOSFET). For these conditions, the electric field along the channel is

very inhomogeneous and reaches its peak value near the drain. In this region, we also have maximum injection of carriers into the gate oxide, which results in a locally confined distribution of injected carriers. To distinguish this sort of measurement from the first category, it is sometimes called channel hot-electron injection. In contrast to the former, the latter's actual field and carrier distribution is very complex, two dimensional and determined not only by choosing the bias conditions but also by an accurate knowledge of the doping profile in the drain region of the device. Because there is at present no way of verifying the correct shape of the doping profile in this region experimentally, this leaves an unresolved uncertainty for calculating the field. However, some indirect methods are available to calibrate the accuracy of the doping profiles obtained from process modeling which are based on measuring the one-dimensional doping profile against its depth into the substrate to verify the junction depth and investigating the threshold voltage dependence on the gate length to verify the sub-diffusion length. A third possibility, which we will merely mention, is the controlled injection of carriers with the aid of a light source. All three methods reveal information about how carriers injected into the oxide can cause damage to the oxide. However, it is not so easy to draw conclusions from one experimental situation about the oxide damage that applies in exactly the same way to another experiment. Although many common features are observed in homogeneous injection experiments and channel hot-electron experiments, it is not generally true that results from the former apply automatically to the latter. It is in particular true that the homogeneous stress experiments probe the better part of the large area in the gate oxide. Channel hot-electron injection experiments are more sensitive to the quality of the gate oxide at the edges or even in the spacer oxide region of an LDD device. Another handicap in degradation studies is that the results are very sensitive to the technology used. In other words, the history of the process to which the investigated device has been subject can have a dramatic effect on its degradation behavior. In this respect, a quantitative analysis of device degradation leaves a lot of open questions and general conclusions drawn from one sample must be handled with care.

Although the situation is very complex and depends on many unknown factors, there are several generally accepted features which make modeling possible. Oxide damage is caused either by fixed charges Q_{ox} trapped in the body of the gate oxide in locally bound oxide traps N_{ox}, or by interface charges Q_{it} trapped in interface states N_{it}. The most significant difference between the former and the latter is that the interface states communicate with the bulk Si while the oxide traps N_{ox} do not. This communication occurs via the quasi-Fermi level ε_F at the interface determined by the potential and density distributions in the Si material at the SiO_2/Si interface. There is some confusion as to how to classify the interface states. In the general non-equilibrium situation of a MOSFET, the quasi-Fermi levels of electrons ε_F^n and that of the holes ε_F^p are not the same. To what Fermi level do the interface states respond and how? There are many ways of answering this question. We found that the following specification is sufficiently general to allow interpretation of most of the experimental data: there are acceptor and donor-like interface states. Acceptor-like interface states respond to the electron's quasi-Fermi level so that states below ε_F^n are occupied with negative charge and states above ε_F^n are neutral. Donor-type interface states communicate with the system of holes such that states above ε_F^p are occupied with positive charge and states below ε_F^p are neutral.

From Fig. 46 we can see that the density of states $D_{it}(\varepsilon)$ of the interface states is important for evaluating the effective charge stored on them. Due to the possible positions of the quasi-Fermi levels, the shape of the distribution function $D_{it}(\varepsilon)$ is important for the energy range between Si valence band and conduction band edge. Its experimental determination will be discussed later. Although the oxide traps also have an energetic distribution in the SiO$_2$ bandgap, device degradation is not sensitive to its special shape because there is no dynamic feedback to the Si material. It is therefore sufficient to consider the trapped charges as an effective fixed charge distributed spatially in the gate oxide. Because of this different coupling to

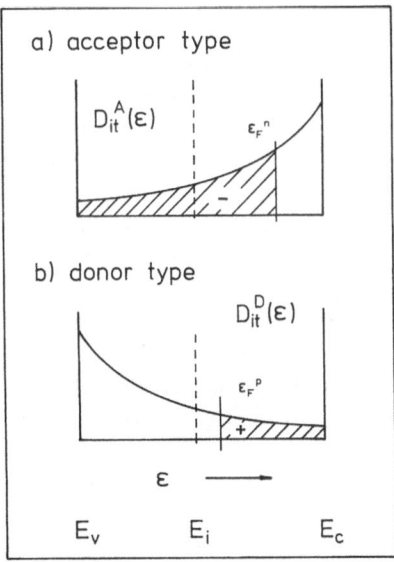

Fig. 46 Definition of interface states: (a) acceptor type interface states are coupled to the quasi Fermi potential of the electrons ε_F^n; (b) donor type interface states are coupled to the quasi Fermi potential of the holes ε_F^p

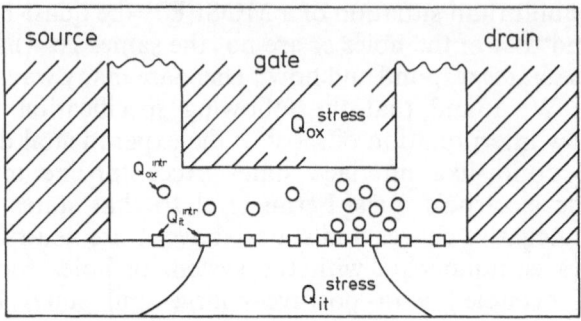

Fig. 47 Distribution of traps in the gate oxide. Open circles—oxide traps; open squares—interface states. There is a process-dependent intrinsic background of oxide traps and interface states prior to the stress induced damage

the carriers in the Si, the oxide traps are often called slow states and the interface states fast states. In Fig. 47, we summarize the trap situation in the gate oxide of a MOSFET.

In the bulk, we have slow traps that can be occupied by negative or positive charges Q_{ox}^- and Q_{ox}^+. At the interface, there are fast acceptor and donor-like interface states filled with negative and positive charge up to the corresponding quasi-Fermi level $Q_{it}^-(\varepsilon_F^n)$ and $Q_{it}^+(\varepsilon_F^p)$, respectively. We will not go into the physics of the traps as such but will stay on this descriptive level. It is of course clear that there is a near-interface transition region where the distinction between fast and slow interface states probably becomes questionable because the oxide traps will also start to communicate with the Si material here. We will avoid this complication for now and strictly adopt the above classification of traps. To answer the question of the physical, or even better, the chemical nature of the traps is a very complicated task which has been pursued for many years. Some concepts are generally accepted whereas others are in permanent flux (Hänsch and Weber 1989). For more details, the interested reader is referred to the original literature on oxide traps and interface states (Di Maria 1978, De Keersmaeker 1983, Grunthener and Grunthener 1986).

Setting up the trap scenario in the oxide with the above model, we can now examine its realization for modeling purposes. It is known from capacity and charge-pumping measurements that the density of states $D_{it}(\varepsilon)$ is not a constant over the Si bandgap. It is better described by an U-shape with maxima at the band edges. It is commonly assumed that the states in the upper half of the bandgap are acceptor-like while those in the lower half are of donor type. If we allow for a constant background over the Si bandgap, the density of states for acceptor-like interface states is given by

$$D_{it}^A(\varepsilon) = D_{max}^A(2\varepsilon/E_G)^\alpha + D_0^A, \quad \varepsilon > 0 \tag{5.1}$$

and for the donor type is

$$D_{it}^D(\varepsilon) = D_{max}^D(-2\varepsilon/E_G)^\beta + D_0^B, \quad \varepsilon > 0 \tag{5.2}$$

Here, the origin of the energy is the mid-gap level. The exponents α and β range from zero to three. Typical values for D_{max}^A and D_{max}^D are $-5\cdot10^{12}/cm^2\,eV$ and $5\cdot10^{12}/cm^2\,eV$, respectively for stressed samples. For a virgin device, these numbers are considerably smaller ($5\cdot10^9/cm^2\,eV$). We obtain the number of charges per unit area in these states by simple integration

$$Q_{it}^-(y,\varepsilon_F^n) = q\Gamma^A(y)\int_{E_v}^{\varepsilon_F^n} d\varepsilon D_{it}^A(\varepsilon) \tag{5.3}$$

$$Q_{it}^+(y,\varepsilon_F^p) = q\Gamma^D(y)\int_{\varepsilon_F^p}^{E_c} d\varepsilon D_{it}^D(\varepsilon) \tag{5.4}$$

Here, we have multiplied the integral by the function $\Gamma(y)$ that models the spatial extent of the damaged region. A reasonable choice for $\Gamma(y)$ is a Gaussian with its maximum at y_0 and a half-width σ. The former corresponds to the location of the damage and the latter to its spread. A typical value for σ is 100 nm. With Eqs. (5.3) and (5.4) we can investigate the influence of locally confined interface state charges on the MOSFET characteristics. We must now specify how to handle the oxide charges Q_{ox}. A rigorous approach would be to place a two-dimensional charge distribution into the oxide region that simulates the oxide charges created during the stress experiment. Because we are not at this stage interested in the correct distribution of the charges in the oxide itself but rather in their influence on the MOSFET characteristic, it is reasonable to replace the oxide charges by an effective charge density at the surface. This then means that we simply have to add another contribution to Eqs. (5.3) and (5.4) for the net negative and positive charge

$$Q^-(y) = Q_{it}^-(y,\varepsilon_F^n) + \Gamma_{ox}^-(y)Q_{ox}^{0-} \tag{5.5}$$

$$Q^+(y) = Q_{it}^+(y,\varepsilon_F^p) + \Gamma_{ox}^+(y)Q_{ox}^{0+} \tag{5.6}$$

Here $Q_{ox}^{0\pm}$ is the maximum of the effective oxide charges and $\Gamma_{ox}^\pm$ describes their distribution along the interface. Equations (5.5) and (5.6) are sufficient to investigate the influence of stress-induced charges on the MOSFET characteristics. Many parameters have to be adjusted for a complete description. In the next section, when we study the dynamic build-up of the traps in the oxide, we will be able to calculate the distribution of the traps in the oxide and interface directly. The position, spread, and number of charges created in a stress experiment are not determined there, but are obtained directly by calculation from the stress experiment. For now, however, we have to consider them as parameters properly chosen to fit the experimental stress data. Depending on the device and stress condition, some parameters are less important than others to fit the experimental results. The apparently large number of parameters therefore reduces to a very small number when the device data is considered. Before we do this, we have to complete our discussion on the possible effects of the oxide damage on the device. With Eqs. (5.5) and (5.6), we have a formulation for an effective interface charge that will alter the potential distribution in the device. It will modify the boundary condition of the Poisson equation at the SiO_2/Si interface. For the oxide charges, this is a straightforward problem. The charge in the interface states depends strongly on the properties of the device itself and it is mandatory to calculate them in a self-consistent cycle. In this cycle, the interface state charge and the quasi-Fermi energy have to be calculated self-consistently. In addition to their influence on the potential distribution, the mobility will also be affected due to the extra Coulomb centers that provide a new scattering channel for the channel electrons or holes. How to incorporate this new scattering channel depends on the mobility model, and in particular on whether the influence of the surface scattering can be explicitly separated. In the mobility model discussed in Chapter 3, we found with Eq. (3.102) a formulation in which the surface mobility appears explicitly. The new scattering channel due to the oxide charges and interface states can be simply taken into account by

$$\frac{1}{\mu_{surf}} \rightarrow \frac{1}{\mu_{surf}} + Q_{tot}(y)\gamma \tag{5.7}$$

$$Q_{tot}(y) = |Q^+(y)| + |Q^-(y)| \tag{5.8}$$

If the center of mass of the oxide charge Q_{ox} is too far away from the interface, its contribution to Q_{tot} must be reduced as it will no longer be an effective scatterer. The constant γ can be determined from the well-defined mobility reduction observed for devices with a known homogeneous oxide charge layer at the SiO_2/Si interface (Sun and Plummer 1980). Here, the potential and mobility effect are decoupled. The potential effect gives a threshold voltage shift which in turn can be used to determine the oxide charge density. The relative change of the drain current can be related directly to γ (compare also the following discussion and Fig. 48). From such an analysis, γ should be $3 \cdot 10^4 \, Vs/c$ and is therefore not an adjustable parameter like those in Eqs. (5.5) and (5.6).

5.2.1 Electrical Characterization

Before we turn to the experimental data, we will discuss the principal features seen in degraded devices and how they are related to the different kinds of disturbance due to the oxide damage. We will consider the effect of oxide charges in NMOS (n-channel) devices first. A layer of homogeneously distributed charge at position x_0, $0 < x_0 < t_{ox}$, in the gate oxide will renormalize the effect of the gate voltage V_G according to

$$\Delta V_G = \frac{Q_{ox}}{\kappa} \cdot \frac{t_{ox} - x_0}{t_{ox}} \cdot t_{ox} \equiv \frac{Q_{ox}^0}{\kappa} \cdot t_{ox} \qquad (5.9)$$

It has exactly the same effect as an interface charge Q_{ox}^0, which corresponds to the approximation of oxide charges made above. For positive charges, we will observe a shift of the threshold voltage to smaller gate voltages and for negative charges a shift to larger gate voltages as shown in Fig. 48.

If we neglect the effect of the mobility, the current characteristics would only experience a parallel shift of $\pm \Delta V_G$ on the V_G axis. If the charge sheet is close enough to the interface, it would contribute to the surface scattering and the current would also decrease somewhat. In the situation that is produced by hot-electron stress, the damage is not homogeneously distributed over the SiO_2/Si interface but confined to a small region near the drain. In the case of a negative charge, there would be a larger threshold voltage for this channel region and the channel could not therefore have been built up unless this threshold was reached. As a result, there is a threshold shift to higher gate voltages for the device. In the case of positive charges, the situation is different. Locally, the threshold voltage is reduced. But to build up the channel, the rest of the device must also form a channel. Therefore there would be no change in the threshold voltage of the device. On the other hand, the positive charge has the effect of reducing the effective channel length of the device, which can in some special cases lead to a decrease of the threshold voltage. It will, however, always give an increased drain current which is proportional to $1/L_{eff}$. This effect can be counteracted by a mobility reduction. The mobility effect will always lead

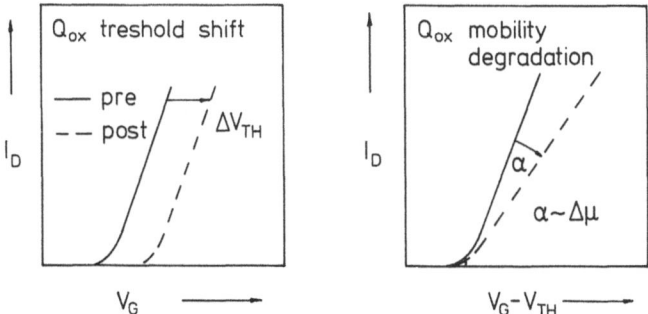

Fig. 48 The influence of a sheet of oxide charge on the MOSFET characteristics. (a) Threshold shift due to modification of the electric potential and (b) mobility degradation due to extra scattering centers located at the interface

to a reduction of the drain current. As already mentioned above, the efficiency of the mobility reduction will depend on the distribution of the oxide charges. Localized negative charge is also provided by acceptor-like interface states. The amount of charge is here proportional to the position of the Fermi level ε_F^n of the electrons in the band gap. It is near mid-gap close to inversion and the charge in the interface state is therefore small. If the device is driven into inversion, the Fermi level will move towards the conduction band edge and the charge in the interface states will increase. The effect on the drain current is shown in Fig. 49. There will be no clear threshold voltage shift as is observed in the case of negative oxide charges but rather a roll-off of the current characteristics which is attributed to a gate voltage dependent threshold voltage shift.

In addition, the build-up of interface charges will also have an influence on the mobility degradation, which will also show a roll-off of the current characteristics. Simple electric current measurements always show a combination of potential effect and mobility degradation. It is therefore very difficult to distinguish between potential and mobility effects on the basis of simple current measurements. As a measure for the maximum electric field, the substrate current is less sensitive to mobility variation than to potential deformations. It can therefore serve as an indicator for the potential deformation due to oxide charges or interface states. A similar situation is encountered in the subthreshold region of the device, where the drain current depends very sensitively on the surface potential and is less sensitive to the detailed features of the mobility.

In Fig. 50 we show the subthreshold characteristic of an n- and p-channel device. The NMOS device has a 1.7 μm effective channel length and an oxide thickness of 42 nm. It was stressed at a drain voltage $V_D = 8$ and a gate voltage of $V_G = 3$ for $5 \cdot 10^4$ s. The p-channel device has an effective channel length of 0.6 μm and a gate oxide of 18 nm. It was stressed at

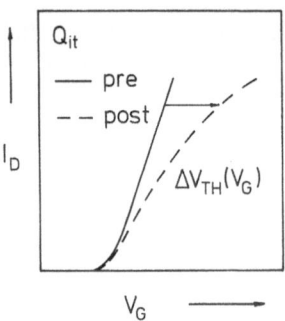

Fig. 49 The influence of interface states on the MOSFET characteristic. Both potential and mobility modification contribute. The charge in the interface state depends on the position of the quasi Fermi level and therefore creates a gate voltage dependent threshold shift

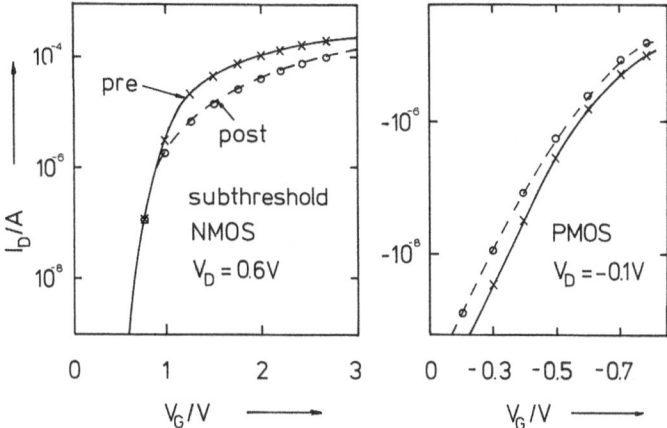

Fig. 50 Measured and simulated subthreshold characteristic for *n*- and *p*-channel MOSFET after stress. Crosses—measured pre-stress data, open circles—measured post-stress data. (a) NMOS degradation due to interface states; (b) PMOS degradation due to negative oxide charge

$V_D = -8\,V$ and $V_G = -2\,V$ for $1.5 \cdot 10^4$ s. Before and after stressing, the devices were characterized with $V_D = 0.6$ and $V_D = -0.1\,V$ for NMOS and PMOS (*p*-channel), respectively. We have also included in that diagram the experimental data.

Different features are observed for the *n*- and *p*-channel devices. For the NMOS device, there is no deviation of post- and pre-stress current for very low gate voltages. Once the threshold voltage is exceeded, however, there is an increasing discrepancy. The PMOS device shows a pronounced difference between post- and pre-stress current, even for gate voltages below threshold. Because we observe a threshold shift towards lower gate voltages, it indicates negative oxide charges. The calculated post-stress values were obtained with $Q_{ox}^{-0} = -3 \cdot 10^{12} q/cm^2$ located at the drain side of the device. The general features of the NMOS device cannot be obtained by using negative oxide charges. The job is best done by acceptor-like interface charges with $D_{max}^A = -2 \cdot 10^{12}/eVcm^2$, $D_0^A = 0$ and $\alpha = 1$. The position x_0 of the damage is obtained by investigating the asymmetry in the output characteristic by swapping the device source and drain, as shown in Fig. 51. This asymmetry for both NMOS and PMOS devices is shown in Fig. 52. For both types of device, the damage is located near the *pn*-junctions at the drain side in the stress configuration. The strong asymmetry of the NMOS device is another indication for the importance of the interface states. In the reverse mode, the damage is located at the source side of the channel. Here, the inversion channel is always formed so that the quasi-Fermi level is high and a maximum number of interface states contain charge. In the normal

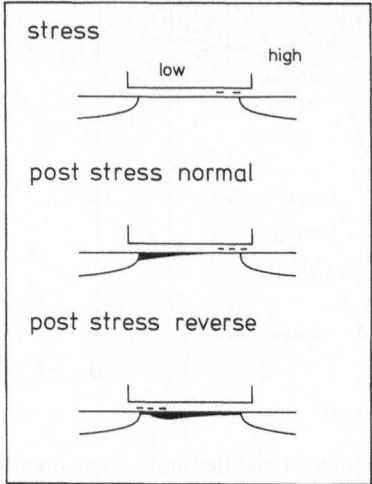

Fig. 51 Characterization of a stressed device. During the stress a low voltage is applied at the gate and a high voltage at the source. Damage is created near the drain *pn*-junction. Post stress normal characterization leaves source and drain in stress configuration. Post stress reverse characterization swaps source and drain

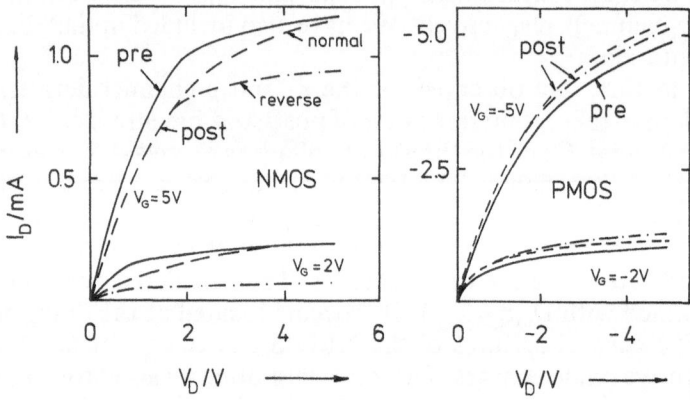

Fig. 52 Pentode regime of *n*- and *p*-channel device in normal and reverse mode. Simulated results assuming interface states for NMOS and negative oxide charges for PMOS

mode, the pinch-off region is formed for high drain voltages. In this region, there is no inversion channel at the surface in the drain region. The quasi-Fermi level therefore moves towards mid-gap and the charge on the interface states decreases. In the case of oxide charges, as seen in PMOS devices, there is only a very weak asymmetry in the forward and reverse

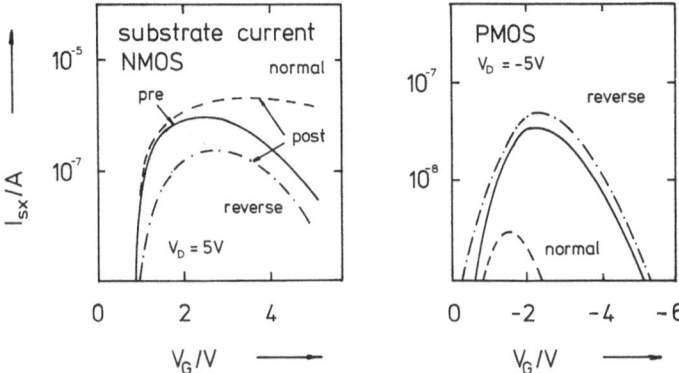

Fig. 53 Substrate current for *n*- and *p*-channel device in normal and reverse mode. Simulated results assuming interface states for NMOS and negative oxide charges for PMOS

modes. Finally, in Fig. 53 we show the substrate currents for both NMOS and PMOS devices. Once again, normal and reverse characterization modes are compared.

We again find distinctive features in both types of device. For the normal mode, the NMOS device shows an increasing deviation of pre- and post-stress currents, which is caused by the interface states. The increasing number of surface charges will give an increasing maximum field resulting in an enhanced substrate current. In the reverse mode, electrons are repelled by the surface charge. This will give a higher channel resistance and thus a larger potential drop at the source end of the channel. The potential drop in the drain region is therefore reduced and gives a lower maximum field. The substrate current is thus significantly reduced for the whole range of gate voltages. For the PMOS device, we observe a considerable reduction of the substrate current after stress for the normal mode because the build-up of negative oxide charge will reduce the maximum electric field. In the reverse mode, the negative charge will cause a channel shortening as discussed above, so we will see an increase in the substrate current. All these features were seen in the experimental data but were fully understood only with the consistent use of simulation (Schwerin *et al.* 1987, Schwerin 1988). To summarize, we can say that for the stress conditions considered, we find that different mechanisms are responsible for degrading NMOS and PMOS devices. The generation of interface states dominates the degradation of the NMOS device while build-up of negative oxide charges causes the PMOS device to degrade. This situation can change for different stress conditions. What really happens is that interface states and oxide charges are formed simultaneously. However, both processes are triggered by mechanisms that respond differently to the electrons and holes injected into the gate oxide. This makes the degradation sensitive to the stress bias, which has a strong influence on the distribution of injected electrons and holes in the gate oxide. We will address this point in the next section, where we will investigate the dynamics of the competing trap mechanisms in more detail.

5.2.2 Charge Pumping

As discussed above, a lot of information can already be drawn from the modification of the terminal currents due to hot-electron stress. A drawback of this approach is that the information we are interested in is a second-order effect on top of the original signal produced by the virgin (pre-stressed) device. For practical purposes, it is sufficient to detect a significant change in the terminal currents due to hot-carrier stress as a measure of the device reliability under operating conditions. To characterize a technology, more information is required than is available with this simple technique. One key factor in successfully manufacturing a MOSFET device is the quality of the gate oxide silicon interface. The original patent for a MOSFET was registered in 1930 (Lillienfeld 1930) and the first operating devices were produced in early 1960s. A major reason for this delay was the large number of interface states at the SiO_2/Si interface that prohibited the device from functioning properly. Even today, as the number of interface states is dramatically reduced, it is still an indicator of the quality of the device, especially with respect to its reliable performance. Another problem is the experimental determination of the density of states $D_{it}(\varepsilon)$, which is impossible with the above-mentioned technique. A very sensitive experiment that gives a signal which is directly proportional to the absolute number of interface states is the charge-pumping method (Burgler and Jespers 1969, Groeseneken et al. 1984). It can therefore be used for very small devices, in contrast to the capacitive approaches, which require large gate areas. In the charge-pumping experiment, the source and drain of the transistor are connected together and held at a specific reverse bias voltage with respect to the substrate. A pulse train of constant amplitude ΔV_{cp} and increasing base level with frequency f_{cp} is applied to the gate. When the surface potential of the transistor is pulsed into inversion, electrons will flow from source and drain into the channel, where some of them will be captured by surface states. While the gate is driving the surface back to accumulation, the mobile charges will return to the source and drain. Owing to their finite lifetime, some of the charges will remain on the interface states until they eventually decay or recombine with the majority carriers from the substrate. Over many periods of the alternating gate pulses, this recombination will give a measurable net charge flow known as the charge pumping current I_{cp}. This current is recorded as a function of decreasing base level of the gate pulse. For a sufficiently high base level, the

device never leaves inversion and no recombination is recorded. If the base
level is lowered, there is strong increase in the charge pumping current until
the base level reaches the flat band voltage V_{FB}. As long as the top pulse
level exceeds the threshold voltage V_{TH}, further lowering of the base level
will give a plateau in the charge pumping current. In this situation, charges
are pumped because the surface potential moves continuously between
inversion and accumulation. Lowering the base level still further, a point
will be reached where the top pulse level will cease to reach the threshold
voltage. The interface states can no longer be filled with channel electrons.
The charge pumping current will therefore drop sharply to zero. The three
phases of the charge pumping experiment are illustrated in Fig. 54.

The physical origin of the charge-pumping current is the finite lifetime $\tau_{it}(\varepsilon)$
of the occupied interface states. During one period in the second phase of
the charge-pumping experiment, the surface potential moves from a value
above the threshold potential through the flat band into accumulation and
then back to inversion. In Fig. 55, we show a schematic example of what
happens to the acceptor-like interface states during this period.

If the surface potential is in inversion, the interface states are filled up to the
quasi-Fermi potential ε_F^n. During the fall time t_f, the pulse moves from
inversion to accumulation, passing the mid gap. The channel electrons now

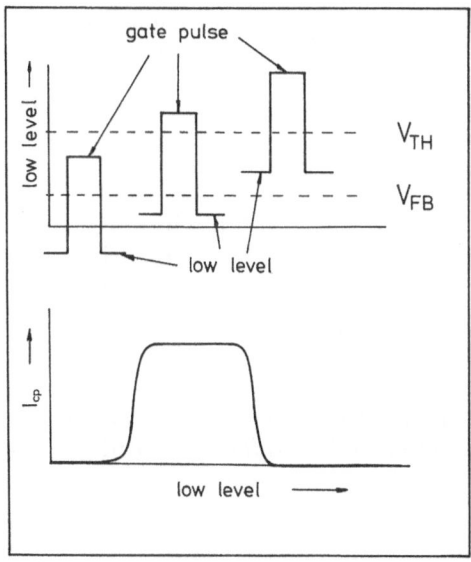

Fig. 54 A pulse train applied to the gate drives the surface subsequently into inversion and
accumulation. If the low level is such that the pulse covers the region from inversion to
accumulation (threshold voltage V_{TH} to flat band voltage V_{FB}) there is a net non-zero charge
pumping current

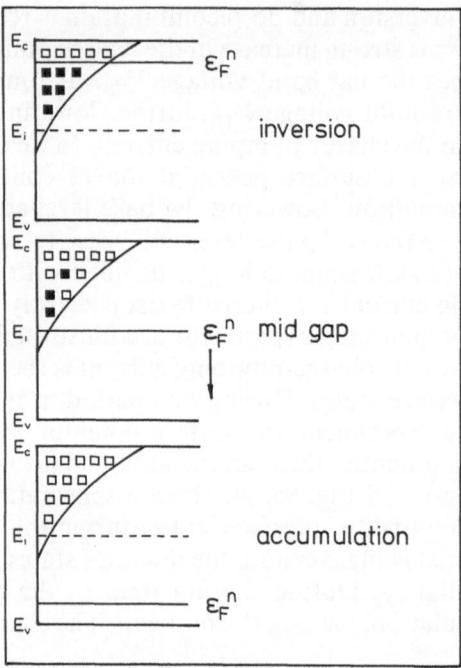

Fig. 55 The three phases of the charge pumping process. I. Inversion: the interface states are in equilibrium with the bulk silicon; II. Midgap: during the fall time of the gate pulse the quasi-Fermi level moves towards mid gap and the interface states are in non equilibrium with the bulk silicon; III. Accumulation: the gate pulse drives the surface into accumulation and not yet decayed interface states recombine with the holes

return to the source and drain, and holes from the substrate are brought to the interface. With negative charge, occupied interface states not yet decayed recombine with the substrate holes. Next, the pulse will move back from accumulation into inversion during the rise time t_r and the interface states will be filled again. A similar argument can be applied to donor-like interface states that are occupied with holes. It is obvious that the charge-pumping current is a function of the shape of the pulse, especially its fall and rise times. To obtain the lifetime $\tau_{it}(\varepsilon)$, we consider the balance equation for the occupied acceptor-like interface states $N_{it}^-(\varepsilon)$, with width $\Delta\varepsilon$, during the inversion period.

$$\dot{N}_{it}^-(\varepsilon) = \sigma_1 \cdot v_{th} n \cdot (\Delta\varepsilon D_{it}(\varepsilon) - N_{it}^-(\varepsilon)) - N_{it}^-(\varepsilon)/\tau_{it}(\varepsilon)$$

The first term describes the capture of mobile carriers in vacant interface states with a capture cross section σ_1 ($\approx 10^{-17}$ cm^2). The second term takes account of the decaying occupied states with a lifetime $\tau_{it}(\varepsilon)$. In equilibrium, the time variation disappears and we obtain the following for the lifetime $\tau_{it}(\varepsilon)$

$$\frac{1}{\tau_{it}(\varepsilon)} = \sigma_1 \cdot v_{th} n \frac{\Delta\varepsilon D_{it}(\varepsilon) - N_{it}^-(\varepsilon)}{N_{it}^-(\varepsilon)} \tag{5.10}$$

In equilibrium, the occupation of the interface states is exactly described by the Fermi distribution function multiplied by their density of states.

$$N_{it}^-(\varepsilon) = \Delta\varepsilon D_{it}(\varepsilon) \frac{1}{1 + \exp\left(\dfrac{\varepsilon - \varepsilon_F^n}{k_B T}\right)} \tag{5.11}$$

and the mobile carrier density is

$$n = n_i \exp\left(\frac{qV - \varepsilon_F^n}{k_B T}\right) \tag{5.12}$$

Here n_i is the intrinsic density of Si. The electric potential V and the Fermi energy have to be taken at the interface. Combining Eqs. (5.11) and (5.12) with Eq. (5.10), we obtain for $\tau_{it}(\varepsilon)$

$$\frac{1}{\tau_{it}(\varepsilon)} = \sigma_1 \cdot v_{th} \exp\left(-\frac{qV - \varepsilon}{k_B T}\right) \tag{5.13}$$

For donor-type interface states, we can derive the energy-dependent lifetime in an analogous way. From Eq. (5.13), we learn that acceptor-like interface states closer to the conduction band edge will decay faster than those in the mid-gap region. This is utilized for determining the density of states from charge pumping measurements (Groeseneken *et al.* 1984). The idea is to compare the charge-pumping currents of two measurements with a slightly different fall time, for instance, but the same base level and amplitude. In Fig. 56a, we show a snapshot of the occupied interface states just before the flat band voltage is reached and the recombination process occurs. Fewer occupied states decay for fast fall times than for slow ones. The difference between both charge distributions, which is shown in Fig. 56b, shows a pronounced maximum at a certain energy $\tilde{\varepsilon}$.
The difference signal is thus proportional to the density of states that belongs to $\tilde{\varepsilon}$. In principle, the complete $D_{it}(\varepsilon)$ can be scanned by appropriately chosen fall times. However, it turns out that only a small energy range is accessible with this technique. The states near the conduction band edge decay too fast and never contribute to the charge-pumping current. Approaching the mid-gap regions $\tilde{\varepsilon} \to qV (= E_i)$, where the decay time is very large, the difference signal will obtain more and more contributions from higher energy states that have not yet decayed so that the signal is no longer proportional to the density of states in a small energy range around $\tilde{\varepsilon}$.

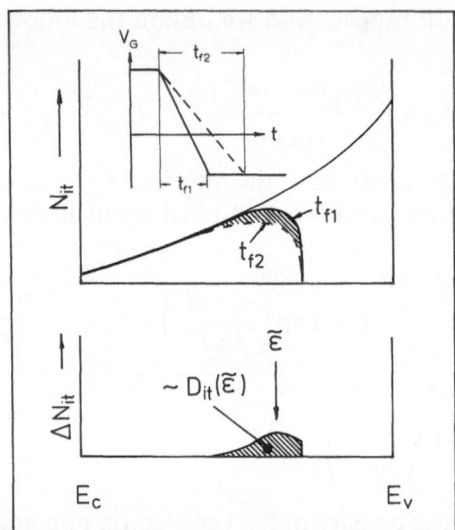

Fig. 56 Scanning the density of states by variation of the fall time t_f. The shaded area is

proportional to the difference $I_{cp}(t_{f1}) - I_{cp}(t_{f2}) \approx \dfrac{\partial}{\partial t_f} I_{cp}(t_f) \cdot \Delta t_f$

To summarize: the charge-pumping method provides a signal which is directly proportional to the number of interface states and the frequency f_{cp} of the pulse train applied to the device gate. We therefore have $I_{cp} \sim f_{cp} \cdot q N_{it}$. It can also be used to determine their density of states by varying the rise and fall times of the pulse. This makes it a very sensitive and useful instrument, especially for investigating very small devices. However, interpretations of the charge-pumping measurement become increasingly difficult in this regime because the device can no longer be treated one dimensionally, which would allow a simple analytical approach. The source and drain regions contribute significantly to the charge-pumping current and the potential and carrier distributions are essentially two-dimensional here. To include the physics of the charge-pumping experiment in a realistic device simulator will help to get a better understanding of the complex experimental data (Hofmann and Hänsch 1989).

A self-consistent calculation of the charge-pumping current has to start from the time-dependent solution of the semiconductor equations in which the effects of the interface states are included. In the first part of this section, we pointed out that even for the stationary case, the interface states require special attention because of their strong coupling to the channel carriers. This coupling must also be maintained in a transient formulation. In addition, however, we must also account for the dynamics of the interface states. So we cannot simply use the time-dependent versions of Eqs. (5.3) and (5.4) to calculate the charge-pumping current. We have to generalize these equations to include the effects of the finite lifetime $\tau_{it}(\varepsilon)$ which is mandatory for obtaining I_{cp}. To find an appropriate equation, we consider the situation where the electron's quasi-Fermi level moves from the conduction band edge towards the valence band edge. This corresponds to the falling pulse in V_G and relates to the contribution of the acceptor-like interface states to the charge-pumping current. We place the origin of time $t_0 = 0$ at the start of the falling gate pulse. For a rectangular pulse shape, we will assume that the plateau was sufficiently long to establish an equilibrium of occupied interface states and channel electrons. We consider a single state with an energy ε_0 which lies somewhere in the upper half of the bandgap and is spatially homogeneously distributed over the active gate area $A_G = W_G \cdot L_G$. At t_0, we suppose this state to be occupied. As the Fermi level sweeps across the bandgap, it will strike the occupied interface state at $t = t_1(\varepsilon_0) > t_0$. This will trigger a clock for the decay of the interface state with a lifetime $\tau_{it}(\varepsilon_0)$. The rate of decay is given by the following for an increasing time t

$$\exp(-\Delta(t, \varepsilon_0)/\tau_{it}(\varepsilon_0)); \quad \Delta(t, \varepsilon_0) = t - t_1(\varepsilon_0) > 0 \tag{5.14}$$

The time interval $\Delta(t, \varepsilon_0)$ measures the time elapsing since the Fermi level struck the energy ε_0. It is therefore obtained from the solution of the following equation

$$\varepsilon_F^n(t - \Delta(t, \varepsilon_0)) = \varepsilon_0 \tag{5.15}$$

The above situation is illustrated in Fig. 57.

Once they have arrived at the accumulation, holes will recombine with the negative charge on the still-occupied interface states $N_{it}^-(t)$. The simple case discussed above is therefore described approximately by the following integral equation

$$N_{it}^-(t) = \Delta\varepsilon\, D_{it}(\varepsilon_0)\exp(-\Delta(t_1\varepsilon_0)/\tau_{it}(\varepsilon_0)) - \sigma_2 \cdot v_{th} \int_{t_0}^{t} d\tau N_{it}^-(\tau)p(\tau) \tag{5.16}$$

The solution of this equation is only physically meaningful for $N_{it}^- \geqslant 0$. For larger times no additional recombination is possible.

The contribution to the charge-pumping current during the fall time t_f is given by the total number of hole captures on occupied interface states during the pulse.

$$I_{cp} = q f_{cp} A_G \sigma_2 v_{th} \int_{t_0}^{t_0+T} d\tau \cdot N_{it}^-(\tau)p(\tau) \tag{5.17}$$

Here σ_2 ($\approx 10^{-16}\, cm^2$) is the recombination cross-section for holes and electrons on acceptor-like interface states and N_{it} is the density of interface

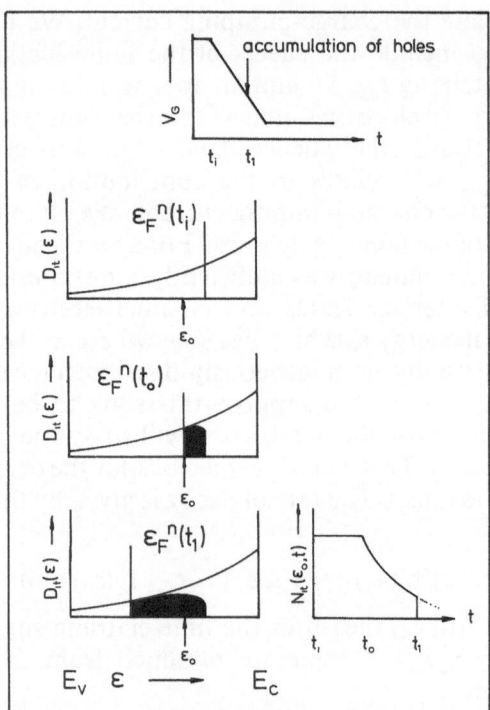

Fig. 57 Dynamics of an interface state with energy ε_0. The decay time of the interface state starts at t_0 and lasts until t_1 when the accumulation phase is reached. The black area at t_1 is proportional to the charge pumping current

states at energy ε_0. For a realistic calculation, we have to allow for a spatially inhomogeneous distribution of interface states and must also consider their continuous distribution in the Si bandgap. For such a continuous distribution we always have, at any instant of time, occupied states below the Fermi energy $\varepsilon_F^n(y, t)$ and still incompletely decayed states above the Fermi level. The states below the Fermi level are in equilibrium with the channel electrons, whereas the states above it are not. The total density of effective interface states therefore separates into an equilibrium part $N_{it}^-(y, t)^{eq}$ and a non-equilibrium part $N_{it}^-(y, t)^{noneq}$

$$N_{it}^-(y, t) = N_{it}^-(y, t)^{eq} + N_{it}^-(y, t)^{noneq} \tag{5.18}$$

The negative charge $Q_{it}^-(y, t)$ assigned to the occupied interface states is given by

$$Q_{it}^-(y, t) = -q N_{it}^-(y, t) \tag{5.19}$$

The equilibrium contribution is obtained by Eq. (5.3), with the time-dependent quasi-Fermi level $\varepsilon_F^n(y, t)$. For the non-equilibrium part, we can generalize the rate equation (5.16)

$$N_{it}^-(y, t)^{noneq} = q \int_{\varepsilon_F^n(t)}^{\varepsilon_F^n(t_0)} d\varepsilon \cdot D_{it}^A(y, \varepsilon) \exp\left(-\Delta(y, t, \varepsilon)/\tau_{it}^A(\varepsilon)\right)$$

$$- \sigma_2 \cdot v_{th} \int_{t_0}^{t} d\tau \cdot N_{it}^-(y, t)^{noneq} p(y, t) \quad (5.20)$$

with

$$\varepsilon_F^n(y, t - \Delta(y, t, \varepsilon)) = \varepsilon \quad (5.21)$$

The charge-pumping current contribution from the acceptor-like interface states is now given by

$$I_{cp} = q f_{cp} W_G \sigma_2 v_{th} \int_{t_0}^{t_0+T} d\tau \int dy \, N_{it}^-(y, \tau)^{noneq} p(y, \tau) \quad (5.22)$$

Equations (5.18)–(5.22) have to be implemented self-consistently within the time-dependent solution of the semiconductor equations. This then allows a direct calculation of the charge-pumping current. To perform the calculation, we must specify the density of states $D_{it}^A(\varepsilon)$ and the spatial distribution of the damage $\Gamma(y)$ as in Eqs. (5.1) and (5.3). A process-related spatially homogeneous background can be included without further problems. So far, we have considered only the influence of acceptor-like interface states. Similar results are applicable to donor-type interface states which are important on the rising pulse of the gate voltage in an NMOS device. Adjusting the density of states and spread of the damage to the experimental data will reveal additional information about the hot-carrier-related oxide damage. Before we turn to this point, we will address the problem of extracting the $D_{it}(\varepsilon)$ from charge-pumping measurements.

Different experimental methods are available to extract the density of states from experimental data. The charge-pumping method is one of them. The key factor in the charge-pumping experiment is to vary the time interval τ_{midgap} in which the surface potential passes from inversion to flatband conditions. Here, the decay of occupied interface states is dominant and the recombination process is still negligible. During this time, only interface states with $\tau_{it}(\varepsilon) < \tau_{midgap}$ release the trapped carriers to source and drain. The remaining trapped carriers contribute to the charge-pumping current I_{cp} in the immediately following accumulation phase. Using a well-defined density of states $D_{it}(\varepsilon)$, the simulation allows us to exactly model the charge-pumping experiment performed to extract this density of states from I_{cp}. A comparison of the result from the experimental procedure and the given density of states will then show the energy window for which the extraction scheme gives reliable results. Experimentally, τ_{midgap} can be adjusted by varying the fall time t_f of the gate pulse (Groeseneken *et al.* 1984). If the gate pulse has an amplitude ΔV_{cp}, then we have the following approximate

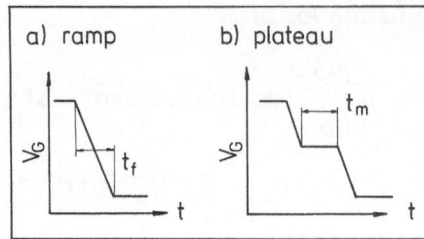

Fig. 58 Different realizations of falling gate pulses: (a) ramp method and (b) plateau method

relation, as shown in Fig. 58

$$\tau_{midgap} \approx \frac{|V_{FB} - V_{TH}|}{\Delta V_{cp}} t_f \qquad (5.23)$$

The interface density of states is then related to the charge-pumping current by

$$D_{it}(\varepsilon) = \frac{t_f}{qA_G k_B T} \frac{\partial}{\partial t_f} Q_{cp} \qquad (5.24)$$

$$\varepsilon(t_f) = -k_B T \cdot \ln(v_{th}^n \sigma_1 n_i \tau_{midgap})$$

Here, $Q_{cp} = I_{cp}/f_{cp}$ is the non-equilibrium charge in the surface states prior to recombination. To scale t_f to τ_{midgap}, the threshold voltage V_{TH} and the flatband voltage V_{FB} must be known. This requires additional experimental effort.

A different experimental method introduces a plateau in the falling pulse with length t_m (Hofmann and Krautschneider 1989). The base level of the gate pulse is varied so that the plateau lies at midgap. Here, I_{cp} shows its minimal value. This set-up requires no extra measurements and the plateau time t_m is easily controlled. We then have $\tau_{midgap} \approx t_m$ and the interface density of states is again extracted from Eq. (5.24). In Fig. 59, we compare the given density of states $D_{it}(\varepsilon) = (2\varepsilon/E_G) \cdot 10^{11}/\text{eV cm}^2$ with the values extracted from the charge-pumping current according to Eq. (5.24). Good agreement is obtained in the range of 0.1–0.35 eV away from midgap for both methods.

At energies greater than 0.35 eV above midgap, there is striking disagreement. In this range, the time constants of the traps are very small; emission of non-equilibrium states occurs almost immediately. These very fast-decaying interface states are not taken into account in the simple analytical approach that leads to Eq. (5.24). The lowest energy which could be determined is approximately $2k_B T$ above midgap. The time constants of the interface states are very long at this energy and below, as discussed above.

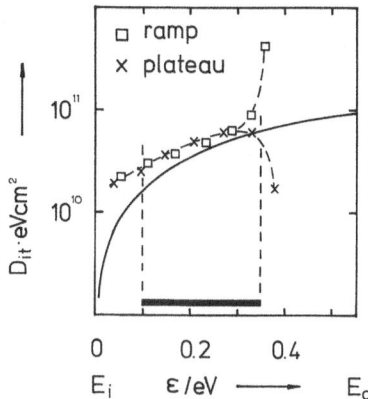

Fig. 59 Extracting a given density of states from calculated charge pumping currents. The extraction was done similarly to the experimental method. Open squares—ramp method; crosses—plateau method. Both methods give similar results in a range of 300 meV (black bar) and fail outside that range

An increasing component in the charge-pumping current can be seen to be attributed to higher energies, as yet undischarged. These states will continue to increase so that $D_{it}(\varepsilon)$ will be overestimated near midgap. In order to obtain more information about the density of states, simple analytical approaches are insufficient. A numerical simulation of $Q_{cp}(t_f)$ in the time domain must be performed. This would correctly cover the decay of the interface states and is very close to the experimental situation.

We will now discuss how modeling the charge-pumping experiment will allow a detailed analysis of a degraded device. We have performed the following experiment as an example: a conventional n-MOSFET ($L_G = 2\,\mu m$, $W = 20\,\mu m$, $t_{ox} = 25$ nm) was stressed with $V_G = 1$ V and $V_D = 8$ V for 1000 s. This stress condition predominantly involves hot-hole injection that builds up fixed positive oxide charges and interface states near the drain. The same device was then stressed with $V_G = 8$ V and $V_D = 8$ V for another 1000 s. The positive charge is now compensated by injected electrons. The experimental findings and simulated results are shown in Fig. 60.

For the simulation of the unstressed device (dashed-dotted line), we used a homogeneous distribution of interface states in the channel with an energetic dependence $D_{it,0}(\varepsilon) = (2\varepsilon/E_G) \cdot 10^{11}/eV\,cm^2$. In order to reproduce the first stress condition, we considered a Gaussian distribution of positive charge Q_{ox}^+ with a maximum $Q_{ox}^{0+} = 2 \cdot 10^{12}$ q cm^{-2} at 50 nm in front of the drain pn-junction and a width of 150 nm. In addition, we placed acceptor-like interface states with a maximum of $5 \cdot 10^{11}/eV\,cm^2$ at the same site. The result of the simulation (solid line) reproduces the experimental features (solid line) very well. If we disregard the influence of the background in

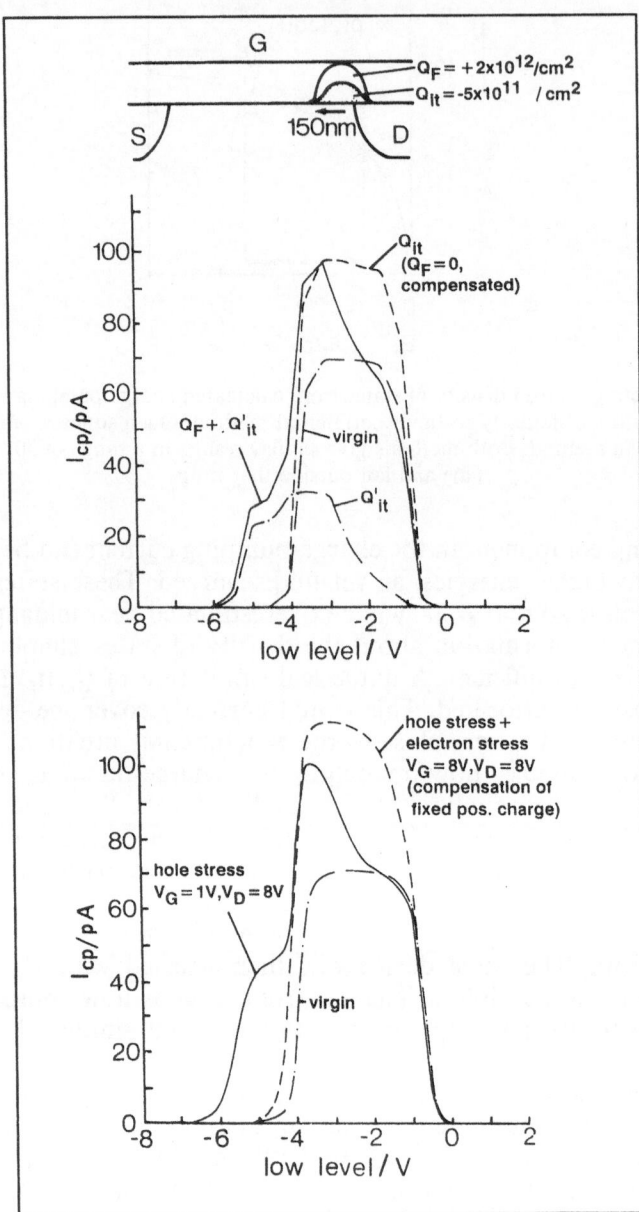

Fig. 60 The charge pumping current before and after hot carrier stress, simulation (left) and experiment (right). Dashed–dotted line—pre stress device; solid line—after hole stress; dashed line—electron stress after hole stress

the unstressed device, which is easily done in the simulation by setting $D_{\text{it},0}(\varepsilon) = 0$, we obtain the dash-dash-dot line which is due to the damaged part of the device alone. The positive oxide charge will locally decrease the threshold voltage so that inversion will already be reached at a lower gate base level. As a result, we observe a shift of the onset in the I_{cp} curve to the left. The complete signal is almost given by superimposing the original and stressed curves. As the unstressed transistor has a homogeneous distribution of interface states in the channel, the original states which are also influenced by the oxide charge now contribute to the I_{cp} signal in the same way as the interface states produced during stress. This means that for the damaged portion of the channel, the contribution to the charge-pumping current moves towards a lower gate base-level voltage. This is also reflected in the lower I_{cp} value of the stress curve for a higher gate base-level voltage. Here, the positive oxide charge prevents accumulation so that no recombination takes place. After the second stress condition, the charge-pumping curve (dashed-dashed line) has almost relaxed to its original position, but with a larger maximum value. This is best described by a compensation of the positive charge, setting $Q_{\text{ox}}^{0+} = 0$ in the simulation. The simulation does not change the maximum value of I_{cp} whereas the experiment does. This indicates that some interface states were also created during the second stress. Furthermore, a close inspection of the experimental curve reveals that the charge compensation is not complete, as assumed in the simulation. This can by seen by the slight shift of the charge-pumping current to the left with respect to the signal of the unstressed device.

We have demonstrated that the charge-pumping technique represents a very sensitive instrument for investigating the degradation of a MOSFET device. The modeling of the charge-pumping current as outlined above provides a very efficient tool for interpreting the complex signal. It is especially important for investigating very small devices which currently have a feature length in the submicron region. Owing to the complicated 2D doping profile in the source and drain region of the channel, edge effects become increasingly important for these devices when interpreting the charge-pumping signal. The self-consistent time-dependent analysis based on the numerical solution of the semiconductor equations for such devices automatically includes these effects. We have implicitly assumed that the traps are filled instantaneously in the rising gate pulse. There is experimental evidence that the finite capture time has an effect on I_{cp}. However, it is argued that this effect is a second-order correction to the charge-pumping signal and can be neglected in a first-order approximation. The model we have presented above can be easily implemented in a device simulator that solves the time-dependent semiconductor equations. A generalization to include a finite capture time during the rising pulse is in principle straightforward, but will inevitably add more complexity to the problem and thus increase computational cost. The key features observed in the experiments are clearly reproduced by our model, whose big advantage is to take the exact device structure into account.

The aim of this section was to discuss how modeling can be used for a quantitative analysis of hot-carrier degraded devices. Our main concern was therefore to verify the influence of locally confined oxide damage on the device characteristics and to determine the nature of the damage. As experimental means, we looked at a simple electrical characterization of the device and the charge-pumping technique. In the next section, we will outline how the degradation process itself can be modeled.

5.3 The Degradation Process

In the previous section, we learned how to model the influence of hot-carrier generated interface states and oxide charges on the I–V characteristics of a MOSFET. If we have enough experimental data available, we can obtain a very detailed description of the damage created in a specific stress experiment. The problem with this approach is that a device design engineer wants to know the degradation behavior of a device in advance in order to design the most reliable device. It is therefore of considerable interest to predict the possible hot-carrier degradation of a MOSFET, which is impossible using the approach presented in the previous section. Here, the position and magnitude of the generated damage are found after performing the stress experiment. A very indirect way to solve this problem is to look at the maximum lateral source-drain field responsible for the injection of carriers into the gate oxide. The bias condition is chosen such that maximum damage occurs. A maximum substrate current is typical for an n-MOSFET and a maximum gate current for a p-MOSFET. To minimize this field as well as the substrate (or gate) current is the goal of drain engineering. Drain engineering means that the source and drain doping profiles of a MOSFET are modified so that optimum performance is achieved with respect to hot-carrier degradation. Unfortunately, it is impossible to optimize a device hot-carrier hardness independently of other important properties such as the maximum current drive, series resistance or features due to the layout. There is no direct experimental access to the maximum source-drain field, but the substrate current as a measure of the avalanche generation rate is easily measurable. It therefore serves as an indicator for the maximum source-drain field. Two-dimensional process and device simulation enables the device engineer to look directly into the device to obtain the desired information. Efficient drain engineering therefore relies on a combination of an accurate 2D process and device simulation. Once the device structure is determined and cast into silicon, a ten-year lifetime is extrapolated from stressing the device for accelerated aging.

As a practical guideline, the substrate current serves as a very good indicator for characterizing a device with respect to hot-carrier hardness (Hu *et al.* 1985). There are, however, exceptions where a device with a larger substrate current shows less damage than a device with lower substrate current. This is because although the former device generates more oxide charges or interface states during the stress experiment, as is expected from the higher field, their influence on the device characteristics is less severe. This observation provides the link between the analysis performed in the previous section and the conventional approach in drain engineering of looking at the maximum source-drain field. The latter predicts the hot-carrier damage from the maximum field strength (substrate current, 2D process and device simulation) and possibly its location (2D process and device simulation), the former gives the amount and location of the stress-induced oxide damage (2D process and device simulation). The dynamics of the degradation process links the two approaches. To be able to calculate the *a priori* unknown position and magnitude of the hot-carrier damage from the initial field and carrier configuration in the device is therefore a major step forward in supporting the device engineer's efforts to find the best possible device.

The calculation of the degradation process can be divided into five steps (we include the chapters devoted to the corresponding problem in parentheses):

   (i) 2D distributions of carrier density and electric field for the realistic device under stress conditions (2 and 3)
   (ii) Hot-carrier injection into the gate oxide (4)
   (iii) 2D distribution of injected carriers in the gate oxide region (5)
   (iv) 2D solution of trap rate equations in the oxide region (5)
   (v) Feedback of stress-created oxide charges to the device (5)

In Chapter 2, we discussed the conventional semiconductor equations and possible extensions for the off-equilibrium case to obtain the correct density and field distributions in the device. The influence of the SiO_2/Si interface on carrier transport was studied in Chapter 3. In Chapter 4, we dealt with the problem of how to calculate the population of high energetic carrier states from first principles, especially in conjunction with impact ionization and gate-oxide injection. At the beginning of Chapter 5 we addressed the problem of the feedback of oxide charges and interface states on the terminal currents of a MOSFET device. To complete our work, the remaining issues involve the trap rate equations and the carrier distribution in the 2D oxide region. Finally, we have to link all components together so that we can self-consistently calculate the temporal development of the oxide damage during the stress experiment.

The complicated interaction of all five components will produce a very complex numerical problem. We will therefore simplify wherever possible without losing the essential core of the degradation process. As we want to monitor the build-up of oxide charges and interface states during the stress experiment, we have an intrinsically time-dependent problem. However, we will only consider a constant bias during the stress experiment. So we will exclude transient effects due to the interface dynamics (Weber 1988) as for example in the charge-pumping experiment. Furthermore, the intrinsic time scale of the device is approximately $\tau_{\text{dev}} \approx L_{\text{eff}}/v_{\text{sat}}$, which is several pico-seconds for a submicron device. This must be compared with the typical time constant for the trapping process, which can be as slow as 1 ms. The device will therefore instantaneously respond to the charges created in the oxide, which means that we can solve the stationary version of the semi-conductor equations, including the oxide charges. The time dependence is completely shifted to the trapping process. This is very convenient for numerical purposes because the time-step control of the problem alone will be given by the time development of the oxide damage. The 2D device simulator will provide the correct carrier and field distributions at any instant of time. The central point of our approach will be to allow these distributions to change in both the active device region and the gate oxide during the time the stress experiment is performed.

5.3.1 Carrier Distributions in the Gate Oxide

The calculation of the carrier distributions in the active device is a subject for device modeling. These distributions are the result of solving the continuity equations, which are coupled to Poisson's equation, taking into account the appropriate boundary conditions. Carriers usually enter and leave the device through contacts, thus producing the terminal currents. Carrier transport through such a contact is a very complicated problem and is not yet fully incorporated in the modeling tools (see also discussion in Chapter 1). Simple approximations are used which do not describe the physics self-consistently. The injection process into the gate oxide also represents a "contact" problem. Carriers are injected from the active device into the gate oxide. Because of the different dielectric constant of Si and SiO_2, the transverse electric field will change abruptly by passing through the interface. In fact, it will increase by a factor of approximately three which is the ratio of the dielectric constants in the different materials. The carriers that come from a stationary distribution in the Si have to adjust to the new field. After a transient length λ_{ox}, they will adjust to a new stationary distribution in SiO_2 due to collisions with phonons or other inelastic channels. From experiments and Monte Carlo studies of high-field transport in SiO_2 (Berglund and Powell 1971, Fischetti et al. 1985, Fischetti et al. 1987) this transient length is derived to be approximately 2–3 nm; which means that for currently used gate oxide thicknesses, a substantial number of carriers can still be found in a stationary state with the field.

Only for those carriers which are in a stationary state with the field can we use the drift–diffusion equation in the oxide. The current density in the oxide comprises a stationary state current density v_{ox} and those carriers that are not yet relaxed towards the stationary state v_{noeq}. Because the only source from which carriers are fed into the oxide is v_{noeq}, we have the following continuity equation in the oxide (Schwerin 1988)

$$\text{div}(v_{ox}(R) + v_{noeq}(R)) = 0 \tag{5.25}$$

If we bring the second term to the right hand side of Eq. (5.25), it can be interpreted as a generation term for the particles in the stationary distribution with the field.

$$\text{div}(v_{ox}(R)) = G_{ox}(R) \tag{5.26}$$

with

$$G_{ox}(R) = -\text{div}(v_{noeq}(R)) \tag{5.27}$$

and

$$v_{ox}(R) = \mu_{ox}(F_{ox}(R))n_{ox}(R)F_{ox}(R) \tag{5.28}$$

Equation (5.28) is the generalization of the conventional drift–diffusion equation (2.170) to the oxide region. The generation term is linked directly to the injected current density, which is given by Eq. (4.161). From Eq. (5.26), we can calculate n_{ox}, the carrier density of the stationary carriers in the force field F_{ox}. The electric field is considered in exactly the same way as in the conventional case and therefore enters the formulation quite naturally as the solution of Poisson's equation in the oxide. Fig. 61 shows the

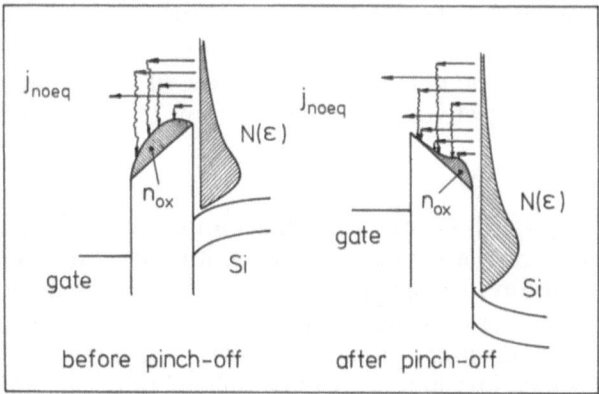

Fig. 61 Carrier injection into the gate oxide. The injected carriers j_{noeq} are in a non-stationary state with the gate field. They relax due to collisions with phonons into a stationary distribution n_{ox} with the oxide field. After pinch-off there is a depth-dependent threshold

typical situation for electrons injected into the gate oxide for different field configurations. Fig. 61a applies to the situation before the pinch-off point and Fig. 61b after the pinch-off. Here, the pinch-off point is defined as the point at which the transverse field is zero, which corresponds to a local flatband situation.

We will now derive the relationship between v_{noeq} and the injected current density from Eq. (4.161). The SiO_2/Si interface is placed at $x = 0$. For $x < 0$ we are in the gate oxide region and for $x > 0$ in the active device. The coordinate y runs parallel to the interface. The current density of injected carriers then has only an x-component and is given by $v_{inj,x}(x = 0, y; \varepsilon_{inj})$ from Eq. (4.161). The non-stationary particle current v_{noeq} is fed from this injected current. The relaxation of carriers into the stationary state is described by an exponential decrease, with a characteristic length scale λ_{ox} of the non-stationary particle density. After the pinch-off point, only particles with an energy greater than the SiO_2/Si barrier can penetrate the inner gate–oxide region. We must therefore assign a position-dependent barrier to these particles. The injected particles will travel predominantly in the direction perpendicular to the interface. Therefore to the SiO_2/Si barrier must be added the potential energy difference

$$q\Delta V(x, y)_{ox} = q[V(x = 0, y) - V(x < 0, y)]\theta[V(x = 0, y)$$
$$- V(x < 0, y)] \tag{5.29}$$

Here, the step function θ ensures that only after pinch-off is $q\Delta V_{ox}$ not zero. With Eq. (5.29) and the exponential relaxation factor, we obtain the following for the non-stationary current density in the oxide region

$$v_{noeq}(x, y) = v_{inj,x}(x = 0, \varepsilon_{inj}^0 + q\Delta V(x, y)_{ox})\exp(-x/\lambda_{ox}) \tag{5.30}$$

Further refinement can be achieved if we include in the effective barrier height ε_{inj}^0 the standard expressions for Schottky barrier lowering through tunneling $\Delta\varepsilon_{inj}^{BL}$ and the barrier reduction caused by the image force $\Delta\varepsilon_{inj}^{IF}$ both of which are effective before the pinch-off region (Ning et $al.$ 1977).

$$\varepsilon_{inj}^0 \rightarrow \varepsilon_{inj}^0 - \Delta\varepsilon_{inj}^{BL} - \Delta\varepsilon_{inj}^{IF} \tag{5.31}$$

With Eqs. (5.25)–(5.30) we can calculate stationary carrier distributions in the gate oxide. For the solution of the continuity equation we still need boundary conditions for the density n_{ox}. For the gate oxide region, we have two kinds of boundaries as shown in Fig. 62.

Interfaces may be insulator–metal or insulator–semiconductor in type. At the insulator–metal interface, n_{ox} will be obliged to be close to equilibrium, which means it assumes its intrinsic value n_i because we have no doping in the system. At the insulator–semiconductor interface, all particles are initially in the non-equilibrium current and we therefore have $n_{ox} \approx 0$ at this interface. Due to the large bandgap of SiO_2, the intrinsic number is very

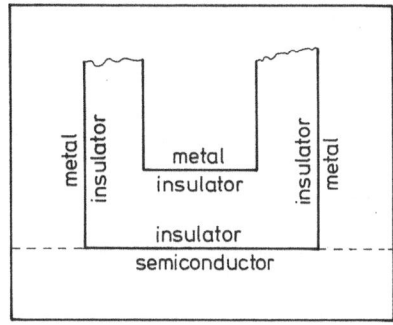

Fig. 62 Boundaries of the gate oxide of a MOSFET

small compared to the particle density n_{ox}. It is therefore justified to use $n_{ox} = 0$ at the insulator–metal interface as well. For the extended oxide regions, we propose $v_{ox,x} = 0$, which completes the boundary conditions for the continuity equation. Although we implicitly assumed electrons, exactly the same approach applies to holes. It remains to specify the mobility of the free carriers in the oxide. Compared to the large amount of material available for the mobility in Si material, only sparse information is available for the mobility in SiO_2 (Hughes 1978). There is experimental evidence that electron transport is very similar to that in Si. It shows saturation at high electric fields in the drift velocity, and the saturation velocity is of the order of 10^7 cm/s. We will therefore use an expression which is very similar to that for bulk silicon.

$$\mu_{ox}^n(F_{ox}^n) = \frac{2\mu_{ox}^{n0}}{1 + (1 + (2\mu_{ox}^{n0}/v_{ox}^{nsat} F_{ox}^n)^\beta)^{1/\beta}} \tag{5.32}$$

The parameters μ_{ox}^{n0}, v_{ox}^{nsat}, and β are adjusted to experimental material found in the work of Hughes (Hughes 1978) which gives $\mu_{ox}^{n0} = 21$ cm^2/Vs, $v_{ox}^{nsat} = 1.5 \cdot 10^7$ cm/s, and $\beta = 4$. The situation is more uncertain for the holes. In the experimental material, there is no indication of a velocity saturation and the low-field mobility $\mu_{ox}^{p0} = 2 \cdot 10^{-5}$ cm^2/Vs is extremely small. The reason for this small value is that holes in the SiO_2 have a strong coupling to the polar optical modes, which results in a very immobile quasi-particle called the small polaron. But the problem is that to form such a polaron the injected hole must have sufficient time to fully establish this interaction on its way through the extremely thin gate oxide. The experiments were performed on relatively thick oxide layers compared to today's gate oxides. So it is not quite clear whether the small polaron picture of hole transport is still valid. The so-called "dry holes" with a low-field mobility $\mu_{ox}^{p0} = 1$ cm^2/Vs seem to indicate such an incomplete formation of a small polaron. At best, we can use a non-saturated hole velocity

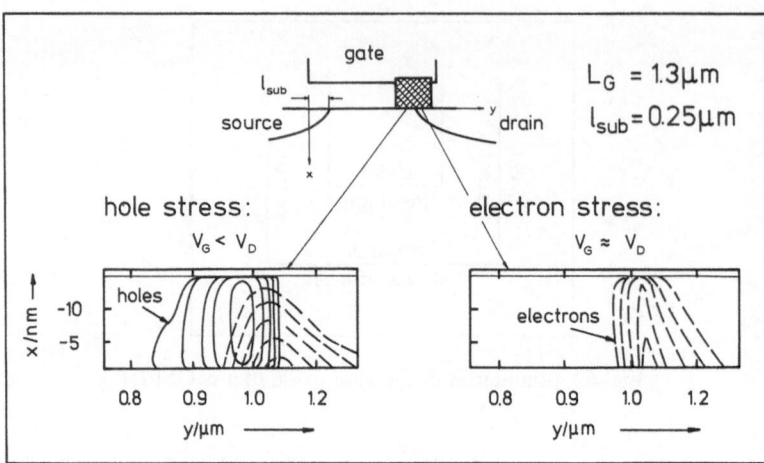

Fig. 63 Stationary distributions of electrons (dashed lines) and holes (solid lines) in the gate oxide for hole stress ($V_G \approx V_D/3$) and electron stress ($V_G \approx V_D$). The drain side pn-junction is located at 1.05 μm

and consider the mobility as an adjustable parameter. Given the electric field and the injected current density, we can now calculate the carrier distribution in the gate oxide. All effects related to the field configuration in the oxide are intrinsically included in our approach. The electron and hole particle densities n_{ox} and p_{ox} relax in a very natural way to a stationary state in the oxide field. In Fig. 63, we show a contour plot for n_{ox} and p_{ox} for different bias conditions of an NMOS device.

For the hole stress condition $V_G < V_D$, we see a deformation of the hole distribution to the left and for the electron distribution to the right. This is caused by the lateral field component in the oxide, which has the tendency to draw the holes to the source and the electrons to the drain side of the device. We also see that the holes penetrate the oxide completely while the maximum of the electron distribution is located near the SiO_2/Si interface. This is a direct consequence of the fact that after the pinch-off point, the transverse oxide field drives the electrons back to the interface. Only a quickly decreasing diffusion tail reaches the gate electrode and still contributes to the gate current. For a higher gate voltage, the oxide field decreases and the electron current density has an increasing drift component which drives the electrons deeper into the oxide, as seen for the electron stress case. The calculation of carrier and current density distributions is the inevitable prerequisite for a physically realistic calculation of the trapping process which we will discuss next.

5.3.2 Trap Equations

In Section 2, we found that the hot-carrier generated oxide damage is a complex interplay between different kinds of oxide traps. There are donor- and acceptor-like interface states which are strongly coupled to the carriers in the active device, and slow oxide traps which can be located throughout the oxide region. The problem of a rigorous investigation of the trap dynamics is to allow all possible trap mechanisms to compete simultan- eously during the occurrence of the oxide damage. Only the bias condition for the stress experiment, which determines the gate injection of carriers and therefore current and particle densities in the gate oxide, should determine how the damage evolves. To this end, we must find a generalized rate equation that takes the time development for the various traps and their interaction into account. This trap equation must then be solved on the 2D gate oxide area as shown in Fig. 62. As a result, we will find the location and number of traps generated during the stress experiment. For convenience, we will work with trap densities N rather than with charge densities Q. The simplest model that covers most of the features observed in a stress experiment will allow for the time evolution of the following components of the generalized trap vector N_{ox}:

N_{ox}^+ slow oxide (hole) traps that captured a single positive charge

N_{ox}^- slow oxide (electron) traps that captured a single negative charge

N_{it}^+ donor-like interface states

N_{it}^- acceptor-like interface states

N_{ox}^{0+} slow oxide (hole) traps (neutral) that can be occupied by a single
        positive charge

N_{ox}^{0-} slow oxide (electron) traps (neutral) that can be occupied by a single
        negative charge

We have omitted the space coordinates for convenience. The first two components N_{ox}^+ and N_{ox}^- describe the formation of positive and negative fixed oxide charges (Di Maria 1978, De Keersmaeker 1983) respectively. N_{it}^+ and N_{it}^- will monitor those traps that eventually turn into fast interface states (Lai 1983, Hofmann et al. 1985, Lyon 1989, Bellens et al. 1988). We will assume that there is a region close to the interface where these states occur and eventually move to the interface to become fast interface states. The last two components N_{ox}^{0+} and N_{ox}^{0-} take account of the fact that a generation of oxide traps (not yet occupied) can take place during the stress experiment (Di Maria and Stasiak 1989, Nissan-Cohen and Gorczyca 1988, Heyns et al. 1989). A more complex trap scenario can be described if we were to allow for multiply occupied traps or if an energetic distribution of the traps in the SiO_2 bandgap were taken into account, for instance. This higher degree of sophistication would not only increase the complexity but would also introduce more uncertainty in the modeling because many of the specific aspects of the trapping process are still not understood on such a detailed level. So we do not use a more complex model. The time evolution of the generalized trap vector N_{ox} is given by the rate equation

$$\dot{N}_{ox}(t) = A(t)N_{ox}(t) + B(t) \tag{5.33}$$

Here A is a 6×6 reaction matrix that must be found by specifying the physical models for the trap dynamics, and the vector B contains the contributions due to oxide trap generation in the strong oxide field. To find the explicit form of these quantities, we must discuss the fundamental processes we will allow to happen during the stress condition. The virgin device, i.e. the device before the stress bias is applied, has a distribution of neutral slow oxide charges $N_{ox}^{0\pm}$ in the gate–oxide area. During the stress experiment, these traps will be filled with charges. We also have a background density of fast interface states which can be parameterized according to Eqs. (5.1) and (5.2). The density of states of this technologically determined base load is sufficiently small for them to have a negligible influence on device performance. They can, however, be detected in the charge-pumping experiment as discussed in the previous section. For this base load, it is justified to assume a homogeneous distribution over the SiO_2/Si interface with a U-shaped distribution in the Si bandgap. During the stress experiment, the density of states of the interface states increases locally. To describe this process, we introduced the volume traps $N_{it}^{\pm}$ that will move to the interface to be transformed into interface states. To generate these interface traps, we will consider either a one- or a two-step process. In the former, a $N_{it}^{\pm}$ is created by electron or hole capture on a neutral oxide trap N_{ox}^{0-} or N_{ox}^{0+}, respectively. The latter will describe the formation of $N_{it}^{\pm}$ by capturing an electron or hole on an already occupied oxide trap N_{ox}^{+} or N_{ox}^{-}, respectively. There is evidence for both one- and two-step processes in published experimental data. A situation often observed is that stress-generated interface states have a similar U-shape of the density of states as the base load. We will therefore describe the effect of hot-carrier stress on the interface state generation by calculating a spatially dependent $D_{max}^{A/D}$ and $D_0^{A/D}$ which parameterize the interface density of states according to Eqs. (5.3) and (5.4). To this end, we assume that the transformation of the $N_{it}^{\pm}$ into interface states is fast compared to the trapping time scale. This allows us to assume that the $N_{it}^{\pm}$ are instantaneously transformed into interface states. Therefore we have the following approximation in the limit $D_0/D_{max} \ll 1$

$$D_{max}^{D/A}(t, y) \approx \pm 2\frac{\alpha + 1}{E_G} \int_0^{-t_{ox}} dx N_{it}^{\pm}(t, x, y) \tag{5.34}$$

which transforms the number of available interface traps $N_{it}^{\pm}$ into an equivalent density of states, which is eventually locally increased and couples to the mobile carriers in the Si according to Eqs. (5.3) and (5.4). Within this model, we cannot cover a drastic change of the density of states (the appearance of extra-local maxima in the upper bandgap, for example) that is sometimes observed in homogeneous stress experiments (Ma 1989). It is not clear at present how relevant these results are for hot channel electron stress experiments where the damage occurs at the gate edges and spacer regions. However, the simplified model is sufficient for a qualitative comparison of competing slow oxide charge formation and interface state generation. Again, we have a trade-off between complexity and physical accuracy. In the beginning of the stress experiment, we assume that the slow oxide traps are not occupied and therefore neutral. When the stress bias is applied, these traps capture electrons and holes provided by the injected current density and new interface states are simultaneously generated. A complicated balance of trapping and detrapping, interface state generation and oxide-trap creation will occur during the stress experiment. Summing up the trap scenario, we have to consider the following basic processes:

(i) Generation of positive oxide charge and interface states due to hole capture on an oxide hole trap

$$p + N_{ox}^{0+} \rightarrow N_{ox}^{+} + N_{it}^{p} \tag{5.35a}$$

(ii) Generation of negative oxide charge and interface states due to electron capture on an oxide electron trap

$$n + N_{ox}^{0-} \rightarrow N_{ox}^{-} + N_{it}^{n} \tag{5.35b}$$

(iii) Generation of interface states and compensation of positive oxide charge due to electron capture on an occupied hole trap

$$n + N_{ox}^{+} \rightarrow N_{ox}^{0+} + N_{it}^{pn} \tag{5.35c}$$

(iv) Generation of interface states and compensation of negative oxide charge due to hole capture on an occupied electron trap

$$p + N_{ox}^{-} \rightarrow N_{ox}^{0-} + N_{it}^{np} \tag{5.35d}$$

We have to add the possible generation of oxide traps $N_{ox}^{0\pm}$ during the stress experiment in the high gate field to Eqs. (5.35). For this process, we introduce the trap generation rates $G_{ox}^{0\pm}$ which we discuss later. From these fundamental processes, we obtain the rate equation for each component of the trap vector N_{ox}.

Positive oxide charge is increased if a vacant hole trap captures a hole. Due to the formation of interface states, however, only a partial number of the total number of vacant hole traps will be available.

$$\frac{\partial}{\partial t} N_{ox}^+ |_+ = \sigma_{ox}^{0p} v_{ox}^p (1 - \delta^p) N_{ox}^{0+} \tag{5.36}$$

Here σ_{ox}^{0p} is the cross-section for holes to be captured in a hole trap and δ^p is the fraction of hole traps that have turned into interface traps. The decrease of positive oxide charge is given by the capture of an electron on an occupied hole trap

$$\frac{\partial}{\partial t} N_{ox}^+ |_- = -\sigma_{ox}^{+n} v_{ox}^n N_{ox}^+ \tag{5.37}$$

Here σ_{ox}^{+n} is the cross-section for an electron to be captured in an occupied hole trap. Combining Eqs. (5.36) and (5.37), we get the rate equation for the positive oxide charge

$$\frac{\partial}{\partial t} N_{ox}^+ = -\sigma_{ox}^{+n} v_{ox}^n N_{ox}^+ + \sigma_{ox}^{0p} v_{ox}^p (1 - \delta^p) N_{ox}^{0+} \tag{5.38}$$

Similarly, we obtain the rate equation for the negative oxide charge

$$\frac{\partial}{\partial t} N_{ox}^- = -\sigma_{ox}^{-p} v_{ox}^p N_{ox}^- + \sigma_{ox}^{0n} v_{ox}^n (1 - \delta^n) N_{ox}^{0-} \tag{5.39}$$

The cross-sections σ_{ox}^{-p} and σ_{ox}^{0n} are for capturing a hole in an occupied electron trap and an electron in a neutral electron trap, respectively. The fraction of electron traps turned into interface traps in the latter process is δ^n. The number of neutral hole traps is increased by electron capture on occupied hole traps. Because some occupied hole traps are simultaneously transformed into interface state traps, only a fraction will become neutral traps. The increase of neutral hole traps is therefore

$$\frac{\partial}{\partial t} N_{ox}^{0+} |_+ = \sigma_{ox}^{+n} v_{ox}^n (1 - \delta^{pn}) N_{ox}^+ + G_{ox}^{0+} \tag{5.40}$$

Here δ^{pn}, is the fraction of occupied hole traps transformed into interface traps and G_{ox}^{0+} is the field-induced generation of neutral hole traps. The decrease of neutral hole traps is due to the hole capture

$$\frac{\partial}{\partial t} N_{ox}^{0+} |_- = -\sigma_{ox}^{0p} v_{ox}^p N_{ox}^{0+} \tag{5.41}$$

The combination of Eqs. (5.40) and (5.41) gives the rate equation for the neutral hole traps

$$\frac{\partial}{\partial t} N_{ox}^{0+} = \sigma_{ox}^{+n} v_{ox}^n (1 - \delta^{pn}) N_{ox}^+ - \sigma_{ox}^{0p} v_{ox}^p N_{ox}^{0+} + G_{ox}^{0+} \tag{5.42}$$

and similarly we get the following for the rate equation for the neutral electron traps

$$\frac{\partial}{\partial t} N_{ox}^{0-} = \sigma_{ox}^{-p} v_{ox}^{p}(1 - \delta^{np}) N_{ox}^{-} - \sigma_{ox}^{0n} v_{ox}^{n} N_{ox}^{0-} + G_{ox}^{0-} \tag{5.43}$$

Here, δ^{np} is the fraction of occupied electron traps that turns into an interface trap due to hole capture. The total rate of interface traps created if

$$\frac{\partial}{\partial t} N_{it} = \sigma_{ox}^{0n} v_{ox}^{n} \delta^{n} N_{ox}^{0-} + \sigma_{ox}^{0p} v_{ox}^{p} \delta^{p} N_{ox}^{0+}$$

$$+ \sigma_{ox}^{-p} v_{ox}^{p} \delta^{np} N_{ox}^{-} + \sigma_{ox}^{+n} v_{ox}^{n} \delta^{pn} N_{ox}^{+} \tag{5.44}$$

We have not yet specified which portion of these traps is donor-like and which is acceptor-like. In a more sophisticated approach, this would certainly be determined by the microscopic description of the trapping mechanism. This is far beyond the semi-empirical level we are following here. As a matter of fact, from the experimental viewpoint it is not even easy to determine which kind of interface states were created. In the charge-pumping current, for example, we see only the net effect of all interface states. If we investigate their density of states with the charge-pumping experiment, we obtain additional information about the change in proportions in the upper and lower bandgap due to hot-carrier stress. By assigning interface states in the lower bandgap to be of donor type and those in the upper bandgap to be of acceptor type, as we have done, we can determine the ratio v of the interface traps which turned into acceptor or donor-like interface states. This allows us to separate donor and acceptor-like interface traps in Eq. (5.44). We obtain the rate equation for donor-type interface traps to be

$$\frac{\partial}{\partial t} N_{it}^{+} = v\sigma_{ox}^{0n} v_{ox}^{n} \delta^{n} N_{ox}^{0-} + v\sigma_{ox}^{0p} v_{ox}^{p} \delta^{p} N_{ox}^{0+}$$

$$+ v\sigma_{ox}^{-p} v_{ox}^{p} \delta^{np} N_{ox}^{-} + v\sigma_{ox}^{+n} v_{ox}^{n} \delta^{pn} N_{ox}^{+} \tag{5.45}$$

and that for acceptor-like interface states is then

$$\frac{\partial}{\partial t} N_{it}^{-} = (1 - v)\sigma_{ox}^{0n} v_{ox}^{n} \delta^{n} N_{ox}^{0-} + (1 - v)\sigma_{ox}^{0p} v_{ox}^{p} \delta^{p} N_{ox}^{0+}$$

$$+ (1 - v)\sigma_{ox}^{-p} v_{ox}^{p} \delta^{np} N_{ox}^{-} + (1 - v)\sigma_{ox}^{+n} v_{ox}^{n} \delta^{pn} N_{ox}^{+} \tag{5.46}$$

With Eqs. (5.38), (5.39), (5.42), (5.43), (5.45) and (5.46), we find the following for the reaction matrix A and the vector B in Eq. (5.33)

$$A = \begin{vmatrix} -\Gamma^{+n} & 0 & 0 & 0 & (1-\delta^p)\Gamma^{0p} & 0 \\ 0 & -\Gamma^{-p} & 0 & 0 & 0 & (1-\delta^n)\Gamma^{0n} \\ v\delta^{pn}\Gamma^{+n} & v\delta^{np}\Gamma^{-p} & 0 & 0 & v\delta^p\Gamma^{0p} & v\delta^n\Gamma^{0n} \\ (1-v)\delta^{pn}\Gamma^{+n} & (1-v)\delta^{np}\Gamma^{-p} & 0 & 0 & (1-v)\delta^p\Gamma^{0p} & (1-v)\delta^n\Gamma^{0n} \\ (1-\delta^{pn})\Gamma^{+n} & 0 & 0 & 0 & -\Gamma^{0p} & 0 \\ 0 & (1-\delta^{np})\Gamma^{-p} & 0 & 0 & 0 & -\Gamma^{0n} \end{vmatrix}$$

(5.47)

$$B = \begin{vmatrix} 0 \\ 0 \\ 0 \\ 0 \\ G_{ox}^{0+} \\ G_{ox}^{0-} \end{vmatrix}$$

(5.48)

Here, we have used the notation $\Gamma^{ij} = \sigma^{ij}v_{ox}^{j}$ with $i = +, 0, -$ and $j = n, p$. The trap kinetics is completely described by Eq. (5.33). The current densities $v_{ox}^{n/p}$ are provided by solving Eqs. (5.26)–(5.28) for the electrons and holes injected into the gate oxide. But the problem is to find appropriate values for the parameters, which are: σ^{ij}, δ^n, δ^p, δ^{pn}, δ^{np}, and v. For the cross-sections σ^{ij} we can distinguish capture in neutral traps ($i = 0$) and capture in charged traps ($i = +, -$). The basic interaction for the latter is Coulomb attraction, which makes it a very efficient interaction compared with that between the charge and neutral traps based on higher dipole moments. We therefore expect the former to have larger cross sections than the latter. For an order of magnitude, a value of $\sigma \approx 10^{-14}\,\mathrm{cm}^2$ is found in the literature for the capture cross-section of charged traps. To determine the capture cross section of neutral traps is far more complicated because there is evidence of a possible field dependence. For an order of magnitude estimate, we show reasonable values for σ^{0n} and σ^{0p} in the table below.

Table 1

$\sigma^{ij}/\mathrm{cm}^2$ i	$+$	0	$-$
j			
n	10^{-14}	10^{-17} (?)	10^{-14}
p	10^{-14}	10^{-15}	10^{-14}

The question mark indicates that a field-independent cross section might not be correct (Nissan-Cohen 1986). Direct interface-state generation by a one-step electron process, Eq. (5.35b), is less significant than by a one-step hole process, Eq. (5.35a). The experimental data seems to indicate that the generation of interface-state traps is very effective if an electron is captured in an occupied hole trap, Eq. (5.35c). The other two-step mechanism where an interface trap is generated by capture of a hole in an occupied electron trap, Eq. (5.35d), seems to have no importance. This gives $\delta^n = \delta^{np} = 0$ and leaves δ^p and δ^{pn} to be freely adjustable parameters. Only in a region close to the interface it is possible to transform bulk traps into interface states. To take this into account, we will introduce an explicit spatial dependence in δ^i that provides a cut-off $x_0 \approx 3\,\text{nm}$ for the transformation of bulk traps into interface states

$$\delta^i(x) = \delta^i(0)\exp\left(-(x/x_0)^2\right); \quad i = p, pn \tag{5.49}$$

The ratio in which the interface traps split up into donor and acceptor-like interface states is a complex issue. Following the D_{it} measurements, an increase is always observed in both. A reasonable choice would be $v = 0.5$. With regard to oxide trap generation, we use

$$G_{\text{ox}}^{0-}(x, y) = v_{\text{ox}}^n(x, y)\alpha_{\text{ox}}^{-n}(F_n) + v_{\text{ox}}^p(x, y)\alpha_{\text{ox}}^{-p}(F_p) \tag{5.50}$$

$$G_{\text{ox}}^{0+}(x, y) = v_{\text{ox}}^n(x, y)\alpha_{\text{ox}}^{+n}(F_n) + v_{\text{ox}}^p(x, y)\alpha_{\text{ox}}^{+p}(F_p) \tag{5.51}$$

Here, $\alpha_{\text{ox}}^{ij} = A_{ij}\exp\left(-B_j/F_j\right)$ has a field dependence which is characteristic for a field-induced process such as impact ionization or field-enhanced emission. Appropriate parameters are (Hsu et al. 1988)

Table 2

$A_{ij}\,\text{cm}$ i	$-$	$+$
j		
n	$4.9 \cdot 10^5$	$9.0 \cdot 10^5$
p	$9.0 \cdot 10^5$	$4.9 \cdot 10^5$

and the critical field is $B_n = 1.75 \cdot 10^7\,\text{V/cm}$ and $B_p = 3.5 \cdot 10^7\,\text{V/cm}$ for electron and hole traps, respectively. A generation term related to the non-stationary current density v_{noeq} may be added to Eqs. (5.50) and (5.51). These ballistic carriers can pick up sufficient energy in the electric field to create a trap. As they travel through the oxide, however, their density decreases exponentially. For very thin oxides ($t_{\text{ox}} \approx \lambda_{\text{ox}}$), these carriers might contribute significantly to the trap generation rate. In that case, we also strike the limits of the drift–diffusion approach for carrier transport in the gate oxide, and the model becomes ambiguous.
There is no doubt that the numbers given above can be contested. Quantitative results about the physics of oxide trapping are very sensitive to sample preparation and are therefore of limited value. In this regard, the values listed above provide a starting point for an iteration procedure to adjust the

trap dynamics to a specific technology. Once a technology is characterized by a set of parameters, the trapping process can be investigated under hot-carrier stress. It is no problem to include general spatial or field-dependent cross-sections in the above formulation if this dependence is known. It is not always clear how to extract this information from the experimental data and whether this data can be applied to the hot-carrier stress experiment on a device. The homogeneous injection experiments, for instance, test the gate area or channel region of the MOS capacitor or MOSFET while hot-carrier stress is more sensitive to source and drain regions where the gate edges are. Assuming that mechanical stress and strain in the peripheral regions of the gate oxide are relaxed towards the channel region, both the oxide quality and the trap dynamics might be different. In small devices, these edge effects might completely determine the gate oxide quality. An independent measurement of trap densities and cross sections in these edge regions is currently not possible because no experimental tool is available which guarantees the required spatial resolution of about a few tens of nanometers to scan the gate edge regions.

Before we turn to examples of how the above model can be utilized, two brief remarks for its implementation in numerical code are in order. The first comment deals with the solution of the self-consistent trap problem. The charges build up in the oxide might eventually modify the electric field so that the carrier injection rate also responds significantly to this change. To be able to monitor this coupling, which is *a priori* unknown, we linearized Eq. (5.33) and developed a time-step control algorithm for the calculation procedure outlined at the beginning of this section. This leaves us with a linear system to solve between two successive time steps t and $t + \Delta t$

$$N_{\text{ox}}(t + \Delta t) = (1 - A(t)\Delta t)^{-1} N_{\text{ox}}(t) + (1 - A(t)\Delta t)^{-1} \Delta t B(t)$$

$$(5.52)$$

If the change of the injected current density v_{noeq} is too large for a chosen time step Δt, then Δt is reduced until the change in v_{noeq} is within the given bounds. By a careful choice of these bounds, a stable solution of the complex numerical problem can be obtained even in the case of strong feedback of the oxide charge onto the field configuration in the active device area. A second problem is to adjust the numerical grid to the trap situation in the oxide. For a virgin device without oxide traps, the optimal grid might be different than the one needed if localized oxide charge is building up. For a numerically accurate evaluation of the charge feedback in particular, it is inevitable to let the grid relax dynamically to the changing situation in the gate oxide. A different approach would be to start with an overkill on grid points. However, this would increase the computational cost dramatically. As an initial condition to solve Eq. (5.33), we must specify the trap distribution of hole and electron traps, N_{ox}^{+0} and N_{ox}^{-0}, throughout the gate

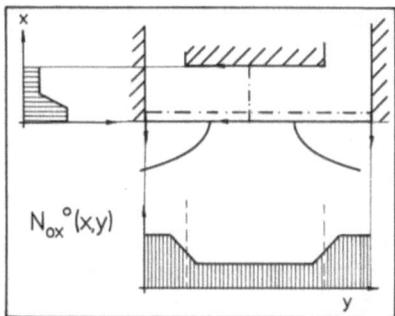

Fig. 64 Two-dimensional profile of neutral oxide traps for the virgin device

oxide area of the unstressed device. In Fig. 64, we show the profile we used
for the following examples.
Due to the effect of mechanical stress at edges and interfaces, we have a
slightly enhanced trap density at these positions. If this enhancement is not
too pronounced, the results of the calculation are not very sensitive to the
detailed shape of the trap profile. The absolute values of these trap densities
can vary considerably if different devices have been differently processed.
They must therefore be determined for a particular technology by adjusting
a stress calculation to a stress experiment. A second required input is the
energy-dependent density of states $D_{it}(\varepsilon)$ for the virgin device according to
Eqs. (5.1) and (5.2). From charge-pumping measurements, we find a reasonable fit for $\alpha = 1.5$. For a device-related technology, the interface state
density for the unstressed device is very small, which means that the choice
of the initial values $D_{max}^{A/D}$ and $D_0^{A/D}$ is not important for the outcome of the
degradation process. The calculations were performed with the device
simulator MINIMOS in which we implemented a program module for a
self-consistent trapping calculation. The profiles were provided by a 2D
process simulation that was calibrated carefully to fit the corresponding
technology. Because the main degradation effects in both NMOS and
PMOS devices originated primarily from the injection of majority carriers
(NMOS: interface state generation is related to hole injection; PMOS:
negative oxide charge is related to electron injection), we must solve the full
set of semiconductor equations, including impact ionization.
    The first example shows a comparison of NMOS and PMOS devices
from a CMOS process. The first step is to verify the drain and substrate
currents for the virgin device by a 2D process and device simulation. We
must then find the trap densities $N_{ox}^{On}\,(t = 0)$ and $N_{ox}^{Op}(t = 0)$ for the virgin
device. The important degradation mechanism for the NMOS device is the
generation of interface states as shown in the previous section. Whether
created by a one or two-step process, both give a generation rate that is

proportional to the number of available hole traps $N_{ox}^{Op}(t = 0)$. This number is therefore obtained by adjusting the calculation to the NMOS stress experiment for a stress condition where interface state generation is dominant (maximum substrate current). The number of electron traps $N_{ox}^{On}(t = 0)$ is found by adjusting the calculation to the PMOS device for a stress condition where the formation of negative oxide charge is important (maximum gate current). For the technology under investigation, we find from the experimental data that $N_{ox}^{Op}(t = 0) \sim 2 \cdot 10^{18}$ cm^{-3} and N_{ox}^{On} $(t = 0) \sim 10^{19}$ cm^{-3}. These numbers are sensitive to technological variations (different S/D complex, especially LDD spacer technique, different gate processing, for instance p^+ poly for n-MOSFET) and must be determined individually for each technology. We find, however, that stress experiments are sufficiently described within a single technology without changing parameters. For the capture cross-sections, we assume σ^{On} $= 10^{-17}$ cm^2 and $\sigma^{Op} = 10^{-15}$ cm^2 in accordance with values found in the literature (Table 1). For trap generation, which turns out to be of significance in the PMOS device, (generation of electron traps) we use a threshold process with a threshold energy 2.3 eV.

If Fig. 65, we show a comparison of the calculated time dependence of DC stress experiments for NMOS and PMOS devices with gate lengths

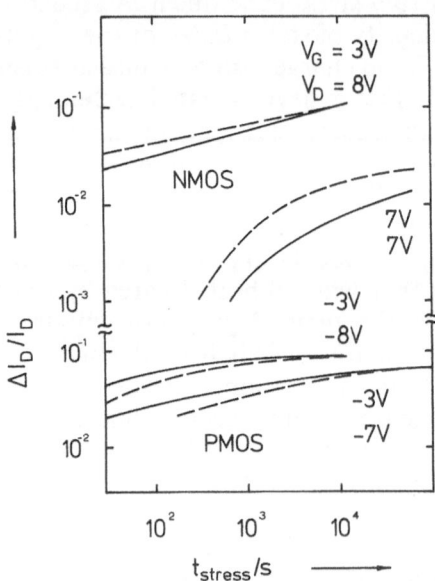

Fig. 65 Stress experiment on NMOS and PMOS devices from the same CMOS process. Solid line—experiment; dashed line—simulation. The change of the drain current was monitored in the linear regime for $V_D = |0.1| V$

$L = 0.65$ μm and oxide thickness $t_{ox} = 16$ nm. The experiments were performed for stress times ranging from a few hours to a day. We characterized the stressed sample in the linear region by monitoring the time dependence of $\Delta I_D / I_D (t = 0)$, where I_D is measured at $V_G = 3$ V and $V_D = 0.1$ V. Other parameters, such as the threshold shift or the substrate current, are naturally also possible.

For a large amount of stress, we find good agreement between experiment and theory for both kinds of devices. There is greater disagreement in the case of low stress level. This can be for technology-related reasons such as latent traps, or is an indication of the accuracy of the surface doping concentration in the LDD regions. A larger surface doping will screen the oxide charge more effectively and has less influence on the channel current. For the NMOS device, the $V_G = 3$ V and $V_D = 8$ V (maximum substrate current) degradation is caused predominantly by the formation of interface states. This is not the case for the $V_G = 7$ V and $V_D = 7$ V conditions. Here, we have a non-negligible amount of fixed negative charge. A comparison of the effective maximum surface charge density for both stress conditions shows that for $V_G = 3$ V and $V_D = 8$ V the contribution of negative oxide charge is less than 3% of the total of $3.4 \cdot 10^{12}$ qcm^{-2}. For the $V_G = 7$ V and $V_D = 7$ V stress, the total maximum of the surface charge density is $4 \cdot 10^{11}$ qcm^{-2}, two-thirds of which is contributed by oxide traps. We also find their position to be at the drain-side gate edge. If the surface doping is sufficiently low in this region, there will be an increased series-resistant effect because the charge is not effectively screened and so will repel electrons from the surface. A saturation in the degradation of LDD devices is often attributed to this increased series resistance. Our samples have seen a high LDD dose of $4 \cdot 10^{13}$ cm^{-2}, which provides a surface doping level in the LDD region that prevents a significant increase in series resistance, at least for the investigated damage. For the PMOS device, degradation is caused by the formation of negative oxide traps. In Fig. 66, we show the distribution of occupied electron traps N_{ox} for two instances of time $t_1 = 66$ s and $t_2 = 10^4$ s.

The figure shows a blow-up of the drain-side gate edge. It is clearly seen how the negative oxide charge is increased during stress time. We also find that the region of damage is confined to a distance of about 100 nm in which the peak maximum is approximately centered. In the previous section, we found the position and spread of the damage by adjusting the parameters of a Gaussian distribution to fit the modified terminal currents. Here, we only adjusted the initial trap densities N_{ox}^{0n} and N_{ox}^{0p}, which are primarily related to the quality of the gate oxide. Comparing the results from n- and p-channel devices, we also observe the very pronounced saturation in the p-channel degradation. In Fig. 67, we show the feedback due to the built-up charge on the maximum lateral field and its position in the PMOS device.

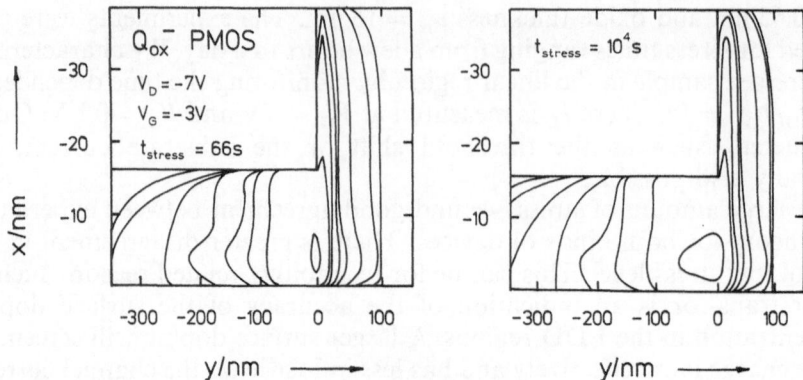

Fig. 66 Distribution of negative oxide charge for stress time $t = 66$ s (left) and $t = 10^4$ s (right). The origin $y = 0$ corresponds to the drain side gate edge. The oxide thickness is 16 nm

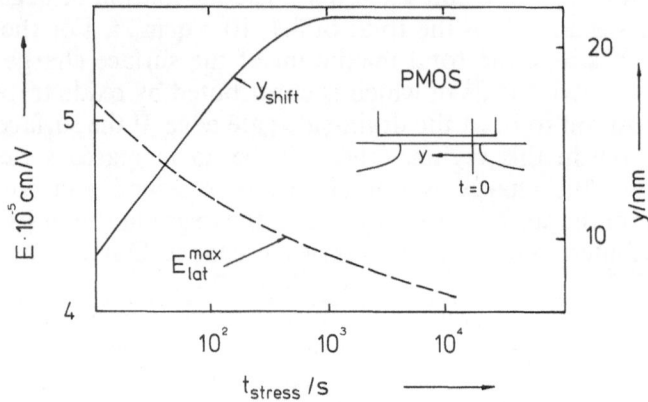

Fig. 67 Feedback of negative oxide charge onto the maximum lateral field in a PMOS device. The shift of the position, measured from its initial position at $t = 0$, corresponds to the channel shortening induced by the negative oxide charge in a PMOS device

The negative charge that accumulates in the oxide during stress reduces the electric field and thus leads to a decrease in the number of injected carriers. To determine this behavior, it is absolutely mandatory to perform a self-consistent calculation of the degradation process. A very interesting feature is that the peak field moves 20 nm from its initial position toward the source end during the stress experiment, which correlates to the channel shortening effect in PMOS devices.

The MOSFET susceptibility to hot carrier degradation is normally monitored by measuring the substrate current characteristics and evaluating the lateral drain field peaks. It is a useful criterion but not enough for S/D optimization. An increase of I_{sx} or of E_{lat}^{max} does not necessarily translate into a lifetime reduction. In this second example, we will consider in some detail how the substrate current correlates to interface state generation and the degradation behavior of n-channel MOSFETs.

Short channel n^+ poly Si gate n-channel MOSFETs with LDD S/D and various channel lengths were used. The gate oxide thickness is 20 nm. The LDD dose and implant energy were varied while keeping the subsequent high dose As implant and thermal budget constant. This covers a large variation of technological parameters within one technology. Unless otherwise specified, the MOSFET stress measurements were performed at the gate voltage V_G corresponding to a maximum substrate current I_{sx}^{max} for a given drain voltage V_{DS}. The stress measurements were performed at room temperature. The device degradation was evaluated by monitoring the relative drain current change $\delta I_{DS}/I_{DS}$ in the linear region. For all the measured cases, the threshold voltage change was below 10 mV which, as was shown in the previous section, indicates that generation of interface states is responsible for the degradation.

The degradation behavior of an LDD MOSFET depends strongly on the stress bias conditions. Increasing the gate voltage at a fixed drain bias reduces the maximum lateral field but increases the drain current. The substrate current usually serves as a monitor for the maximum lateral field and is therefore used to optimize the device for hot-carrier stability. We performed stress experiments for three different operating points with the same initial substrate current as shown in Fig. 68: the usual stress condition at peak substrate current for a given drain bias ($\#2$), higher drain voltage at a lower ($\#1$) and higher ($\#3$) gate voltage.

The stress conditions, the experimentally determined lifetimes τ and the simulated maximum lateral fields E_{Lat}^{max} for these stress conditions are listed

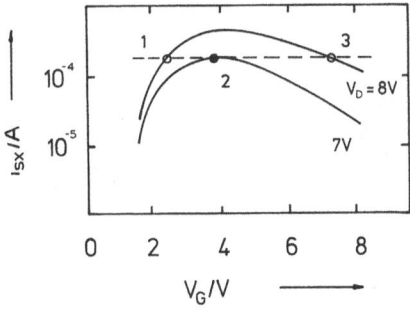

Fig. 68 Equal substrate current stress. A NMOS sample is stressed under equal substrate current I_{sx} for different bias conditions. Devices with $L_G = 1\ \mu m$, $t_{ox} = 20$ nm and LDD with $4 \cdot 10^{13}\ cm^{-2}$ at 60 keV

in Table 3. The difference of device lifetime τ for these conditions is dramatic (Table 3). From #1 to #3 it varies by two orders of magnitude. From the simulation, we obtained the maximum lateral electric field as listed in Table 3. For case #1, it shows a 25% higher field than for cases #2 and #3. The maximum field for the latter two is equal, although they differ by a factor 5 in their lifetimes. With our simulation tool, we calculated the device degradation for the corresponding DC stress experiment. In Fig. 69 we present $\delta I_D/I_D$, evaluated in the linear regime for $V_{GS} = 3$ V and $V_{DS} = 0.1$ V, which corresponds to the experimentally characterized samples.

The calculated lifetimes found for a 3% change in drain current are in good qualitative agreement with the experimental values. The deviation for the case of #1 comes from uncertainties in the channel doping profile that have a strong influence on the substrate current for low gate bias. The slope of the curves for the samples #2 and #3 is very flat, so that an uncertainty in the drain current of 1— can easily account for the disagreement. However, the qualitative behavior of the experimental situation is reproduced very well. The situation of the injected electrons and holes for the different stress conditions #1, #2 and #3 is compared in Figs. 70a–70c.

The electron injection maximum decreases by two orders of magnitude when we go from #1 to #3. In contrast, the maximum hole injection

Table 3

stress condition	1	2	3
lifetime for $\Delta I_{DS}/I_{DS} = 3\%$ [sec]	6.6×10^2	1.8×10^4	9×10^4
E_{lat}^{max} simulation [V/cm]	5.1×10^5	4.3×10^5	4.4×10^5

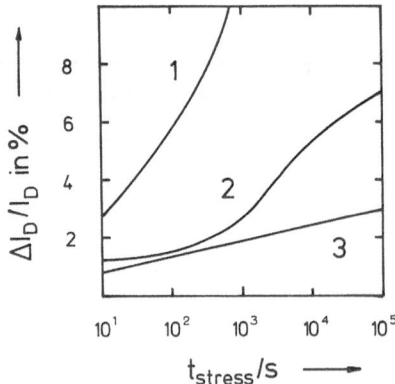

Fig. 69 Time evolution of the maximum interface state charge during the stress experiment for equal substrate current stress. Numbers refer to stress conditions in Fig. 68

reduces by as much as seven orders of magnitude. Hole injection is responsible for the formation of interface states. The degradation behavior reflects this fact in both measurements and simulation. Our simulation tool correctly describes the formation of interface states for different bias

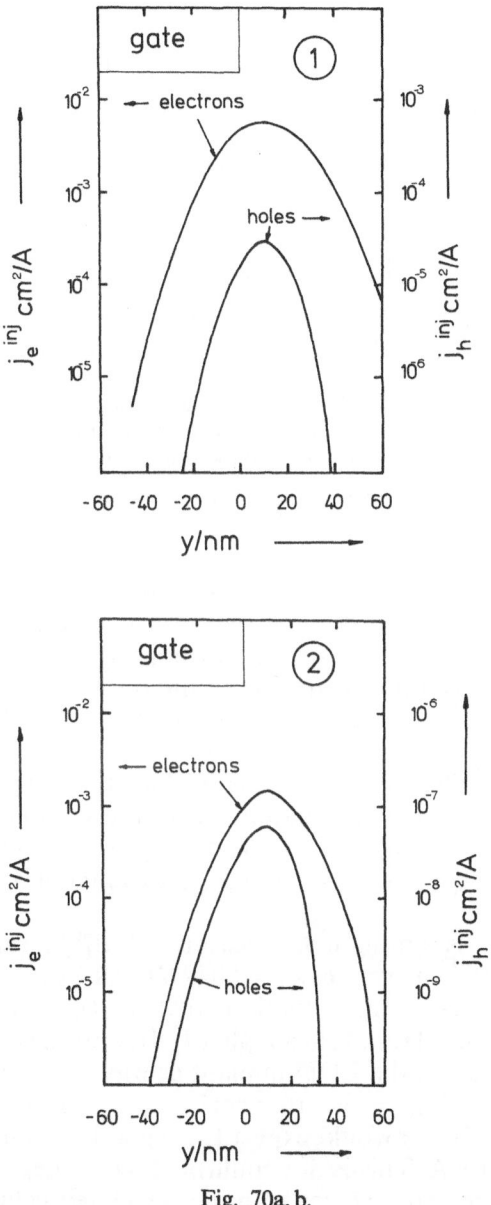

Fig. 70a, b.

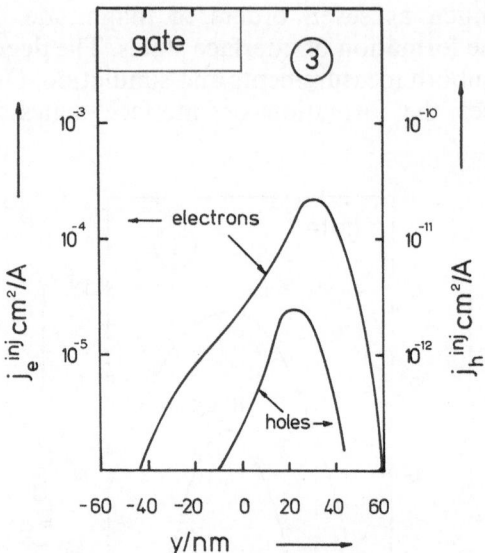

Fig. 70 Spatial distribution of injected electrons and holes along the Si/SiO$_2$ interface. (a) stress condition 1, (b) stress condition 2, (c) stress condition 3. Please note the change of scale for the injected hole current density while going from 1 (hole stress) to 3 (electron stress)

conditions. The simulation allows us to study the generation of interface states separately by either a one or two-step process. In the former, the interface states are created directly by injected holes. The latter initially assumes hole trapping and subsequently trapping of an injected electron to transform the occupied hole trap into an interface state. Both mechanisms give the same results in the degradation behavior of the device and cannot be distinguished by electrical characterization.

Next, we consider different LDD types for which the device lifetime τ results do not correlate with the expectations from the I_{sx} and I_{sx}/I_{DS} measurements. We specify the samples by A, B and C. Fig. 71 shows the effect of the LDD dose D_{LDD} and energy E_{LDD} on I_{sx}/I_{DS} as a function of the polySi gate length L_G.

The I_{sx}/I_{DS} ratio is an indirect measure of E_{Lat}^{max}. Taking LDD type B ($D_{LDD} = 3 \times 10^{13}$ cm^{-2}; $E_{LDD} = 60$ keV) as the reference LDD construction, we have the situation that a D_{LDD} increase (case A: $D_{LDD} = 4 \times 10^{13}$ cm^{-2}) results in a higher I_{sx}/I_{DS} ratio, i.e. E_{lat}^{max} increases. In contrast, an increase of the LDD implant energy E_{LDD} reduces the I_{sx}/I_{DS} ratio as in case C ($D_{LDD} = 3 \times 10^{13}$ cm^{-2}; $E_{LDD} = 80$ keV). According to the results of Fig. 71, we would expect the highest τ results for case C and the lowest τ for case A. This is not confirmed by the degradation measurements, which are shown together with the calculated values in Fig. 72. We

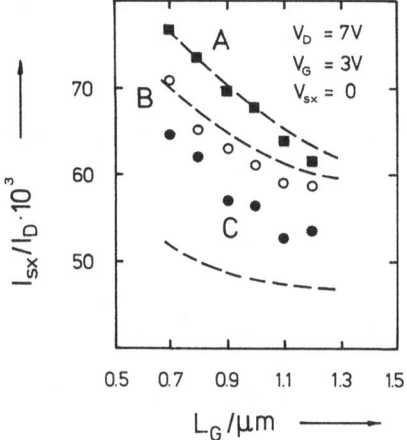

Fig. 71 Substrate current versus drain current for different LDD type devices at different gate length L_G. Filled squares—A: LDD with $4 \cdot 10^{13}$ cm^{-2}/60 keV; open circles—B: LDD with $3 \cdot 10^{13}$ cm^2/60 keV; filled circles—C: LDD with $3 \cdot 10^{13}$ cm^{-2}/80 keV Symbols—experiment; dashed lines—simulation

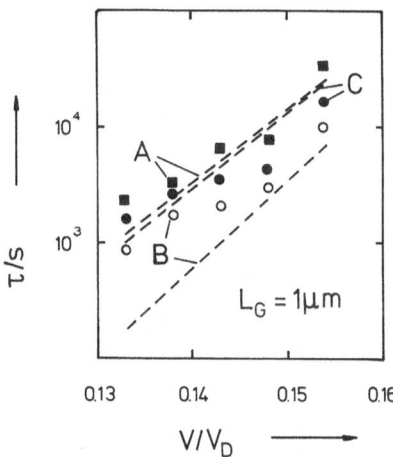

Fig. 72 Lifetime of different LDD type devices for $L_G = 1$ μm. The lifetime was determined by a 2% change of the drain current in the linear mode. Symbols—experiment; dashed lines—simulation; same notation as in Fig. 71

find that τ is higher for case C than for case B in agreement with the I_{sx}/I_{DS} results. But the results for case A are unexpected. In spite of the high I_{sx}/I_{DS} ratio, case A has the longest lifetime.

The simulation describing these experiments is in good agreement with the measured data. There is a deviation for case C, the calculated I_{sx}/I_{DS} is about 20% too small, thus shifting the lifetime estimate to larger τ values. The origin of this discrepancy is not clear at present. It could, however, be related to an inaccurate description of the lateral profiling of S/D or of the effective channel length.

To analyze the degradation behavior of the three samples, we calculated the time development of interface state generation during the stress experiment. In Fig. 73, we show the temporal evolution of the spatial maximum of the interface states occupied during characterization for all three samples.

For cases B and C, we find the maximum to be located outside the gate, and for A it is well below the gate. The surface doping at the location of the maximum interface states is greatest for sample A ($5 \times 10^{17} \, \text{cm}^{-3}$) and least for sample B ($1.4 \times 10^{17} \, \text{cm}^{-3}$) and C ($3.1 \times 10^{17} \, \text{cm}^{-3}$). For long times, the number of occupied interface states correlates with the substrate current measurements. However, they have completely different effects on device performance. Although A has the greatest number of interface states, their effect is screened by the high doping level. Samples B and C require the same number of interface states for a 2% change in the drain current. This level is reached later in sample C, which again corresponds to the substrate current behavior and lifetime results. The saturation (log–log scale!) indicated for A has its origin in the tendency of the acceptor-like interface states to turn themselves off. Moving the Fermi level toward the conduction band edge produces a higher negative interface charge, which in turn reduces the electron density and moves the Fermi level toward the middle of the bandgap. A self-imposed saturation therefore occurs for a high density of interface states.

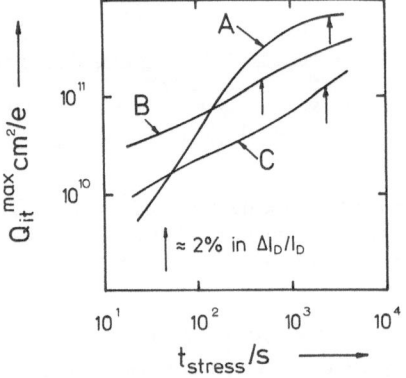

Fig. 73 Time evolution of the maximum interface state charge during the stress experiment for the LDD type devices A, B and C as defined in Fig. 72. The arrows refer to the time when 2% change in drain current in the linear mode was reached

In the foregoing example we have gained some insight into how profiling of the S/D regions has an influence on the hot carrier stability of an LDD MOSFET. In the third and last example we will investigate how the choice of the gate material will modify the hot carrier susceptibility of N- and PMOS devices. Usually the gate material is chosen to be of n-type material, for instance heavily doped poly silicon. The substrate doping is p and n-type for NMOS and PMOS, respectively. To obtain a comparable threshold voltage, in magnitude, for n- and p-channel device the extra volt in work function difference for the PMOS compared to the NMOS has to be compensated by a shallow p-type implant at the surface. As a result of this compensation the channel in the PMOS moves deeper into the substrate than in the uncompensated NMOS device where it is located at the surface. The buried channel PMOS device is more sensitive to punch through for decreasing gate length than the surface channel NMOS device. This has especially consequences for the degradation of short channel PMOS where through the hot carrier induced channel shortening the threshold voltage shift can be significant. If instead of n-type gate material one would use a p-type gate material the PMOS would be a surface channel device and the NMOS a buried channel device with a n-type compensation doping. By the choice of the proper gate material, therefore, both NMOS and PMOS can be designed to be surface channel devices.

We will now study the hot carrier degradation of NMOS and PMOS with different gate material. The devices we have investigated have an oxide thickness of 16 nm and an effective channel length of 0.8 μm. Except for the gate definition sequence the processing was identical for both types of devices especially the source/drain profiling was identical. For the n^+ gate devices conventional phosphorus doped poly silicon was used. The p^+ gate version consisted of boron doped $TaSi_2$/poly silicon bilayers (polycide). The silicide films were prepared by cosputtering Ta and Si from separate targets. A LTO-cap was used to mask the gates against counterdoping from S/D implants. The hot carrier stress of the NMOS device was performed at maximum substrate current, whereas the PMOS was stressed at maximum gate current. The devices were characterized in the linear mode at $V_D = 0.1$ V. In addition we have performed charge pumping measurements to evaluate the number of interface traps prior to and after stress.

In Fig. 74 we show the measured and calculated life time data for the NMOS devices. Sample A is the surface channel device and sample B a buried channel type with an arsenic $1.4 \cdot 10^{12}$ cm^{-2}/100 keV compensation doping. The life time of the surface channel device is significantly shorter than that of the buried channel device. The life time of the devices was determined by a 2% change of the linear region drain current.

The degradation is due predominantly to acceptor type interface state generation. In Fig. 75 we show the calculated maximum interface charge in the characterization mode versus stress time. One has to bear in mind that

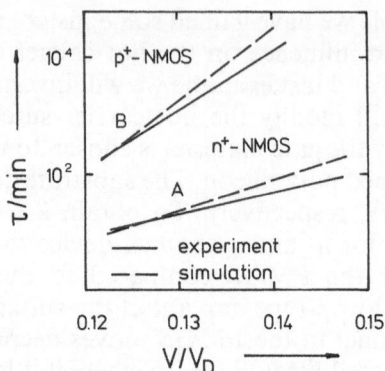

Fig. 74 Comparison of the lifetime of p^+ and n^+ NMOS devices with a comparable effective channel length $L_{\mathrm{eff}} = 0.8 \ \mu$m. Solid lines—experiment, dashed line—simulation. A is a surface channel LDD and B a buried channel device with a As compensation doping of $1.4 \cdot 10^{12} \ \mathrm{cm}^{-2}/100$ keV and the same source drain complex as A. The hole trap density N_{ox}^{0+} of the n^+ NMOS was $5 \cdot 10^{18} \ \mathrm{cm}^{-2}$. To obtain the steeper slope for the p^+ NMOS $N_{\mathrm{ox}}^{0+} = 1 \cdot 10^{19} \ \mathrm{cm}^{-2}$ had to be used

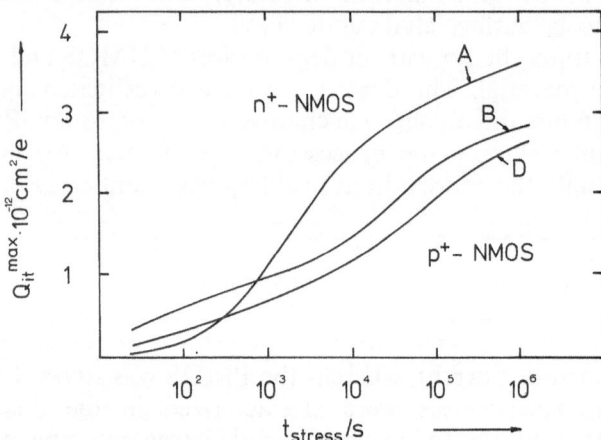

Fig. 75 Time evolution of the maximum interface state charge during the stress experiment for NMOS devices with n^+ and p^+ poly gate material. A and B are the devices specified in Fig. 74. The device D is a buried channel device with an As compensation of $1.8 \cdot 10^{12} \ \mathrm{cm}^{-2}/160$ keV which gives a deeper channel junction than device B

in the characterization experiment only a fraction of the total number of hot-carrier-induced interface states is populated because only those interface states are occupied with charge that are below the electrons Fermi level ε_F^n. A is the surface channel device, B and D are buried channel devices with

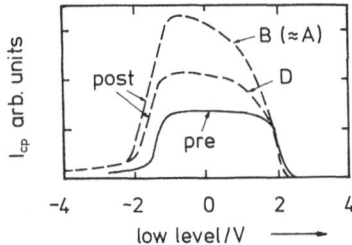

Fig. 76 Charge pumping current for the buried channel devices B and D. Solid line—pre-stress; dashed line—post-stress. The surface channel device A coincides with the device B after hot-carrier stress. The pre-stress data is identical for all three devices

shallow and deep channel junctions, respectively. The device D has an arsenic $1.8 \cdot 10^{12}$ cm^{-2}/160 keV compensation doping which gives a deeper channel junction than device B.

We find that for larger stress times the amount of effective interface charge correlates with the depth of the channel junction and is consequently largest for the surface channel device. The charge pumping signal is directly proportional to the total number of interface states. In Fig. 76 we show the result of a charge pumping measurement prior and after the stress experiment was performed. The pre stress data is the same for all three devices. The total number of interface states that are created during the stress experiment is comparable for the surface channel device A and the buried channel device B and less for device D which has a deeper channel junction than B. The reduced degradation in the buried channel device is, however, the consequence of a lower interface minority carrier density which pulls the quasi Fermi level towards midgap and henceforth reduces the effective interface charge as clearly demonstrated in Fig. 75. The different slope observed in the experimental life time data in Fig. 74 can only be obtained if we increase the initial hole trap density N_{ox}^{0+} for the p^+-NMOS device by a factor of two compared to the n^+-NMOS. This is related to the different process required for the formation of the p^+-gate. A calculation with the same initial N_{ox}^{0+} gives a parallel shift of the n^+-NMOS lifetime curve without a change in slope.

In Fig. 77 we show the change of drain current due to stress induced oxide charge for the PMOS devices in the linear regime. The buried channel PMOS device is more affected than the surface channel device. Its lifetime is approximately 80% reduced compared to the surface channel device. In Fig. 78 we show the calculated change of linear drain current for the PMOS device versus stress time. We find the same trends as observed in the experiments. The negative oxide charge leads to an increase of the drain current as discussed in the previous section. However, mobility degradation will counteract this increase because of the extra scattering.

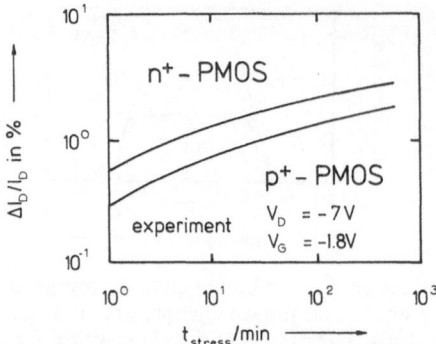

Fig. 77 Time evolution of drain current during hot-carrier stress for surface and buried
channel PMOS. Experimental data

In the previous section we have introduced a model how to incorporate
mobility degradation by oxide charges, compare Eq. (5.7) and the discussion thereafter. In that simplified treatment we assumed that an effective
oxide charge can be projected onto an interface charge which then leads to
mobility decrease. This approach is reasonable in the case of analyzing
degraded samples where only the effective oxide charge is important. In our
more detailed calculation we calculate the distribution of oxide charge in
the entire gate oxide region explicitly. We include this 2D charge distribution in the solution of the Poisson equation and therefore obtain the correct
modification of the electric potential distribution. However, we do not
know from first principles how such a 2D oxide charge distribution acts on
the mobility of the carriers in the silicon. It is reasonable to assume that
charge close to the interface is a more efficient scatterer than charge that is
located deeper in the oxide. This would suggest a projection of the oxide
charge on to the interface with a weight function that is rapidly decreasing
for charge contributions that are located deeper in the oxide removed from
the interface. The scattering strength is related to the overlap of the mobile
carrier wave function with that of the oxide trap. Therefore the intrinsic
cutoff parameter of the interface projection is, due to the absence of long
range order, related to the range of the oxide traps wave function which
should be of the order of the interatomic distances of SiO_2 which is 2–3 Å. If
we include the mobility degradation due to oxide charges, which always
leads to a decrease of current, degradation is slowed further down counter-
acting the increase of current due to the potential effect. In Fig. 78 we have
included the effect of mobility degradation based on the projection of the
2D oxide charge distribution onto an effective interface charge which then
allows to apply the approach from the previous section. Comparing experimental results in Fig. 77 and calculated ones in Fig. 78 we find that the
mobility effect leads to different features than that shown in the experi-

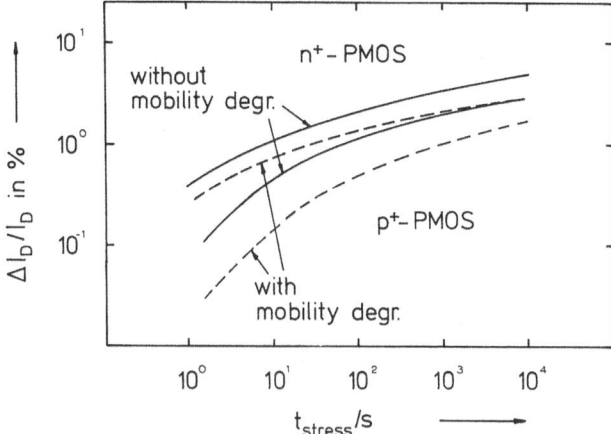

Fig. 78 Time evolution of drain current during hot-carrier stress for surface and buried channel PMOS. Simulated data. Solid line—mobility degradation is switched off; dashed lines—maximum mobility degradation is taken into account. Please note the different time scale.

mental data. Especially for increasing stress time we see a more pronounced influence on the buried channel device (n^+-PMOS) than in the surface channel PMOS (p^+-PMOS). This is directly traced back to the spatial distribution of oxide charges which are shown in Figs. 80 and 81. We have to point out that the problems discussed above are only connected with the effect of oxide charge, distributed in the gate oxide, on the mobility. Charge on interface states is located directly at the interface and therefore the approach discussed in the previous section is justified. In Fig. 79 we show the measured substrate and gate currents for the considered PMOS devices. We see that the order of substrate and gate current is reversed for surface and buried channel device. Both currents are a measure for the maximum electric field in the device as discussed above. Therefore it is surprising that their magnitude is reversed for surface and buried channel PMOS. To find the reason for this behavior we investigate in more detail how the carriers are heated up in the electric field. In Figs. 80 and 81 we show the temperature distribution along the Si/SiO$_2$ interface for minority and majority carriers in the surface channel and buried channel PMOS, respectively. These temperatures were calculated with the carrier temperature model developed in Chapter 3. The calculation fully self consistently accounts for minority carrier and majority carrier temperatures including avalanche generation.

Comparing surface and buried channel device we find that the holes have a considerably higher temperature in the former than in the latter which correlates with the substrate current. For the electrons we find the opposite

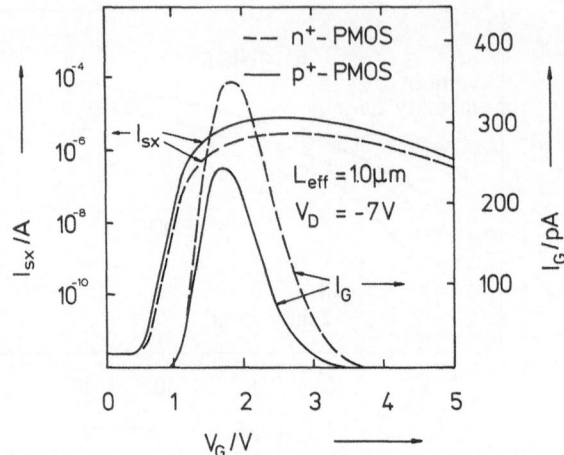

Fig. 79 Substrate and gate currents for n^+ and p^+ poly gate PMOS. Experimental data

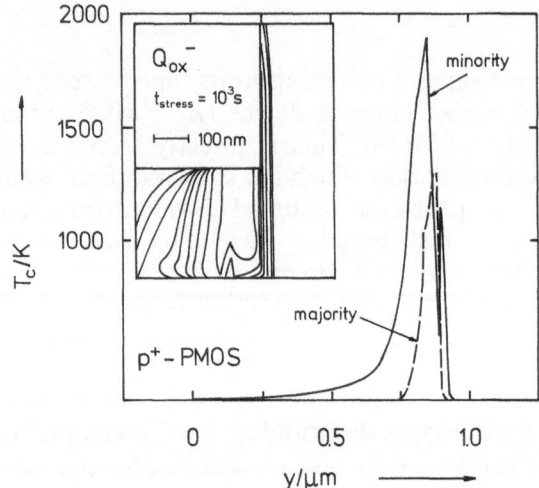

Fig. 80 Minority (solid line) and majority (dashed line) carrier temperature along the Si/SiO$_2$ interface of a surface channel PMOS device. The insert shows the distribution of negative oxide charge after 10^3 s stress time. The stress condition was $V_D = -7$ at maximal gate current

result which correlates with the measured gate current. The physical reason for this behavior is that minority carriers are heated up in the lateral electric field to generate majority carriers which gives the substrate current. The majority carriers are heated up in the transverse electric field and are injected into the gate oxide to give the gate current. Because the position of

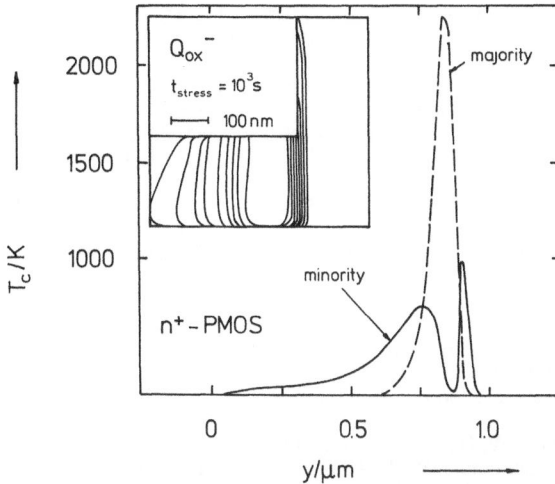

Fig. 81 Minority (solid line) and majority (dashed line) carrier temperature along the Si/SiO$_2$ interface of a buried channel PMOS device. The insert shows the distribution of negative oxide charge after 10^3 s stress time. The stress condition was $V_D = -7$ at maximal gate current

maximum impact ionization is deeper in the buried channel device the electrons can pick up more energy in the transverse field and this gives a higher gate current in the buried channel PMOS. We also show as an insert in Figs. 80 and 81 the distribution of negative oxide charge after 10^3 s stress time. There is a characteristic difference in the surface (p^+-gate) and buried (n^+-gate) channel PMOS. Due to the enhanced injection of holes, negative charge near the interface is compensated in the former. That moves the center of charge into the gate oxide. As a result it less effects the drain current which we discussed in connection with the mobility degradation. Dual work function engineering, as discussed in this last example, provides the device engineer in addition to S/D-profiling with an alternative for the optimization of NMOS and PMOS devices. The advantage of an improved short channel behavior by only using surface channel devices is, however, compensated by a higher process complexity due to two-gate formation processes and a significantly reduced hot carrier hardness of the surface channel NMOS.

References

Bellens R., Heremans, P., Groeseneken, G., Maes H. E. (1988): IEEE Electr. Device Lett. *EDL9*, 232.
Berglund C. N., Powell R. J. (1971): J. Appl. Phys. *42*, 573.

Burger J. S., Jespers R. G. A. (1969): IEEE Trans. Electr. Devices *ED16*, 297.
De Keersmaeker R. F. (1983): In: Insulating Films on Semiconductors. (ed. J. F. Verweij and
    D. Wolters) North Holland, Amsterdam p 85.
Di Maria D. J. (1978): In: The Physics of SiO_2 and Its Interfaces. (ed. S. T. Pantelides).
    Pergamon, New York p 160.
Di Maria D. J., Stasiak J. W. (1989): J. Appl. Phys. *65*, 2342.
Fischetti M. V., Di Maria D. J., Brorson S. D., Theis T. N., Kirtley J. R. (1985): Phys. Rev.
    B31, 8124.
Fischetti M. V., Di Maria D. J., Dori L., Batey J., Tiernney E., Stasiak J. (1987): Phys. Rev.
    B35, 4404.
Grunthener F. J., Grunthener P. J. (1986): Materials Science Reports 1, North Holland,
    New York p 65.
Groeseneken G., Maes H. E., Beltran N., De Keersmaeker R. F. (1984): IEEE Trans. Electr.
    Devices *ED31*, 42.
Hänsch W., Weber W. (1989): The Physics of Hot Carrier Degradation in Si-MOSFET's: a
    Satellite Workshop in Connection With INFOS 89 München. Appl. Surf. Sci. *39*, 511.
Hänsch W., Orlowski M., Weber W. (1988): J. Physique Colloque *C4*, 597.
Heyns M. M., Krishna Rao D., De Keersmaeker R. F. (1989): Appl. Surf. Sci. *39*, 327.
Hofmann F., Hänsch W. (1989): J. Appl. Phys. *66*, 3092.
Hofmann F., Krautschneider W. H. (1989): J. Appl. Phys. *65*, 1358.
Hofmann K. R., Wrener Ch., Weber W., Dorda G. (1985): IEEE Trans. Electr. Devices *ED
    32*, 691.
Hsu C. C., Nishida T., Sah C. T. (1988): J. Appl. Phys. *63*, 5882.
Hu C., Tam C., Hsu F. C., Ko P. K., Chan T. Y., Terril K. W. (1985): IEEE Trans. Electr.
    Devices *ED 32*, 375.
Hughes R. C. (1978): Solid State Electr. *21*, 251.
Lai S. K. (1983): J. Appl. Phys. *54*, 2540.
Lillienfeld J. E. (1930): US Patent 1,745,175.
Lyon S. A. (1989): Appl. Surf. Sci. *39*, 552.
Ma T. P. (1989): In: Semiconductor Science and Technology Vol. 4, 1061.
Ning T. H., Osburn C. M., Yu H. N. (1977): J. Appl. Phys. *48*, 286.
Nissan-Cohen Y., Shappir J., Frohmann-Bentchkowsky D. (1986): J. Appl. Phys. *60*, 2024.
Nissan-Cohen Y., Gorczyca T. (1988): IEEE Electr. Device Lett. *9*, 287.
Orlowski M. K., Werner Ch., Klink J. P. (1989a): IEEE Trans. Electr. Devices *ED36*, 375.
Orlowski M. K., Werner Ch. (1989b): IEEE Trans. Electr. Devices *ED36*, 382.
Schwerin A. (1988): Oxidegradation von MOS transistoten durch heiße Ladungsträger.
    Thesis. Leopold Franzens Universität Innsbruck Austria.
Schwerin A., Hänsch W., and Weber W. (1987): IEEE Trans. Electr. Devices *ED 34*, 2493.
Sun S. C., Plummer J. D. (1980): IEEE Trans. Electr. Devices *ED27*, 1497.
Takeda E., Suzuki N. (1983): IEEE Electr. Device Lett. *EDL4*, 111.
Weber W. (1988): IEEE Trans. Electr. Devices *ED35*, 1476.

Appendix I: Perturbation Theory and Diagram Technique

In Chapter 1, we introduced Green's functions to calculate some properties of many-particle systems. We considered the equilibrium case in Chapter 1.2 and the non-equilibrium case in Chapter 1.3. In both cases we found similar concepts of obtaining Green's function as the solution of the Dyson equation with a properly determined self-energy operator. The aim of the many-body perturbation theory is to formalize a procedure to calculate the self-energy in ascending order of interaction strength. The equilibrium Green's function was at $T = 0$, a ground state expectation value of a time-ordered product of field operators. At $T \neq 0$, it was a grand canonical ensemble average of the same time-ordered product. Both cases require a different perturbational approach, however. Whereas we can develop a perturbation theory in the physical space-time domain for $T = 0$, this is not possible for $T \neq 0$. Here, a thermodynamic perturbation theory was developed. The basic ideas for $T = 0$ carry directly over to the Keldysh formulation for the non-equilibrium Green's function even at finite temperatures. We need only take account of the modified time-ordering path. So we will discuss in detail the basic concepts of the perturbation theory in real space-time that was developed for $T = 0$ (Abricosov 1963, Fetter and Walecka 1971). From Eq. (1.80), we have the following for Green's function

$$G(x\tau, x'\tau') = -i \frac{\langle 0, \infty | T_\tau S \, \Psi_{\mathrm{I}}(x\tau) \Psi_{\mathrm{I}}^+ (x'\tau') | - \infty, 0 \rangle}{\langle 0, \infty | S | - \infty, 0 \rangle} \qquad (A1.1)$$

with the S-matrix from Eq. (1.81)

$$S = T_\tau \exp\left(-i \int_{-\infty}^{\infty} d\tau' H_{\mathrm{I, int}}(\tau') \right) \qquad (A1.2)$$

The interaction is given either by a coupling to an external potential $V(x)$, which could be an applied field or a field created by impurities, or an electron coupling to the phonon field. For the first kind of interaction, we get

$$H_{\text{I, int}}(\tau) = \int dx\, V(x) \Psi_{\text{I}}^{+}(x\tau) \Psi_{\text{I}}(x\tau) \qquad \text{(A1.3)}$$

and the second is

$$H_{\text{I, int}}(\tau) = g \int dx\, \Phi_{\text{I}}(x\tau) \Psi_{\text{I}}^{+}(x\tau) \Psi_{\text{I}}(x\tau) \qquad \text{(A1.4)}$$

Here, g is the coupling strength of the electron–phonon coupling and $\Phi(x\tau)$ is the phonon field operator, which is of boson-type and obeys a commutation rule according to Eq. (1.4). Equation (A1.4) is Eq. (1.16) expressed in field operators. It is worthwhile noting that the electron–phonon interaction contains two electron field operators and one phonon field operator. This will be of importance later. We will not include electron–electron interaction, and will therefore omit its interaction term. We can express the electron field operators $\Psi(x\tau)$ according to Eq. (1.6) through the complete basic set of the one-particle states s.

$$\Psi(x\tau) = \sum_{s} \varphi_{s}(x) a_{s}(\tau) \qquad \text{(A1.5)}$$

$$\Psi^{+}(x\tau) = \sum_{s} \varphi_{s}^{*}(x) a_{s}^{+}(\tau)$$

and for the phonon field we have

$$\Phi(x\tau) = \sum_{q} \Omega_{q}[b_{q}(\tau)\exp(iqx) + b_{q}^{+}(\tau)\exp(-iqx)] \qquad \text{(A1.6)}$$

with $\Phi(x\tau) = \Phi^{+}(x\tau)$. The function Ω_{q} is determined by the specific coupling mechanism. Fermion-like and boson-type creation operators a_{s} and b_{q} obey the anti-commutation or commutation rules of Eq. (1.4), respectively.

To obtain an expression for the Green's function, Eq. (A1.1) in ascending order of the interaction, we formally expand the S-matrix

$$G(x\tau, x'\tau') = \frac{-i}{\langle 0, \infty|S|-\infty, 0\rangle} \sum_{n=0}^{\infty} \frac{(-i)^{n}}{n!} \int_{-\infty}^{\infty} d\tau_{1} \cdots \int_{-\infty}^{\infty} d\tau_{n}$$

$$\cdot \langle 0, \infty|T_{\tau}H_{\text{I, int}}(\tau_{1}) \ldots H_{\text{I, int}}(\tau_{n})\Psi_{\text{I}}(x\tau)\Psi_{\text{I}}^{+}(x'\tau')|-\infty, 0\rangle \qquad \text{(A1.7)}$$

If we write down the first two contributions explicitly, we have

$$G(x\tau, x'\tau') = \frac{1}{\langle 0, \infty|S|-\infty, 0\rangle}$$

$$\cdot [-i \langle 0, \infty|T_{\tau}\Psi_{\text{I}}(x\tau)\Psi_{\text{I}}^{+}(x', \tau')|-\infty, 0\rangle +$$

$$-i\frac{(-i)^{1}}{1} \int_{-\infty}^{\infty} d\tau_{1} \langle 0, \infty|T_{\tau}H_{\text{I, int}}(\tau_{1})$$

$$\cdot \Psi_{\text{I}}(x\tau)\Psi_{\text{I}}^{+}(x'\tau')|-\infty, 0\rangle + \ldots] \qquad \text{(A1.8)}$$

In any order of the interaction, we must therefore calculate the expectation value of a time-ordered product of field operators. Owing to some remarkable properties of the field operators, which are directly related to their commutation rules, this can be done in a very easy way. The key concept is the "pairing" of field operators. We consider the time-ordered product of two fermion field operators $\Psi_I^+(x\tau)$ and $\Psi_I(x'\tau')$, which is as follows with Eq. (A1.5)

$$T_\tau\{\Psi_I^+(x'\tau')\Psi_I(x\tau)\} = \sum \varphi_{s'}^*(x')\varphi_s(x)\exp(i\varepsilon_{s'}\tau' - i\varepsilon_s\tau)a_{s'}^+ a_s\theta(\tau' - \tau)$$
$$- \sum \varphi_{s'}^*(x')\varphi_s(x)\exp(i\varepsilon_{s'}\tau' - i\varepsilon_s\tau)a_s a_{s'}^+ \theta(\tau - \tau')$$

$$(A1.9)$$

We now consider their normal ordered product, which is defined as an operation that rearranges the field operators so that the creation operator Ψ_I^+ is moved to the left of the annihilation operator Ψ_I.

$$N\{\Psi_I^+(x'\tau')\Psi_I(x\tau)\} = \Psi_I^+(x'\tau')\Psi_I(x\tau) \qquad (A1.10)$$
$$N\{\Psi_I(x\tau)\Psi_I^+(x'\tau')\} = -\Psi_I^+(x'\tau')\Psi_I(x\tau)$$

and, applied to the ground-state vector of the non-interacting system it is identically zero by virtue of

$$\Psi_I(x\tau)|0, -\infty\rangle = 0 \qquad (A1.11)$$

Using the number representation of the field operators, we obtain

$$N\{\Psi_I^+(x'\tau')\Psi_I(x\tau)\} = \sum \varphi_{s'}^*(x')\varphi_s(x)\exp(i\varepsilon_{s'}\tau' - i\varepsilon_s\tau)a_{s'}^+ a_s\theta(\varepsilon_s - \mu)$$
$$- \sum \varphi_{s'}^*(x')\varphi_s(x)\exp(i\varepsilon_{s'}\tau' - i\varepsilon_s\tau)a_s a_{s'}^+ \theta(\mu - \varepsilon_s)$$

$$(A1.12)$$

Here μ is the Fermi energy up to which the states are occupied. In Eq. (A1.12) we used the fact that the conjugate of Eq. (A1.11) is also true. The minus sign of the second term comes from the anti-commutation property of the number operators a_s and a_s^+ and the diagonal term $s = s'$ is zero because all states below the Fermi energy μ are already occupied. For the difference between time-ordered and normally ordered product, we obtain a c-number

$$T_\tau\{\Psi_I^+(x'\tau')\Psi_I(x\tau)\} - N\{\Psi_I^+(x'\tau')\Psi_I(x\tau)\} = -ig_0(x\tau, x'\tau')$$

$$(A1.13)$$

which is easily verified by virtue of

$$g_0(x\tau, x'\tau') = -i\langle -\infty, 0|T_\tau\Psi_I(x\tau)\Psi_I^+(x'\tau')|0, -\infty\rangle \qquad (A1.14)$$

A "pairing" of two field operators is defined by the difference between its time and normally ordered products. Although many different combinations can be paired, only a few give a non-zero result. For fermion field operators $\Psi_1(x\tau)$, only "pairings" of destruction and creation operators $\Psi_1(x\tau)$ and $\Psi_1^+(x\tau)$ give a non-zero contribution. Pairings between fermion and boson field operators $\Psi_1(x\tau)$ and $\Phi_1(x\tau)$, respectively, always disappear. The "pairing" of two phonon field operators gives the zero-order phonon Green's function $d_0(x'\tau', x\tau)$.

$$d_0(x\tau, x'\tau') = -i\langle -\infty, 0|T_\tau\Phi_1(x\tau)\Phi_1(x'\tau')|0, -\infty\rangle \qquad (A1.15)$$

Therefore the only "pairings" that contribute are

$$T_\tau\{\Psi_1(x\tau)\Psi_1^+(x'\tau')\} - N\{\Psi_1(x\tau)\Psi_1^+(x'\tau')\} = ig_0(x\tau, x'\tau') \qquad (A1.16)$$

and

$$T_\tau\{\Phi_1(x\tau)\Phi_1(x'\tau')\} - N\{\Phi_1(x\tau)\Phi_1(x'\tau')\} = id_0(x\tau, x'\tau') \qquad (A1.17)$$

The clue to evaluate the time-ordered products in Eq. (A1.7) is to employ the concept of "pairing". As an example let us consider the second term of Eq. (A1.8). If we assume an interaction of the type in Eq. (A1.3), we must calculate the ground state expectation value $\langle 0, \infty|T_\tau\Psi_1^+(x_1\tau_1)\Psi_1(x_1\tau_1)\Psi_1(x\tau)\Psi_1^+(x'\tau')|-\infty, 0\rangle$. For the time order $\tau_1 > \tau > \tau'$, we obtain the following by repeatedly utilizing Eq. (A1.16)

$$\langle 0, \infty|\Psi_1^+(x_1\tau_1)\Psi_1(x_1\tau_1)\Psi_1(x\tau)\Psi_1^+(x'\tau')|-\infty, 0\rangle =$$
$$- \langle 0, \infty|\Psi_1^+(x_1\tau_1)\Psi_1(x_1\tau_1)\Psi_1^+(x'\tau')\Psi_1(x\tau)|-\infty, 0\rangle$$
$$+ \langle 0, \infty|\Psi_1^+(x_1\tau_1)\Psi_1(x_1\tau_1)|-\infty, 0\rangle ig_0(x\tau, x'\tau') =$$
$$+ \langle 0, \infty|\Psi_1^+(x_1\tau_1)\Psi_1^+(x'\tau')\Psi_1(x_1\tau_1)\Psi_1(x\tau)|-\infty, 0\rangle$$
$$+ ig_0(x\tau, x_1\tau_1)ig_0(x_1\tau_1, x\tau') - ig_0(x_1\tau_1, x_1\tau_1)ig_0(x\tau, x'\tau')$$
$$= ig_0(x\tau, x_1\tau_1)ig_0(x_1\tau_1, x\tau') - ig_0(x_1\tau_1, x_1\tau_1)ig_0(x\tau, x'\tau')$$

The ground state expectation value, with all the annihilation operators to the right, disappears because of its normal ordering. Had we chosen a different time ordering to begin with, the result would have been the same, as can be easily verified. We can therefore write the following for the first-order time-ordered product

$$\langle 0, \infty|T_\tau\Psi_1^+(x_1\tau_1)\Psi_1(x_1\tau_1)\Psi_1(x\tau)\Psi_1^+(x'\tau')|-\infty, 0\rangle =$$

$$ig_0(x\tau, x_1\tau_1)ig_0(x_1\tau_1, x\tau') - ig_0(x_1\tau_1, x_1\tau_1)ig_0(x\tau, x'\tau') \qquad (A1.18)$$

Wick's theorem generalizes our finding to a time-ordered product of arbitrary order:

To calculate the ground state expectation value of a time-ordered product of field operators, we have to calculate the sum of all possible non-vanishing complete "pairings" of this product. The overall sign of a single contribution is determined by the number of permutations of Fermi field operators required to bring into a "pairing" the creation operator Ψ_1^+ on the right side of the annihilation operator Ψ_1.

We have marked the possible "pairings" in Eq. (A1.18). The top markings will give the second contribution, which has a minus sign because an odd permutation of fermion operators is required to obtain the field operators in correct order. The bottom markings give the first contribution. To be utilized in the perturbation expansion of the Green's function, we must multiply Eq. (A1.18) by $(-i)$ and integrate over the internal variables x_1 and τ_1, which is found by inspecting the S-matrix expansion in Eq. (A1.7). For a low-order contribution, a direct calculation of the "pairings" is possible. However, the higher the order, the more complicated it becomes to find all possible contributions. The Feynman diagram technique gives a systematic graphical way to find all possible completely "paired" products of a given order. We first demonstrate this in Eq. (A1.18). The physical interpretation of the first contribution is that a particle starts at a space-time point $x'\tau'$ and propagates to $x_1\tau_1$ where an interaction event takes place. It then proceeds to $x\tau$. The second term describes a particle propagating freely from $x'\tau'$ to $x\tau$ and another one with a self-scattering contribution at $x_1\tau_1$. In a graphical representation, this is translated into the following picture where we represent the scattering event by a ●

There is a distinct difference between the two graphs. In the first one, the external and internal variables are topologically connected whereas in the second they are not. The former is called a connected diagram and the latter an unconnected one. In any order, both types of diagrams will be created. It can be shown that only the connected diagrams contribute to the Green's function. The unconnected diagrams, which have no external labels, provide a contribution that is exactly the expectation value of the S-matrix and are therefore canceled by the normalization of the Green's function. An nth order contribution in the S-matrix expansion will generate n internal space and time labels $x_i\tau_i$. Because these are integration variables, a permutation

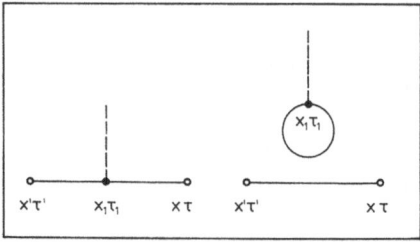

Fig. AI.1 Graphical representation of the pairings in Eq. (A1.17). Part one—left side; part two—right side

will not change the result. For a given diagram, we must therefore multiply its contribution by $n!$. This factor is canceled by the same number in the series expansion. For an interaction of external-field type we obtain the following set of rules for the nth order contribution to the Green's function.

(i) Draw all topological non-equivalent connected diagrams with two external points $x\tau$ and $x'\tau'$ and n interaction vertices and label each vertex with an internal space-time pair $x_i\tau_i$. Connect all labeled points by an oriented line starting at space–time point $x'\tau'$.

(ii) A propagator $ig_0(x\tau, x'\tau')$ must be assigned to each line running from $x'\tau'$ to $x\tau$.

(iii) Integrate over all internal variables and multiply the result by $(-i)(-i)^n$.

The only types of diagram that contribute to an external interaction are those shown in Fig. A1.2.

These are also the basic diagrams for randomly distributed impurities. Here, we must also calculate the configuration average over the distribution. We will discuss this later.

The second type of interaction is the coupling of the electrons to the phonons. As already pointed out in the discussion about the "pairings", there must be an even number of phonon field operators to have a non-zero contribution. This means that only $n = 2, 4, 6, \ldots$ terms in the S-matrix expansion relating to the electron–phonon interaction will contribute to the Green's function. The rules for obtaining the total set of diagrams that

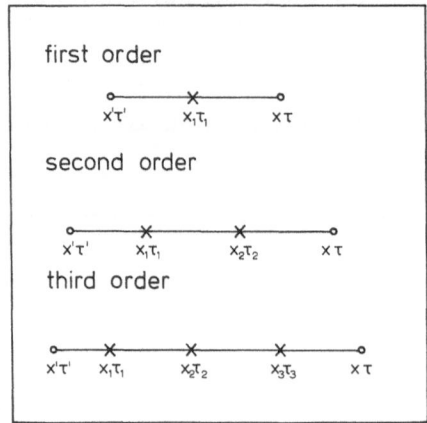

Fig. A1.2 S-matrix expansion for an external perturbation. The open circles stand for the external space time points $x\tau$ and $x'\tau'$. Each cross is a interaction matrix element with the internal label $x_i\tau_i$. Orientation of graphs are omitted for convenience

represent the completely "paired" T_τ-products are:

(i) For the nth order, draw all connected topological non-equivalent diagrams with $2n$ vertices and external labels $x'\tau'$ and $x\tau$. Each vertex is labeled with an internal space-time pair $x_i\tau_i$. Connect all labels by oriented lines starting at label $x'\tau'$. One incoming and one outgoing electron line and a phonon line (arbitrarily directed) are attached to each vertex.

(ii) To each electron line running from $x'\tau'$ to $x\tau$, assign an electron propagator $ig_0(x\tau, x'\tau')$ and to each phonon line running from $x'\tau'$ to $x\tau$ the phonon propagator $id_0(x\tau, x'\tau')$.

(iii) Integrate over all internal variables and multiply by $(-i)(-ig)^{2n}$ $(-1)^m$. Here, m is the number of closed fermion loops.

In Fig. A1.3, we show some examples of first- and second-order diagrams. Only in diagram (e) do we have a Fermi loop which is a topologically connected Fermi line contour within a diagram. Therefore this diagram has $m = 1$. In the second-order contributions (b) and (c) we recognize a part that is similar to the first-order contribution (a).

In Chapter 1, we derived the Dyson equation (1.56) for the Green's function by introducing the concept of the self-energy Σ. The Dyson equation is an integral equation for the Green's function whose kernel is the self-energy. We will now rediscover the Dyson equation and find a way of calculating the self-energy operator. For simplicity, we will continue only for the electron–phonon system where we found a graphical representation for a perturbation expansion of the Green's function G in the electron–phonon

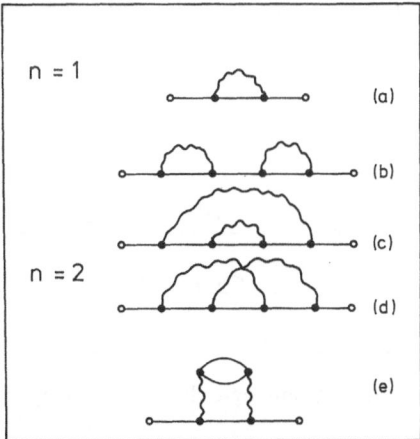

Fig. A1.3 Lowest order electron–phonon interaction contributions to the Green's function. Orientation of graphs are omitted for convenience

interaction strength g^2. Up to second order in the electron–phonon interaction strength, which is g^4, we can write down the Green's function symbolically by the following graphical representation.

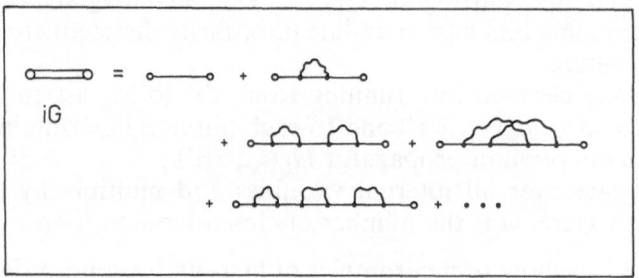

Fig. A1.4 The exact Green's function of the electron–phonon system as a result of an S-matrix expansion

For the Green's function G we have used a double line to distinguish it from the zero-order propagator g_0. Following the rules given above, we can find the corresponding analytical expressions. To be consistent, we must replace the overall factor $(-i)(-ig)^{2n}(-1)^m$ by $(-ig)^{2n}(-1)^m$ and translate the double line for G by iG in accordance with the ig_0 for the single lines. We have already observed that some of the second-order diagrams in Fig. A1.3 have parts that are identical to the first-order diagram. This will carry through to infinite order. Some of these simple low-order diagrams can be summed to infinite order, such as diagram (a) for example. Considering only subsets that are created by using this diagram recursively, for instance, we obtain the following for the Green's function

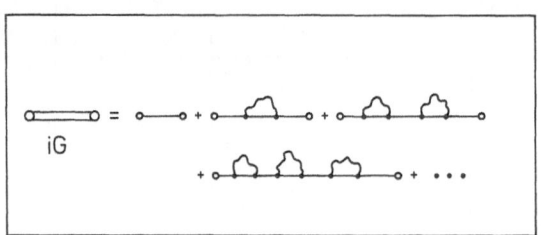

Fig. A1.5 Summing-only reducible diagrams

The infinite sum shown in Fig. A1.5 can alternatively be created by the following recursive relationship

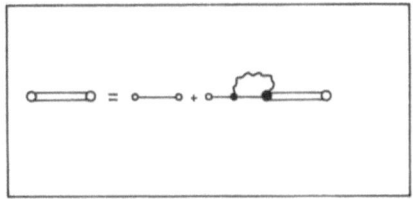

Fig. A1.6 Dyson's equation as a recursive relationship to generate an infinite sum of reducible diagrams

Translated into an analytical formula with the given rules, this gives us the following integral equation

$$G = g_0 + g_0 \Sigma G \tag{A1.19}$$

Here, we have omitted the variables for convenience. Comparing with the Dyson equation (1.59), we find the self-energy to be

$$\Sigma \approx \Sigma^{(a)} = ig^2 d_0 g_0 \tag{A1.20}$$

The self-energy $\Sigma^{(a)}$ is a first-order diagram. By iterating this simple first-order diagram as shown in Fig. A1.5, we can obtain an approximation of the Green's function to infinite order. The idea behind many-particle perturbation theory is now to consider only self-energy contributions that cannot be obtained by an iteration procedure. These are also known as irreducible self-energy diagrams. The solution of the Dyson equation then provides an iteration of these irreducible self-energy contributions to infinite order. The self-energy is obtained by removing the external lines from the diagram giving the Green's function. Irreducible self-energy diagrams are those self-energy contributions that cannot be separated by cutting a single electron line. Diagram (b) from Fig. A1.3 will not give an irreducible diagram because it is the first iteration of the first-order contribution. Although it is an irreducible diagram according to the definition given above, diagram (c) can also be considered as a result of an iteration: the internal line in the first-order diagram (a) is replaced by the next-higher order correction of the electron Green's function. The iteration procedure will result in the following graphs for the self-energy

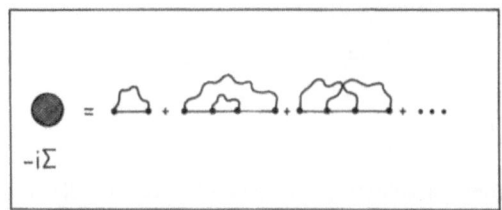

Fig. A1.7 The self-energy as kernel for Dyson's equation

In the context of the Dyson equation for a self-consistent calculation in the first-order irreducible self-energy operator, we would have the following equation

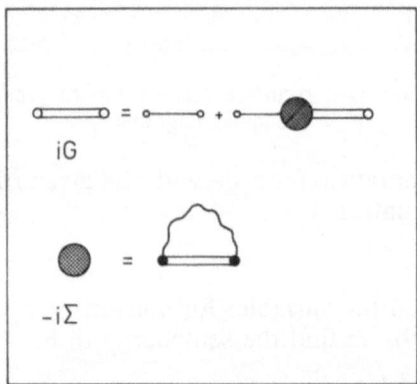

Fig. A1.8 A fully self-consistent set of Dyson's and self-energy equation in lowest order of the electron phonon coupling

Which means that we must replace the zero-order propagator g_0 by the full Green's function G.

$$\Sigma \approx \Sigma^{(1)} = ig^2 d_0 G \qquad (A1.21)$$

The diagrams (d) and (e) of Fig. A1.3 are not yet included in the solution of the Dyson equation. They contribute to higher-order corrections of the irreducible self-energy. A formal theory that studies the influence of such higher corrections is beyond the scope of this introduction. They are included in renormalized vertex contributions in the lowest-order self-energy diagram. Its general form would be as in Fig. A1.9.

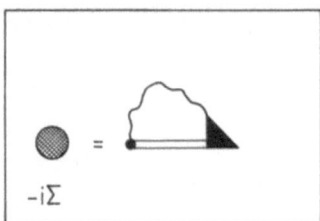

Fig. A1.9 Higher-order corrections are lumped into a renormalized electron–phonon interaction and called vertex corrections. Only one side of the self-energy is affected by the vertex correction to avoid double counting of diagrams

A reader interested in the more formal aspects is referred to the many-body textbooks quoted in Chapter 1.

Wick's theorem also holds true for the Keldysh time-ordered products. We can therefore adopt the complete analysis for the Green's function, Dyson equation and self-energy operator as presented above. The only problem is that we have to take into account the different time-ordering contour for the Keldysh propagator. Times can proceed forwards and backwards in physical time. As discussed in Chapter 1, this will generate additional features, and will especially cast the Green's function into a 2 × 2 matrix according to the different locations of τ and τ' on the time-ordering path. This can easily be taken into account by assigning a 2 × 2 matrix structure to the vertex functions. A graphical representation is again helpful. The fundamental graph for an external perturbation is shown in Fig. A1.10

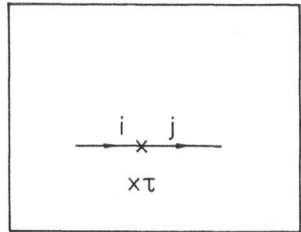

Fig. A1.10 External perturbation, including the time labels for the Keldysh path

In addition to the rules stated above, we must assign an index for both the incoming and outgoing propagators, which indicates the position on the time path. Thus we replace the interaction V

$$V \to V_{ij} = V\sigma_{ij}^3 = \begin{cases} +V; t \text{ on forward path } (+) \\ -V; t \text{ on backward path } (-) \end{cases} \tag{A1.22}$$

with

$$\sigma^3 = \begin{vmatrix} 1 & 0 \\ 0 & -1 \end{vmatrix} \tag{A1.23}$$

The electron–phonon interaction has the fundamental vertex

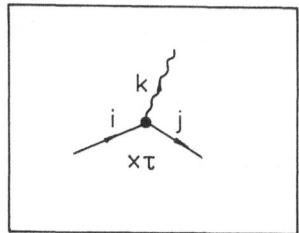

Fig. A1.11 Electron–phonon interaction, including the time labels for the Keldysh path

All lines attached to the vertex must have a label for the Keldysh path. Here we must modify the scattering strength g

$$g \to g_{ij}^k = g\delta_{ij}\sigma_{jk}^3 = \begin{cases} +g; \ i = j = 1 \ k = 1, \ t \text{ on forward path } (+) \\ -g; \ i = j = 2 \ k = 2, \ t \text{ on bacward path } (-) \end{cases}$$

(A1.24)

With Eq. (A1.21) and the modified vertex Eq. (A1.24), we can immediately write down the lowest-order irreducible self-energy contribution for the Keldysh formalism.

$$\Sigma_{ij} = ig_{ii'}^k d_{0,kk'} G_{i'j'} g_{j'j}^{k'}$$

(A1.25)

Here we use Einstein's sum convention that repeated indices must be summed over. Performing the summation, we obtain the self-energy matrix

$$\Sigma = ig^2 \begin{vmatrix} d_{0,11}G_{11} & -d_{0,12}G_{12} \\ -d_{0,11}G_{21} & d_{0,22}G_{22} \end{vmatrix}$$

(A1.26)

The off-diagonal elements of this matrix are the self-energy contributions $\Sigma^{</>}$ of Eq. (1.164). Higher-order contributions can be calculated directly from the corresponding irreducible self-energy diagram. Due to the matrix structure, all the components on the Keldysh time path will mix in a complex way the more complicated the diagram is. Self-energy contributions for diagrams of type (d) in Fig. A1.3 can be found in the literature (Hänsch 1985).

To conclude this section, we come back to scattering by randomly distributed impurities. The general nth-order contribution to the Green's function is shown in Fig. A1.2. It is convenient for what follows to assume that plane waves are an appropriate basic set which permits us to work in $k\omega$-space rather than in $x\tau$-space. For every vertex ($\times$) we must use the interaction matrix element according to Eq. (1.12). For the first low-order contributions we then obtain

$$G_{kk'}(\omega) = g_{0,k}(\omega) + g_{0,k}(\omega) V_{kk'} g_{0,k'}(\omega)$$

$$+ \sum_{k''} g_{0,k}(\omega) V_{kk''} g_{0,k''}(\omega) V_{k''k'} g_{0,k'}(\omega) + \dots$$

(A1.27)

with the interaction matrix element $V_{kk'}$

$$V_{kk'} = V_{\text{imp},kk'} \sum_1 \exp(iR_1(k - k'))$$

(A1.28)

The Green's function will depend implicitly on the distribution of the scattering centers. However, if they are distributed randomly we can calculate a meaningful configuration average $G_{kk'} \to \{G_{kk'}\}$. The first term on the right hand side is the zero-order Green's function. It does not depend on the position of the scatterers and is therefore not affected by the configuration average. The second term contains the scattering potential. Using Eq. (A1.28) we get

$$\{V_{kk'}\} = V_{\text{imp}, kk'} \left\{ \sum_{1} \exp(iR_1(k - k')) \right\} = N_{\text{imp}} V_{\text{imp}, kk'} \delta_{kk'} \quad \text{(A1.29)}$$

Due to the randomness of the distribution of positions R_i for $k \neq k'$, the sum is zero. Only if $k' = k$ will it provide a contribution identical to the number of impurities N_{imp}. Similarly, we obtain for the third term of the right-hand side of Eq. (A1.27)

$$\{V_{kk''} g_{0,k''}(\omega) V_{k''k'}\} = V_{\text{imp}, kk''} g_{0,k''}(\omega) V_{\text{imp}, k''k'}$$

$$\cdot \left[\left\{ \sum_{1=1'} \exp(iR_1(k - k')) \right\} \right.$$

$$\left. + \left\{ \sum_{1 \neq 1'} \exp(iR_1(k - k')) \exp(iR'_1 \cdot (k - k')) \right\} \right]$$

$$= N_{\text{imp}} V_{\text{imp}, kk''} g_{0,k''}(\omega) V_{\text{imp}, k''k'} \delta_{kk'}$$

$$+ N_{\text{imp}} (N_{\text{imp}} - 1) V_{\text{imp}, kk''} \delta_{kk''} g_{0,k''}(\omega) V_{\text{imp}, k''k'} \delta_{k''k'}.$$

$$\text{(A1.30)}$$

The physical interpretation of Eq. (A1.30) is that the first term describes two successive scattering events occurring at the same impurity and is therefore linear in N_{imp}. The second contribution describes two scattering events at different impurities. While in the former momentum is conserved for the complete process, the latter conserves momentum at each scattering event separately. These scattering events can therefore be interpreted as being uncorrelated. However, the momentum conservation is a result of the configuration average and not a property of the single physical scattering event. To obtain a higher-order contribution, the average procedure can be represented in graphical form. On the left side of Fig. A1.12 we show the fundamental diagrams up to third order before the configuration average was calculated. On the right hand side we show the graphical representation of averaged diagrams.

The diagrams of order one and two have just been calculated. Comparing Eqs. (A1.29) and (A1.30) with the corresponding diagrams, we find that each dashed line stands for a matrix element $V_{\text{imp}, kk'}$ and the cross indicates whether we have successive scattering at a single impurity or not. If such successive scattering occurs, incoming and outgoing momentum is conserved at the impurity. We can already construct a large number of diagrams for order three. Those diagrams in Fig. A1.12 that are labeled (a)

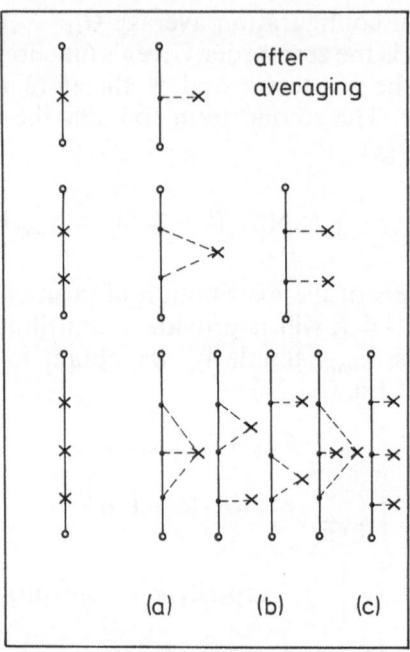

Fig. A1.12 Graphical representation of configurational average of randomly distributed impurities. The number of interaction lines (dashed lines) in the averaged diagrams corresponds to the number of crosses in the fundamental diagrams (left side)

are linear in the number of impurities, those labeled (b) are quadratic in N_{imp} ($N_{\mathrm{imp}} \gg 1$), and those labeled c are of order N_{imp}^3.

Taking the configuration average gives a diagonal Green's function $\{G_{kk'}(\omega)\} \to G_k(\omega)$ as well as a diagonal self-energy $\{\Sigma_{kk'}(\omega)\} \to \Sigma_k(\omega)$. By inspection of Fig. A1.12, we find as in the electron–phonon case that some of the higher-order diagrams contain elements of lower-order ones. Thus, diagram (c) is an iteration of the first-order contribution and the second diagram in group (b) is an iteration of a first and second-order one. We can find the irreducible self-energy of order n by removing all diagrams containing elements of order lower than n and replacing g_0 by G. These are shown in Fig. (A1.25) for single-impurity scattering events which are proportional to N_{imp}.

Configurationally averaged Green's function and self-energy are linked together by a Dyson equation. For the lowest-order self-energy contribution, we have the following according to Fig. A1.13

$$\Sigma_k(\omega) = N_{\mathrm{imp}} V_{\mathrm{imp}, kk'} \delta_{kk'} + N_{\mathrm{imp}} \sum_{k'} |V_{\mathrm{imp}, kk'}|^2 G_{k'}(\omega) \qquad (A1.31)$$

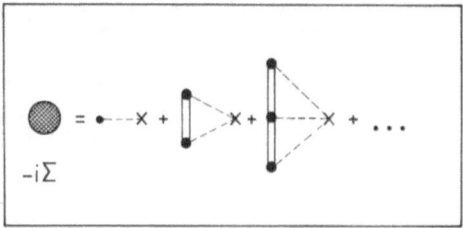

Fig. A1.13 Irreducible self-energy for electrons scattered by randomly distributed impurities. Contribution linear in the density of impurities

Because the first term in Eq. (A1.31) contributes only to a simple constant shift in the one-particle energy, it is usually omitted. The connection to the Keldysh operators is obtained if we go back into physical space-time. Because of the static nature of the interaction (it carries momentum but no energy away), we must simply replace the equilibrium operators by Keldysh matrices. However, this works only for the simple lowest-order diagrams. In higher orders, the correct Keldysh vertex as discussed above has to be taken into account. The self-energy contribution that we found with Eq. (A1.31) is appropriate for low-impurity densities. For higher densities, we must account for interfering scattering centers. Such diagrams would contain crossing interaction lines and would be of at least order N_{imp}^2. For more detailed information, we refer the reader to the literature (Mahan 1981, Rammer and Smith 1986).

Appendix 2: Inversion Channel Particle-Density Distribution in Equilibrium

In Chapter 3, we investigated the effect of the finite barrier between Si and SiO_2 on the particle-density distribution of a MOSFET in strong inversion. We found that due to the boundary condition of the wave function, the particle density will have its maximum away from the interface, in contrast to the classical result. This effect was correctly obtained by replacing the conventional drift diffusion equation by Eq. (3.86). In this equation, the drift term contains a built-in field that repels the carriers from the surface. In this appendix, we will compare this semi-classical approach with the exact calculation for the equilibrium case. In contrast to the usual treatment of this problem, where the coupled Poisson and Schrödinger equation is solved to obtain the carrier distribution in a self-consistent field, we chose a different approach. It is based on the observation that to a first-order approximation, the quantum mechanical corrections can be directly included in the current equation. A generalization of Eq. (3.86) would read

$$j(R) = -qn\mu_{ch}(x)\nabla_R V^*(R) + qD_{ch}(x)\nabla_R n(R) \tag{A2.1}$$

$$V^*(R) = V(R) - u_0 \ln \gamma(x) \tag{A2.2}$$

Here, γ is a still unknown built-in potential. In a first-order approximation, it is given by Eq. (3.87). The continuity equation, with the current equation (A2.1) and Poisson's equation, comprise the conventional set of equations giving the field and density distribution in the device. To include quantum corrections as well, we must add Schrödinger's equation, which in its simplest form is

$$-\frac{1}{2m}\frac{\partial^2}{\partial x}\varphi_n(x) + qV(x)\varphi_n(x) = E_n\varphi_n(x) \tag{A2.3}$$

We will neglect the variation of the potential along the channel, which is weak in the case of strong inversion. Mistakes are made near the *pn*-junctions of the source and drain regions. Because we are only interested in comparing the approximate solution Eq. (3.86) by an exact calculation, it is sufficient to stay in the middle of the channel where, under strong inversion, the one-dimensionality of the potential is a good approximation. The energy eigenvalues E_n cover both bound states in the potential well and the continuous spectrum. The solution of the Schrödinger equation in a given potential $V(x)$ will give a complete set of eigenfunctions. This set will contain a finite number N of bound states $\varphi_n(x)$, for which we have $qV_\infty > E_n > qV_0$ ($V_0 < V_\infty$) and the continua wave functions $\varphi_k(x)$ with $E_k = k^2/2m + qV(x) > qV_\infty$. With Eq. (3.22) we obtain the density distribution for a non-degenerate electron gas

$$
n_{\text{sch}}(x) = \frac{1}{2\pi}(mk_B T)\sum_{n=1}^{n=N}\exp\left(-\frac{\varphi_n - q\varepsilon_F}{k_B T}\right)|\varphi_n(x)|^2\theta(V_\infty - V(x))
$$

$$
+ n_\infty \exp\left(\frac{qV(x) - qV_\infty}{k_B T}\right)\cdot\left[\theta(V(x) - V_\infty)\right.
$$

$$
+ \theta(V_\infty - V(x))\frac{2}{\sqrt{\pi}}\int_{\frac{q(V_\infty - V(x))}{k_B T}}^{\infty} dy\sqrt{y}\exp(-y)\bigg] \qquad (A2.4)
$$

We have introduced the label to distinguish $n_{\text{sch}}(x)$ from the solution obtained from the modified semiconductor equations. The first term in Eq. (A2.4) contributes only if bound states are possible, $V_\infty > V_0$. In the limit $x \to \infty$, the wave functions from the bound states are zero and interface-related features have disappeared. Hence the normalization of the continuums states is chosen so that deep in the substrate, where the bound states have died out, n_{sch} matches the density distribution obtained from the classical solution n_∞. The number of bound states N can be estimated within the semi-classical quantization, which is justified in the limit $N \to \infty$.

$$
N \approx \frac{1}{\pi}\sqrt{(2m/\hbar^2)}\int_0^\infty (qV_\infty - qV(x))^{1/2} \qquad (A2.5)
$$

For a typical potential in a MOSFET, N is of the order of a hundred. The iteration procedure is now to solve the ordinary semiconductor equations with $\gamma = 1$ in the first step. This provides a potential and carrier distribution, V^1 and n^1, respectively. In a second step, we solve the Schrödinger equation in the potential V^1. This gives a density n_{sch}^1 and we define $\gamma = 1 \cdot n_{\text{sch}}^1/n^1$. The modified semiconductor equations are solved and the potential is fed into the Schrödinger equation. The new correction is found to be $\gamma = 1 \cdot n_{\text{sch}}^1/n^1 \cdot n_{\text{sch}}^2/n^2$. The iteration is repeated until we have $n_{\text{sch}}^i/n^i \approx 1$. For the equilibrium situation, this iteration scheme is equivalent to the Poisson–Schrödinger system usually considered, which is an

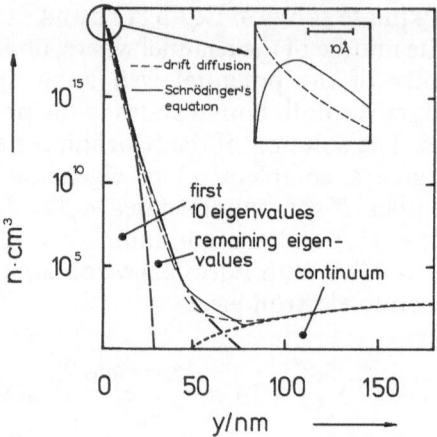

Fig. A2.1 Carrier density in the inversion channel. Thin dashed line—conventional drift diffusion equation; thin solid line—exact quantum mechanical solution; thick dashed line—contribution from the first 10 eigenvalues; thick dashed–dotted line—contributions from the remaining eigenvalues; thick dashed line—continuums states

eigenvalue problem in a self-consistent field. Our proposed iteration scheme is coupled directly to the continuity and current equations and is therefore very naturally suited to the transport problem.

For the evaluation of Eq. (A2.4), we considered the complete spectrum of the Schrödinger equation. Usually, only the contributions of the first few bound states are considered. This is sufficient if we are interested in the carrier density near the surface, because the corresponding wave functions decay rapidly once they reach the classical turning points in the potential well. Wave functions in higher bound states contribute more significantly far away from the interface. In the limit $x \to \infty$, only the continua states contribute. These different contributions are shown in Fig. A2.1.

Only if we consider the complete spectrum of the quantum mechanical problem does the carrier density coincide, deep in the bulk, with the conventional one. In fact, a significant difference is only observed very close to the surface where the contribution from the first few bound states is most important. Taking into account the higher bound states and the continua states as well guarantees a correct connection to the bulk region, where the classical behavior is unchanged. This is an important result if quantum corrections near the surface are incorporated in the drift diffusion approximation to cover the non-equilibrium case as well. Otherwise, we would eventually obtain a non-physical carrier density which would result in a mismatch of the quantum and classical regimes in the form of non-physical current density contributions. A more detailed comparison of the conventional case, the first-order approximation of Eq. (3.87), and the exact solution for the equilibrium case are shown in Chapter 3, Fig. 26.

Author Index

Subject Index

Henk C. de Graaff and François M. Klaassen

Compact Transistor Modelling for Circuit Design

(Computational Microelectronics)

1990. 184 figures. XII, 351 pages.
Cloth DM 186,-, öS 1300,-
ISBN 3-211-82136-8

Prices are subject to change without notice

This book describes analytical compact transistor models that can be used in circuit simulation programs like SPICE. It provides the reader with a thorough knowledge of many aspects of compact models. The book starts with the necessary device physics: Boltzmann transport equation, continuity equations, Poisson equation and physical modelling of mobility, recombination, bandgap narrowing, avalanche multiplication and noise.

Then a systematic treatment of the analytical formulas that describe the device behaviour in d.c., a.c. and transient situations, is given for both bipolar and MOST devices. The book contains complete sets of model equations for various models, including some new ones, and special attention is paid to the numerical problems of analytical continuity.

Separate chapters are devoted to parameter determination, the parameter temperature dependence as well as their relation to process variables, the statistical correlations between parameters, the scaling rules for submicron devices, the side wall effects and the parasitics.

The book thus contains all the relevant aspects of compact transistor modelling for integrated circuit design and serves as a state-of-the-art description in compact modelling.

Springer-Verlag Wien New York

Carlo Jacoboni, Paolo Lugli

The Monte Carlo Method for Semiconductor Device Simulation

(Computational Microelectronics)

1989. 228 figures. X, 356 pages.
Cloth DM 186,–, öS 1300,–
ISBN 3-211-82110-4

Prices are subject to change without notice

The subject of the book is the application of the Monte Carlo method to the simulation of semiconductor devices. It introduces the reader to the Monte Carlo technique as applied to the study of transport in semiconductors, and to the modelling of semiconductor devices. The book is at a tutorial level, where all the details of a Monte Carlo algorithm are discussed. Since the use of the Monte Carlo technique requires an accurate knowledge of the physical system under investigation, a general overview of the basis of the physics of transport in semiconductors is also provided. A review of the Monte Carlo simulation of actual devices is also presented, together with possible vectorization schemes.

The Monte Carlo technique is a fairly new tool in the area of device modelling, traditionally dominated by simulators based on drift-diffusion or on balance-equation models. A comparison of the characteristics of the different methods is presented, pointing out the areas and limits of applicability of each of them.

The book allows the reader (even with a limited knowledge of transport in semiconductor devices) to become accustomed to the Monte Carlo simulation and to quickly set up a simulator for a specific problem.

Springer-Verlag Wien New York